Lecture Notes in Computer Science   7652

Commenced Publication in 1973
Founding and Former Series Editors:
Gerhard Goos, Juris Hartmanis, and Jan van Leeuwen

Lecture Notes in Computer Science

Armin Haller   Guangyan Huang
Zhisheng Huang   Hye-young Paik
Quan Z. Sheng (Eds.)

Web Information Systems Engineering – WISE 2011 and 2012 Workshops

Combined WISE 2011 and WISE 2012 Workshops
Sydney Australia, October 12-14, 2011 and Paphos,
Cyprus, November 28-30, 2012
Revised Selected Papers

 Springer

Volume Editors

Armin Haller
CSIRO ICT Centre, Canberra, ACT, Australia
E-mail: armin.haller@csiro.au

Guangyan Huang
Victoria University, Melbourne, VIC, Australia
E-mail: guangyan.huang@vu.edu.au

Zhisheng Huang
Vrije Universiteit Amsterdam, The Netherlands
E-mail: huang@cs.vu.nl

Hye-young Paik
The University of New South Wales, Sydney, NSW, Australia
E-mail: hpaik@cse.unsw.edu.au

Quan Z. Sheng
The University of Adelaide, SA, Australia
E-mail: qsheng@cs.adelaide.edu.au

ISSN 0302-9743   e-ISSN 1611-3349
ISBN 978-3-642-38332-8   e-ISBN 978-3-642-38333-5
DOI 10.1007/978-3-642-38333-5
Springer Heidelberg Dordrecht London New York

Library of Congress Control Number: 2013937715

CR Subject Classification (1998): H.3, H.4, H.5, J.1, I.2, C.2

LNCS Sublibrary: SL 3 – Information Systems and Application, incl. Internet/Web
and HCI

Typesetting: Camera-ready by author, data conversion by Scientific Publishing Services, Chennai, India

Printed on acid-free paper

Springer is part of Springer Science+Business Media (www.springer.com)

Preface

Welcome to the proceedings of the Web Information Systems Engineering - Combined WISE 2011 and WISE 2012 Workshops. The international conference series on Web Information Systems Engineering (WISE) aims to provide an international forum for researchers, professionals, and industrial practitioners to share their knowledge in the rapidly growing area of Web technologies, methodologies and applications. The 13th WISE (WISE 2012) was held in Paphos, Cyprus, in November 2012. Previous WISE conferences were held in Hong Kong, China (2000), Kyoto, Japan (2001), Singapore (2002), Rome, Italy (2003), Brisbane, Australia (2004), New York, USA (2005), Wuhan, China (2006), Nancy, France (2007), Auckland, New Zealand (2008), Poznan, Poland (2009), Hong Kong, China (2010), and Sydney, Australia (2011).

The seven workshops of WISE 2011-2012 have reported the recent developments and advances in the related fields of: Advanced Reasoning Technology for e-Science (ART 2012), Cloud-Enabled Business Process Management (CeBPM 2012), Engineering the Semantic Enterprise (ESE 2012), Social Web Analytics for Trend Detection (SoWeTrend 2012), Big Data and Cloud (BDC 2012), Personalization in Cloud and Service Computing (PCS 2011), and User-Focused Service Engineering, Consumption and Aggregation (USECA 2011).

Many colleagues helped toward the success of the above mentioned workshops. We would especially like to thank the Program Committee members and reviewers for a conscientious reviewing process. We are also grateful to the WISE Society for generously supporting our workshops. We greatly appreciate Springer for publishing the WISE workshop proceedings in their LNCS series.

December 2012

Armin Haller
Guangyan Huang
Zhisheng Huang
Hye-young Paik
Quan Z. Sheng

Preface

Organization

Executive Committee

Workshops Co-chairs

Armin Haller	CSIRO, Australia
Zhisheng Huang	Vrije University of Amsterdam, The Netherlands
Quan Z. Sheng	University of Adelaide, Australia

Publication Co-chairs

Guangyan Huang	Victoria University, Australia
Hye-young Paik	University of New South Wales, Australia

Advanced Reasoning Technology for e-Science (ART 2012)

Workshop Co-chairs

Alexey Cheptsov	University of Stuttgart, Germany
Jeff Z. Pan	University of Aberdeen, UK
Zhisheng Huang	Vrije University Amsterdam, The Netherlands

Workshop on Cloud-enabled Business Process Management (CeBPM 2012)

Workshop Co-chairs

Dana Petcu	West University of Timisoara, Romania
Vlado Stankovski	University of Ljubljana, Slovenia

Program Chair

András Micsik	MTA SZTAKI, Hungary

Engineering the Semantic Enterprise (ESE 2012)

Workshop Co-chairs

Maciej Dabrowski	DERI Galway, Ireland
John Breslin	DERI Galway, Ireland
Alexandre Passant	Seevl.net
Eric Gordon Prud'hommeaux	W3C

Social Web Analytics for Trend Detection (SoWeTrend 2012)

Workshop Co-chair

Athena Vakali	Aristotle University of Thessaloniki, Greece
Hakim Hacid	Bell Labs, France

International Workshop on Big Data and Cloud (BDC 2012)

Workshop Co-chair

Surya Nepal	CSIRO, Australia
Athman Bouguettaya	RMIT University, Australia

The 5th International Workshop on Personalization in Cloud and Service Computing (PCS 2011)

Workshop Co-chairs

Chengfei Liu	Swinburne University of Technology, Australia
Yanbo Han	Institute of Computing Technology, CAS, China
Bernhard Holtkamp	Fraunhofer ISST, Germany
Bing Li	Wuhan University, China

Program Co-chairs

Jian Yu	Swinburne University of Technology, Australia
Hong-Linh Truong	Vienna University of Technology, Austria

User-Focused Service Engineering, Consumption and Aggregation (USECA 2011)

Workshop Co-chairs

Hye-young Paik	University of New South Wales, Australia
Ingo Weber	University of New South Wales, Australia
Marek Kowalkiewicz	SAP Research Brisbane, Australia

Program Committees

ART 2012 PC Members

Stefan Wesner	High Performance Computing Center Stuttgart, Germany
Marko Grobelnik	Jozef Stefan Institute in Ljubljana, Slovenia

Volker Tresp	Ludwig Maximilian University of Munich, Germany
Michael Witbrock	Cycorp, USA
Dongmo Zhang	University of Western Sydney, Australia
Frank van Harmelen	Vrije University Amsterdam, The Netherlands
Kewen Wang	Griffith University, Australia
Guilin Qi	Southeast University, China
Grigoris Antoniou	University of Huddersfield, Greece
Jianfeng Du	Guangdong University of Foreign Studies, China

CeBPM 2012 PC Members

Ciprian Dobre	Politehnica University of Bucharest, Romania
Enn Õunapuu	Tallinn University of Technology, Estonia
Georgina Gallizo	High Performance Computing Center Stuttgart, Germany
Huy Tran	University of Vienna, Austria
Jānis Grabis	Institute of Information Technology, Riga Technical University, Latvia
Jorge Ejarque	Barcelona Supercomputing Center, Spain
José Luis Vázquez-Poletti	Universidad Complutense de Madrid, Spain
László Kovács	MTA SZTAKI, Hungary
Matjaž Branko Jurič	University of Ljubljana, Slovenia
Matthias Book	University of Duisburg-Essen, Germany
Parastoo Mohagheghi	NTNU-IDI, Norway
Pawel Czarnuł	Gdansk University of Technology, Poland
Ricardo Jardim Goncalves	UniNova, Portugal
Tomáš Pitner	Masaryk University, Brno, Czech Republic

ESE 2012 PC Members

Soren Auer	Universität Leipzig, Germany
John Breslin	NUI Galway, Ireland
Maciej Dabrowski	DERI Galway, Ireland
Renaud Delbru	Sindice Tech
Keith Griffin	Cisco Systems
Conor Hayes	DERI Galway, Ireland
Bart van Leeuwen	Netage.nl, The Netherlands
Vincenzo Loia	University of Salerno, Italy
Fabrizio Orlandi	DERI Galway, Ireland
Adrian Paschke	Freie Universität Berlin, Germany
Alexandre Passant	Seevl.net
Eric Gordon Prud'hommeaux	W3C, USA
Yves Raimond	BBC, UK
Shawn Simister	Google, USA

SoWeTrend 2012 PC Members

Ludovic Denoyer	University of Paris 6, France
Arne Barre	Sintef, Norway
Ernestina Menasalvas	Universidad Politecnica de Madrid, Spain
George Pallis	University of Cyprus, Cyprus
Hye-young Paik	University of New South Wales, Australia
John Garofalakis	University of Patras, Greece
Shengbo Guo	Xerox Research Centre Europe, France
Grigoris Tsoumakas	Aristotle University, Greece
Julien Velcin	ERIC Lab, University of Lyon, France
Tetsuya Yoshida	Hokkaido University, Japan

BDC 2012 PC Members

Xumin Liu	Rochester Institute of Technology, USA
Qi Yu	Rochester Institute of Technology, USA
Suraj Pandey	IBM Research Australia
Xuyun Zhang	University of Technology Sydney, Australia
Wanita Sherchan	IBM Research Australia
Julian Jang	CSIRO Australia
Zaki Malik	Wayne State University, USA
Shiping Chen	CSIRO Australia
Dongxi Liu	CSIRO Australia

PCS 2011 PC Members

Mahammad Ali Babar	IT University of Dennmark, Denmark
Jinjun Chen	University of Technology Sydney, Sydney
Marco Comerio	University of Milano-Bicocca, Italy
Paolo Falcarin	University of East London, UK
Hans-Georg Fill	Stanford University, USA
G.R Gangadharan	IBM India, India
Ligang He	University of Warwick, UK
Martin Hirsch	Dortmund University of Applied Sciences and Arts, Germany
Hien Le	Norwegian University of Science and Technology, Norway
Xitong Li	MIT, USA
Jiangang Ma	University of Adelaide, Australia
Fulvio Mastrogiovanni	University of Genoa, Italy
Maurizio Morisio	Politecnico di Torino, Italy
Tran-Vu Pham	Ho Chi Minh City University of Technology, Vietnam
Fangzhong Qi	Zhejiang University, China
Stephan Reiff-Marganiec	University of Leicester, UK
Kurt Sandkuhl	University of Rostock, Germany

Aviv Segev   KAIST, Republic of Korea
Jun Shen   University of Wollongong, Australia
Michael Sheng   The University of Adelaide, Australia
Philippe Thiran   University of Namur, Belgium
Guiling Wang   Institute of Computing Technology, CAS,
                            China

Manfred Wojciechowski   Fraunhofer ISST, Germany
Jinhua Xiong   Institute of Computing Technology, CAS,
                            China

Jian Yang   Macquarie University, Australia
Muhammod Younas   Oxford Brooks University, UK
Xiaohui Zhao   Unitec Institute of Technology, New Zealand

USECA 2011 PC Members

Boualem Benatallah   University of New South Wales, Australia
Fabio Casati   University of Trento, Italy
Jinjun Chen   University of Technology Sydney, Australia
Alan Colman   Swinburne University of Technology, Australia
Florian Daniel   University of Trento, Italy
Keith Duddy   Queensland University of Technology, Australia
Alexander Dreiling   SAP Research, Australia
Karl M. Goeschka   Vienna University of Technology, Austria
Hakim Hacid   Alcatel-Lucent Bell Labs, France
Christian Janiesch   Karlsruhe Institute of Technology, Germany
Tomasz Kaczmarek   Poznan University of Economics, Poland
Dimka Karastoyanova   University of Stuttgart, Germany
Ryszard Kowalczyk   Swinburne University of Technology, Australia
Hamid Motahari   HP Labs, USA
Surya Nepal   CSIRO, Australia
Anne Ngu   Texas State University, San Marcos, USA
Fethi Rabhi   University of New South Wales, Australia
Chris Smith   University of New Castle, UK
Vladimir Tosic   NICTA, Australia
Jian Yu   Swinburne University of Technology, Australia
Julien Vayssiere   Germany

Table of Contents

Engineering the Semantic Enterprise (ESE 2012)

Social Web Analytics for Trend Detection (SoWeTrend 2012)

International Workshop on Big Data and Cloud (BDC 2012)

The 5th International Workshop on Personalization in Cloud and Service Computing (PCS 2011)

User-Focused Service Engineering, Consumption and Aggregation (USECA 2011)

Preface to the Advance Reasoning Technology for eScience (ART-2012) Workshop

Zhisheng Huang[1], Alexey Cheptsov[2], and Jeff Z. Pan[3]

[1] Free University of Amsterdam, The Netherlands
huang.zhisheng.nl@gmail.com
[2] High-Performance Computing Center Stuttgart, Germany
cheptsov@hlrs.de
[3] University of Aberdeen, UK
jeff.z.pan@abdn.ac.uk

Reasoning is widely recognized as the most challenging application area of the modern Semantic Web Technology in terms of performance and scalability demands, involving exciting research on ontology languages, inference logic, data management, scale-up techniques, etc. Reasoning is widely adopted by the social media, bioinformatics, smart cities, and many other domains working at the leading edge of the science and technology. The availability of the large-scale computing infrastructures and software platforms targeting them (such as LarKC - the Large Knowledge Collider, www.larkc.eu) has enabled the application of reasoning to solving the emerging problems of the modern eScience, such as Big Data.

Following the discussions at the European and International Semantic Web Conferences from 2009, 2010, and 2011, the main goal of the ART workshop was to elaborate a strategy and a roadmap for the traditional reasoning approaches (e.g. complete and deductive reasoning) to be advanced by the novel techniques (e.g. automated and streaming reasoning) in order to take up the dominating position on the Intelligent Information Management system market, leveraging the on-demand infrastructures like Supercomputers and Cloud.

The workshop proved a successful event in terms of both the quality of presentations and the interest attracted by the visitors. The presentations included in the workshop spawned over a wide range of topics from the application of context-aware systems to the tools facilitating porting the reasoning applications to supercomputing infrastructures. To the workshop's highlights can be referred the invited talks held by Prof. Yanchun Zhang from the Victoria University on "Real-time and Self- Adaptive Stream Data Analysis" and by Dr. Jeff Z. Pan from the University of Aberdeen on "How Approximate Reasoning Helps in Centralised and Distributed OWL Reasoning".

The ART workshop was a first standalone event organized on the reasoning technology in the cooperation with the WISE conference. However, this first experience was very successful, also in terms of enhancing the main conference value. The good acceptance of the workshop will surely motivate us to continue the collaboration with the WISE conference and organize similar events in the future.

We would like to acknowledge the work of the Program Committee, invited speakers as well as the organizers of the hosting WISE'2012 conference. We would also like to thank to the large audience of the workshop's participants for their involvement and participation.

The ART'2012 Workshop's Organizing Committee

Real-Time and Self-Adaptive Stream Data Analysis

(Invited Talk)

Yanchun Zhang

Centre for Applied Informatics, Victoria University, Melbourne, Australia
yanchun.zhang@vu.edu.au

In recent years, with the advances in hardware technology, abundant medical surveillance data streams can be easily collected using various kinds of medical devices and sensors. Accurate and timely detection of abnormalities from these physiological data streams is in high demand for the benefit of the patients. However, the state-of-the-art data analysis techniques face the following challenges: First, the raw data streams are in sheer volume, which is attributed to both the number and the length of the data streams. Massive data streams pose a challenge to storing, transmitting, and analysing them. Second, multiple physiological streams are often heterogeneous in nature. These data streams collected from different devices have different value ranges and meanings. Domain knowledge is required for fully understanding them. Third, multiple physiological data streams are not independent. As a matter of fact, they often exhibit high correlations. Abnormalities can be evidenced not only in individual stream but also in the correlation among multiple data streams.

Our work embraces the above challenges and aims at increasing the accuracy and efficiency of abnormality detection via mining multiple correlated heterogeneous time series. In this presentation, we put forward our recent work towards this direction.

1) We developed a novel abnormality detection framework for multiple heterogeneous yet correlated time series. In this framework, we transform heterogeneous data streams from various feature spaces into a uniform trend space. A clustering based compression algorithm is proposed to reduce the number of time series and summarize the information provided by original time series. Consequently, the data size is significantly reduced. We employed the eigenvalues of the correlation matrix of multiple time series to characterize the data and detect emerging abnormalities. Experimental result indicates that our framework is more effective and efficient than its peers.

2) A novel algorithm is proposed to detect abnormal period patterns from multiple physiological streams. In our preliminary experiment, we observe that physiological time series frequently exhibit periodic patterns. Thus, we proposed a novel clustering algorithm to facilitate the extraction of periodic features of these time series. Consequently, the original data streams can be represented in a concise manner with much smaller size. Based on such periodic feature, abnormalities can be efficiently detected. The efficiency and effectiveness of our method is demonstrated by experiments on a real-world massive physiological database which comprises 250 patients' streams with a total size of 28GB.

A. Haller et al. (Eds.): WISE 2011 and 2012 Combined Workshops, LNCS 7652, pp. 2–3, 2013.
© Springer-Verlag Berlin Heidelberg 2013

3) In addition, we present the current version of Victoria University Patient Health Monitor (VUPHM) system, which is an accurate and efficient real-time abnormality detection system over co-evolving stream data. The proposed system can lower mortality and reduce the critical response time during the surgical operation. Moreover, it provides a friendly use interface and data visualization.

An Approach for Distributed Parallelization of Large-Scale Semantic Web Reasoners Based on MPI

Alexey Cheptsov

High-Performance Computing Center Stuttgart, Nobelstr. 19,
70569 Stuttgart, Germany
cheptsov@hlrs.de

Abstract. High performance is a key requirement for a number of Semantic Web applications. Reasoning over large and extra-large data is the most challenging class of Semantic Web applications in terms of the performance demands. For such problems, parallelization is a key technology for achieving high performance and scalability. However. development of parallel application is still a challenge for the majority of Semantic Web algorithms due to their implementation in Java – a programming language whose design does not allow the porting to the High Performance Computing Infrastructures to be trivially achieved. The Message-Passing Interface (MPI) is a well-known programming paradigm, applied beneficially for scientific applications written in the "traditional" programming languages for High Performance Computing, such as C, C++ and Fortran. We describe an MPI based approach for implementing parallel Semantic Web applications and evaluate the performance of a pilot Semantical Statistics application - random indexing over large text volumes.

Keywords: Semantic Web, Reasoning, Java, Parallelization, MPI.

1 Introduction

The tremendous amount of inter-connected and linked data annotated with expressive ontology models, such as RDF, offers several challenging scenarios for their efficient processing in large-scale reasoning engines, such as provided by OWLIM [1], or LarKC [2]. Recent advantages in the Semantic Web require its Java applications to be scaled up to the requirements of rapidly increasing amount of data, such as coming from automated reasoning engines or sensor networks. Given the large problem sizes that are addressed by the applications implemented on top of those engines, and given the tremendous growth of the data sets addressed by the Semantic Web, it seems natural to explore the benefits of applications parallelization and thus porting to High Performance Computing (HPC) platforms for this domain as well.

Several forms of parallelism are recognized for reasoning applications, such as inter-querying (running more than one query in parallel), intra-query (running subqueries in parallel and pipelining operators), or intra-operation (distribution single operations for concurrent execution). Moreover, parallelization can be applied for the operational chains of application workflows (Fig. 1).

A. Haller et al. (Eds.): WISE 2011 and 2012 Combined Workshops, LNCS 7652, pp. 4–12, 2013.

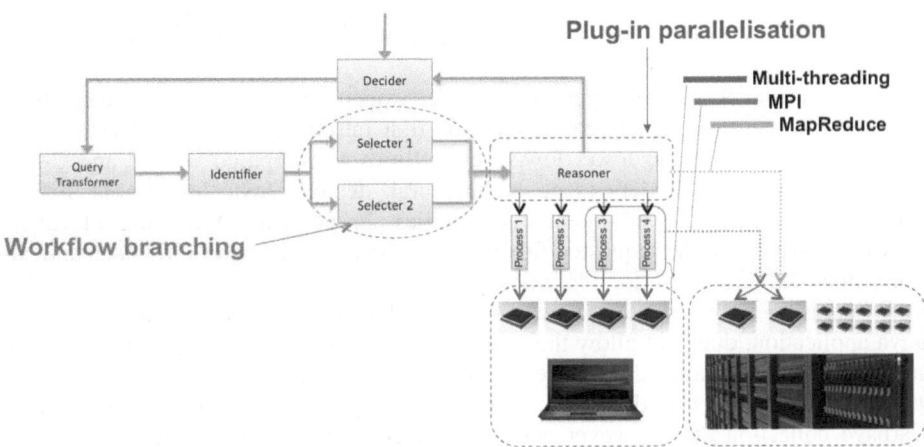

Fig. 1. Parallelization in LarKC application's workflow

However, there are still certain difficulties presented that appear when applying those approaches in practice; they are mainly related to considerable implementation efforts of the parallel processing in the existing applications.

Whereas the thread-based strategies, such as *Multi-Threading* [3], are considered to be very efficient and easy to implement, they don't allow the application to achieve high performance speed up due to resource limitation of the currently existing compute architecture (the total number of CPU cores provided by such system is usually not higher then 8). On the contrary, the process-based strategies, such as *Map-Reduce* [4] or *Message-Passing Interface* [5] enable distributed compute architectures for application execution, including clusters, HPC, or Grid systems.

Map-Reduce is a software framework for distributed computing on large data sets on clusters of workstations. Although Map-Reduce has proved it efficiency for processing large datasets on certain kinds of distributable problems, this technique requires considerable re-think of the application algorithms in order to confirm to Map and Reduce steps.

Opposite to Map-Reduce, MPI doesn't require such considerable changes in the application code as being a utility library supporting information exchange among the parallel application instances and is thus much more attractive to be adopted by the already existing applications. The main drawback of a wide-spread use of MPI in the Semantic Web application community lies in the fact that the majority of applications are written in Java programming language, that has prevented for a long time implementation of distributed memory code parallelism.

In this paper we introduce and discuss solutions for implementation of Java applications with MPI (Section 2). Then we introduce the case study application which performs Semantic Random Indexing (Section 3) and propose a generic parallelization algorithm for data intensive application (Section 4). We evaluate the performance of the parallel implementation in Section 5. The conclusions as well as consideration about further research objectives are collected in the end.

2   Parallelization of Java Applications by Means of MPI

Java is an object-oriented, general-purpose, concurrent, class-based, and object-oriented programming language, which was first introduced in 1995. Thanks to the simpler object model and fewer low-level facilities as compared with C and C++ as well as its platform-independent architectural design, Java has found a wide application in many development communities. In Semantic Web, use of Java is necessary to achieve the requested flexibility in processing and exploiting semantically annotated data sets.

Whereas the multi-threading mechanism, which is relatively easy to implement for a Java application, does not allow the application performance to scale well beyond the number of cores available, the use of the distributed-memory approaches has been beyond the scope of Java due to the design features pertained to this language, such as garbage collection etc. However, since the emerge of Grid and Cloud technologies offering a virtually unlimited resource pool for the execution of particular applications, the interest in Java computing is shifting also towards distributed-memory programming, which allows the use of high-performance resources.

Among the sustainable parallelization approaches, used over the last years in a wide range of software projects, the Message-Passing Interface (MPI) has become de-facto a standard in the area of parallel computing. MPI is a wide-spread implementation standard for parallel applications, introduced in many programming languages. As the acronym suggests, MPI is a process-based technique, whereby processes communicate by means of messages transmitted between (a so called "point-to-point" communication) or among (involving several or even all processes, a so called "collective" communication) the nodes. Normally, one process is executed on a single computing node, as shown in Fig. 2. If any of the processes needs to send/receive data to/from other processes, it should call a corresponding MPI communication function. Both point-to-point and collective communications available for MPI processes are documented in the MPI standard [5,6].

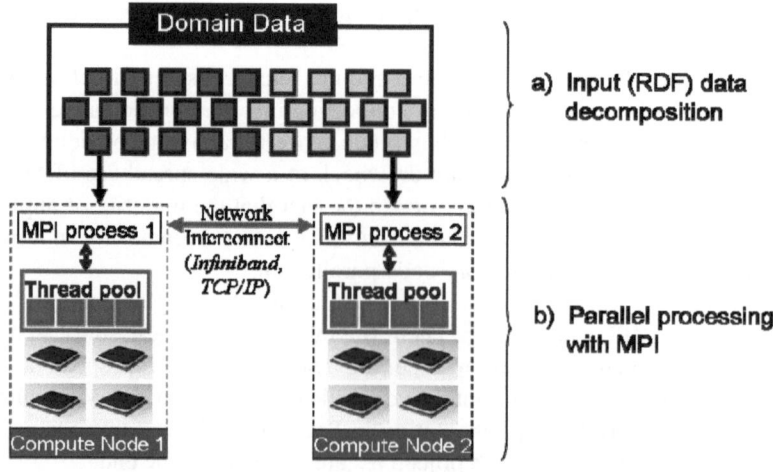

Fig. 2. Execution of MPI processes on HPC system's distributed nodes

There have been several initiatives striving to provide support for Java in HPC environments. One of the most successful MPI implementations is considered to be mpiJava[1] [6], which came out of the HPJava project and is being developed in the frame of the Large Knowledge Collider (LarKC). The key feature of mpiJava is that it wraps the calls of the native C library, which mpiJava is installed on top of (e.g. MPICH or Open MPI). This allows mpiJava to substitute MPI operations with calls to the native library (thanks to Java Native Interface - JNI), that ensures better communication performance as in case of the "Java-only" realization, as was for example done in MPJ-Express[2] [7] library (see Fig. 3).

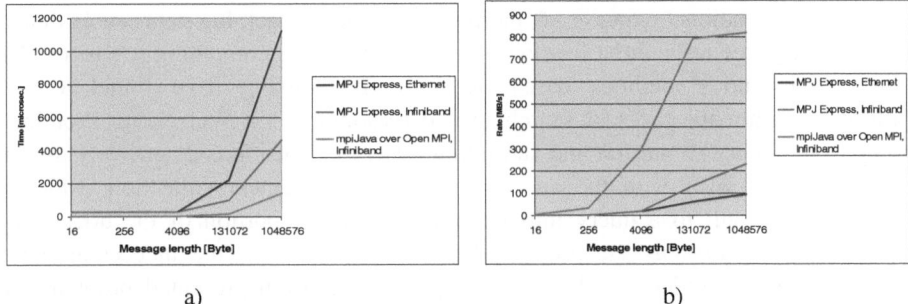

a)   b)

Fig. 3. Comparison of time (a) and bandwidth (b) characteristics of MPI communication for different Java MPI libraries

The applications implemented by means of MPI follow a process-oriented parallel computing paradigm. Each process is identified by means of a rank, which is unique within a group of processes involved in the execution (Listing 1). Among others, the rank allows every process to identify a part of the data to be processed.

```
int my_rank = MPI.COMM_WORLD.Rank();
int comm_size =
        MPI.COMM_WORLD.Size();
System.out.println("Hello from Process " +
        my_rank + " out of " comm_size);
```

Listing 1. Requesting rank of the process and number of the involved processes, following the Java specification of MPI

---

[1] https://sourceforge.net/projects/mpijava/
[2] http://mpj-express.org/

3   Application Use Case

In the recent years, a tremendous increase of structured data sets has been observed on the Web, in particular in the government domain, as for example promoted by the Linked Open Data[3] project. The massive amount of data, in particular described by RDF (Resource Description Framework) – a standard model for data interchange on the Web, is a key challenge for many Semantic Web applications. As a reaction to this challenge, a new technique – Random Indexing – has emerged, that is a vector-based approach for extracting semantically similar words from the co-occurrence statistics of the words in textual data [8]. The technique can be applied for data sets of very big dimensionality, e.g. Linked Life Data[4], Wikipedia[5], and other global data repositories. The high computational expense of finding similarities in such big data sets is thus of great challenge for efficient utilization of high-performance computing resources. The statistical semantics methods based on Random Indexing have found a wide application within the tasks of searching and reasoning on a Web scale. Prominent examples are query expansion and subsetting. Query expansion is extensively used in Information Retrieval with the aim to expand the document collection, which is returned as a result to a query thus covering the larger portion of the documents. Subsetting (also known as selection), on the contrary, deprecates the unnecessary items from a data set in order to achieve faster processing. Both presented problems are complementary, as can change properties of a query process to best adapt it to the search needs of the agent, and are quite computationally expensive.

The pilot use case considered in this paper is a random indexing application based on the Airhead Semantic Spaces library [9]. The library is used for processing text corpora and mapping of semantic representations for words onto high dimensional vectors. The main challenge of the application is that the computation time increases linearly with the size of the word base and is extremely high for the real world data involving several billions of entries. Moreover, the requirements to the hardware resources (e.g., RAM, disc space etc.) increase according to the data set's dimensionality as well. The latter mostly prevents the efficient processing of large data sets on the currently available non-parallel computing architectures. For example, search over the LLD repository, which consolidates over 4 billion RDF statements for various sources covering the biomedical domain, can take up to months of CPU time.

Application of distributed-memory parallelization techniques (such as MPI) is thus straightforward for leveraging large data sources for Semantic Web applications performing Random Indexing.

---

[3] http://linkeddata.org
[4] http://linkedlifedata.com
[5] http://wikipedia.org

4   Parallelization Approach

The main idea of parallelization for Semantic Web Applications lies in decomposing the data access and processing operations into fragments, each of them is executed in parallel and concurrently. In case of MPI, each fragment is handled by a separate process. The more computation intensive is the parallelized algorithm and the more processes are involved, the higher is the application performance improvement thanks to the parallelization.

In case of a Random Indexing application, its most computing intensive part is a search, which is performed over all elements of the vector spaces. The search is performed over the entries of the semantic vector space according to the schema depicted in Figure 4a. All the vectors are processed independently and concurrently. The trivial parallelization can be achieved by mapping the contiguous sets of vectors in the vector space to a parallel block, each running on a compute node. The execution on single nodes is followed then by a synchronization to wait for other parallel blocks and gather the partial results of all the blocks. The division of the vectors among the parallel blocks is specified by the domain decomposition (Figure4b). The domain decomposition ensures the optimal load balancing among the compute nodes and therefore the highest performance of the parallelized algorithm.

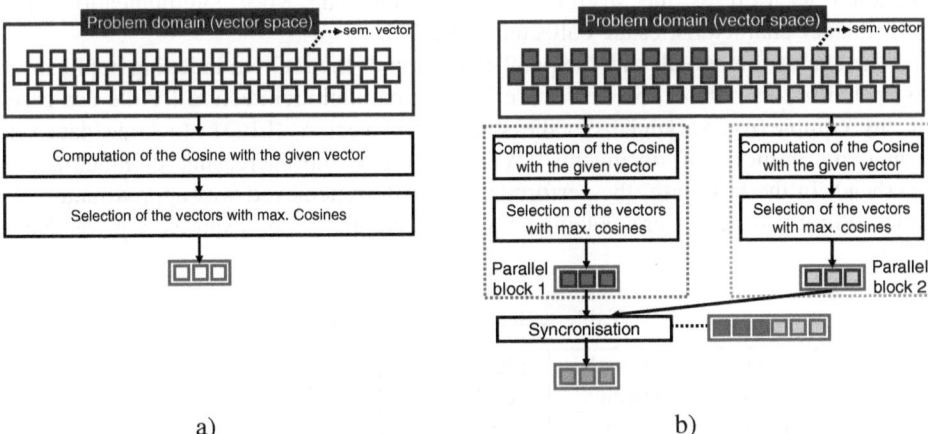

a)   b)

Fig. 4. The sequential (a) and parallelized by means of domain decomposition (b) search algorithm

The results of the search in the part of the vector space, assigned to the parallel block/process, are stored in the block's memory space and can not be accessible from another block. However, this is needed to perform the final selection among the results of each of the blocks. For this purpose, all the partial outputs might be gathered, in one of the blocks (the "root" one), where then the final selection is performed. The necessity of passing the results (n selected words) from each block to the root one as well as the following final selection prevents the parallelized application performance from super-linear scalability. Nevertheless, the optimal realization of the synchronization allows the parallel algorithm to minimize the computation overhead of this operation in total execution time.

In order to synchronize the data, MPI processes communicate by means of messages transmitted between (a so called "point-to-point" communication) or among (involving several or even all processes, a so called "collective" communication). In case of Figure 4, a collective "gather" operation can be applied for gathering all the partial outputs produced by each of the processes in one (the root) process, which then performs the final selection over the gathered values. More detailed information about point-to-point and collective communications available for MPI processes are documented in the MPI standard for Java [10].

5   Performance Impact of Parallelization

The evaluation of the actual application performance was on the Intel Nehalem cluster[6] of High Performance Computing Center Stuttgart. Configuration of 1, 2, 4, 8, and 16 distributed compute nodes were benchmarked to evaluate the scalability of the parallel realization. We also varied the size of the analyzed data sets to evaluate the algorithmic scalability as well as stability of the parallelized version (Table 1).

During the evaluation, we were concentrating on the total execution time, the time of loading the vector space from the file on the disk, the duration of the search operation as well as the overhead of the inter-node MPI communication. The performance characteristics are collected in Table 2.

The evaluation reveals that the parallel version of the random indexing application, suggested in this publication, scales well on the parallel architecture for the use cases of any complexity, varying from the sparse term vectors (LLD1) to large data sets containing millions of documents (Wiki2), despite the increasing communication overhead. In the best case, the performance speed-up achieved was approximately 27 times. Similar results were obtained for other compute architectures, e.g. on Xeon CPUs.

Table 1. Benchmarked data sets

Vector space	Nr. of entries	Size on the disk, GB	Description
LLD1	0,064 M	0,082	A subset of Linked Life Data
LLD2	0,5 M	0,65	A subset of Linked Life Data
Wiki1	1 M (low density, terms only)	1,6	A term set from Wikipedia articles
Wiki2	1 M (high density, entire documents)	16	A document set from Wikipedia articles

---

[6] http://www.hlrs.de/systems/platforms/nec-nehalem-cluster/

Table 2. Performance characteristics for the parallelized algorithm

Vector Space	Number of compute nodes	Time, s.				Speed-up
		Loading	*Search*	*MPI comm.*	*Total*	
LLD1	1	-	-	-	2	1
	2	0,75	0,5	0,04	1,61	1,25
	4	0,4	0,4	0,05	1,2	1,7
	8	0,23	0,32	0,1	1,01	1,98
	16	0,17	0,29	0,16	0,94	2,13
LLD2	1	12	6	-	19,5	1
	2	4	3,3	0,03	7,9	2,47
	4	2,4	1,8	0,23	4,6	4,24
	8	1,2	1	0,16	2,9	6,72
	16	0,6	0,7	0,2	2	9,75
Wiki1	1	18	4	-	22	1
	2	8,9	3,8	1	13,3	1,65
	4	4,6	2	0,08	7,4	2,97
	8	2,3	1,3	0,23	4,4	5
	16	1,2	0,75	0,52	2,8	7,86
Wiki2	1	309	83	-	395	1
	2	59	27	0,58	88	4,5
	4	35	13	16	59,1	6,7
	8	20	8	4	32,2	12,3
	16	10	3,7	0,16	14,6	27

6   Conclusions

The Message-Passing Interface is the most efficient technique of implementation of parallel applications, also introduced in Java. Nevertheless, for a long time this technique was underestimated in use for Java developments due to many reasons; perhaps main of them is complexity of applying process based programming model. This paper is an attempt to close the gap between Java and MPI. Presenting a common parallelization strategy, which is based on domain decomposition, we implemented the parallel version of the search operation from the Airhead library with MPI, based on the sequential code. The described technique allows any other Java developer to apply parallelism to his/her application with the minimum knowledge about MPI. For the tested application, we achieved a speed-up of almost 27 times already on 16 compute nodes, as compared with the sequential version. Moreover, the parallel implementation allowed us to perform a complex experiment on the resource, whose capacities were not enough to run the sequential version of the application. With our experience we would like to encourage other researchers to apply the MPI-based parallelization for their Java applications as well.

References

1. Kiryakov, A.: OWLIM: balancing between scalable repository and light-weight reasoner. Presented at the Developer's Track of WWW 2006, Edinburgh, Scotland, UK, May 23-26 (2006), http://www2006.org/programme/item.php?id=d15
2. Fensel, D., van Harmelen, F., Andersson, B., Brennan, P., Cunningham, H., Della Valle, E., Fischer, F., Huang, Z., Kiryakov, A., Kyung-il Lee, T., School, L., Tresp, V., Wesner, S., Witbrock, M., Zhong, N.: Towards LarKC: a Platform for Web-scale Reasoning. IEEE Computer Society Press, Los Alamitos (2008)
3. Multithreading fundamentals in Java. An on-line tutorial, http://www.geekpedia.com/tutorial289_Multithreading-Fundamentals-in-Java.html (retrieved)
4. Ranger, C., Raghuraman, R., Penmetsa, A., Bradski, G., Kozyrakis, C.: Evaluating MapReduce for Multi-core and Multiprocessor Systems. In: HPCA 2007 (2007), http://www.willowgarage.com/evaluating-mapreduce-multi-core-and-multiprocessor-systems
5. The MPI standard, http://www.mcs.anl.gov/research/projects/mpi/standard.html
6. Carpenter, B., Getov, V., Judd, G., Skjellum, T., Fox, G.: MPJ: MPI-like Message Passing for Java. Concurrency: Practice and Experience 12(11) (September 2000)
7. Baker, M., Carpenter, B., Shafi, A.: MPJ Express: Towards Thread Safe Java HPC. In: IEEE International Conference on Cluster Computing (Cluster 2006), Barcelona, Spain, September 25-28 (2006)
8. Sahlgren, M.: An introduction to Random Indexing. In: Methods and Applications of Semantic Indexing Workshop at the 7th International Conference on Termonology and Knowledge Engineering, TKE 2005, Citeseer (2005)
9. Jurgens, Stevens: The S-Space Package: An Open Source Package for Word Space Models. In: System Papers of the Association of Computational Linguistics (2010)
10. The Java API for MPI, http://www.hpjava.org/theses/shko/thesis_paper/node33.html

Making Web-Scale Semantic Reasoning More Service-Oriented: The Large Knowledge Collider

Alexey Cheptsov[1] and Zhisheng Huang[2]

[1] High-Performance Computing Center Stuttgart, Nobelstr. 19, 70569 Stuttgart, Germany
`cheptsov@hlrs.de`
[2] Free University of Amsterdam, De Boelelaan 1081, 1081 HV Amsterdam, The Netherlands
`huang@cs.vu.nl`

Abstract. Reasoning is one of the essential application areas of the modern Semantic Web. Nowadays, the semantic reasoning algorithms are facing significant challenges when dealing with the emergence of the Internet-scale knowledge bases, comprising extremely large amounts of data. The traditional reasoning approaches have only been approved for small, closed, trustworthy, consistent, coherent and static data domains. As such, they are not well-suited to be applied in data-intensive applications aiming on the Internet scale. We introduce the Large Knowledge Collider as a platform solution that leverages the service-oriented approach to implement a new reasoning technique, capable of dealing with exploding volumes of the rapidly growing data universe, in order to be able to take advantages of the large-scale and on-demand elastic infrastructures such as high performance computing or cloud technology.

Keywords: Semantic Web, Reasoning, Big Data, Distribution, Parallelization, Performance.

1 Introduction

The large- and internet-scale data applications are the primary challenger for the Semantic Web, and in particular for reasoning algorithms, used for processing exploding volumes of data, exposed currently on the Web. Reasoning is the process of making implicit logical inferences from the explicit set of facts or statements, which constitute the core of any knowledge base. The key problem for most of the modern reasoning engines such as Jena [1] or Pellet [2] is that they can not efficiently be applied for the real-life data sets that consist of tens, sometimes of hundreds of billions of triples (a unit of the semantically annotated information), which can correspond to several petabytes of digital information. Whereas modern advances in the Supercomputing domain allow this limitation to be overcome, the reasoning algorithms and logic need to be adapted to the demands of rapidly growing data universe, in order to be able to take advantages of the large-scale and on-demand infrastructures such as high performance computing or cloud technology. On the other hand, the algorithmic principals of the reasoning engines need to be reconsidered as well in order to allow for very large volumes of data. Service-oriented architectures

A. Haller et al. (Eds.): WISE 2011 and 2012 Combined Workshops, LNCS 7652, pp. 13–26, 2013.

(SOA) can greatly contribute to this goal, acting as the main enabler of the newly proposed reasoning techniques such as incomplete reasoning [3]. This paper focuses on a service-oriented solution for constructing Semantic Web applications of a new generation, ensuring the drastic increase of the scalability for the existing reasoning applications, as elaborated by the Large Knowledge Collider (LarKC)[1] EU project.

The paper is organized as follows. In Section 2, we collect our consideration towards enabling the large-scale reasoning. In Section 3, we discuss LarKC – a service-oriented platform for development of fundamentally new reasoning application, with much higher scalability barriers as by the existing solutions. In Section 4, we introduce some successful applications implemented with LarKC, such as BOTTARI – the Semantic Challenge winner in 2011. In Section 5, we discuss our conclusions and highlight the directions for future work in highly scalable semantic reasoning.

2   Semantic Reasoning on the Web Scale

2.1   From Web to the Semantic Web

The Web as it is seen by the users "behind the browser" has traditionally been one of the most successful examples of the SOA realization. The possibility to transform the application's business logic into a set of the linked services supplied with the transparent access to those services over standardized protocols such as HTTP was a key asset for tremendous wide-spread of the Internet worldwide. However the possibility to organize business relationship between the data located on several hosts had been extremely poor. The research seeking for a concept of applying a data model on the Web scale resulted in the Semantic Web – the later advance of the Web, which offers a possibility to extend the Web-enabled data with the annotation of their semantics, thus making the context in which the data is used meaningful for the applications [4]. Nowadays, there are several existing well-established standards for annotation of data web-wide, such as for example Resource Description Framework (RDF)[2] schema.

The practical value of the Semantic Web is that it enables development of applications that can handle complex human queries based not only on the value of the analyzed data, but also on its meaning. Promotion of such platforms as (Friends-of-a-Friend) FOAF[3] at the early stages of the Semantic Web has forced a lot of data providers to actively expose and interlink their data on the Web, which resulted in many problem-oriented data repositories, as for example Linked Life Data (LLD)[4], which is a collection of the data for biomedical domain; alone the LLD dataset comprises over one billion web resources presented in RDF. On the other hand, social networks like Twitter or Facebook encourage people to upload there personal data as well, thus drastically increasing the weight of the digital information on the Web.

---

[1] http://www.larkc.eu/
[2] http://www.w3.org/RDF/
[3] http://www.foaf-project.org/
[4] http://linkedlifedata.com/

2.2   Semantic Reasoning

Thanks to the ability to offer the structured data as the Web content, the Semantic Web has become de-facto an indispensable aspect of the human's everyday life. The application areas of the modern Semantic Web spawn a wide range of domains, from social networks to large-scale Smart Cities projects in the context of the future internet However, data processing in such applications goes far beyond a simple maintenance of the collection of facts; based on the explicit information, collected in datasets, and simple rule sets, describing the possible relations, the implicit statements and facts can be acquired from those datasets. For example, supposed that bulldogs are dogs, and cats hate dogs, cats must also hate bulldogs, which is however not explicitly stated but rather inferred from the content.

Many data collections as well as application built on top of them allow for rule-based inferencing to obtain new, more important facts. The process of inferring logical consequences from a set of asserted facts, specified by using some kinds of logic description languages (e.g., RDF/RDFS and OWL[5]), is in focus of semantic reasoning. The goal is to provide a technical way to determine when inference processes is valid, i.e., when it preserves truth. This is achieved by the procedure which starts from a set of assertions that are regarded as true in a semantic model and derives whether a new model contains provably true assertions.

2.3   Big Data Challenge and New Reasoning Approaches

The latest research on the Internet-scale Knowledge Base technologies, combined with the proliferation of SOA infrastructures and cloud computing, has created a new wave of data-intensive computing applications, and posed several challenges to the Semantic Web community. As a reaction on these challenges, a variety of reasoning methods have been suggested for the efficient processing and exploitation of the semantically annotated data. However, most of those methods have only been approved for small, closed, trustworthy, consistent, coherent and static domains, such as synthetic LUBM sets. Still, there is a deep mismatch between the requirements on the real-time reasoning on the Web scale and the existing efficient reasoning algorithms over the restricted subsets.

Whereas unlocking the full value of the scientific data has been seen as a strategic objective in the majority of ICT- related scientific activities in EU, USA, and Asia [5], the "Big Data" problem has been recognized as the primary challenger in semantic reasoning [6][7]. Indeed, the recent years have seen a tremendous increase of the structured data on the Web with scientific, public, and even government sectors involved. According to one of the recent IDC reports [8], the size of the digital data universe has grown from about 800.000 Terabytes in 2009 to 1.2 Zettabytes in 2010, i.e. an increase of 62%. Even more tremendous growth should be expected in the future (up to several tens of Zettabytes already in 2012, according to the same IDC report [8]).

---

[5] http://www.w3.org/TR/owl-ref/

The "big data" problem makes the conventional data processing techniques, also including the traditional semantic reasoning, substantially inefficient when applied for the large-scale data sets. On the other hand, the heterogeneous and streaming nature of data, e.g. implying structure complexity [9], or dimensionality and size [10], makes big data intractable on the conventional computing resource [11]. The problem becomes even worse when data are inconsistent (there is no any semantic model to interpret) or incoherent (contains some unclassifiable concepts) [12].

The broad availability of data coupled with increasing capabilities and decreasing costs of both computing and storage facilities has led the semantic reasoning community to rethink the approaches for large-scale inferencing [13]. Data-intensive reasoning requires a fundamentally different set of principles than the traditional mainstream Semantic Web offers. Some of the approaches allow for going far beyond the traditional notion of absolute correctness and completeness in reasoning as assumed by the standard techniques. An outstanding approach here is interleaving the reasoning and selection [14]. The main idea of the interleaving approach (see Figure 1a) is to introduce a selection phase so that the reasoning processing can focus on a limited (but meaningful) part of the data, i.e. perform incomplete reasoning.

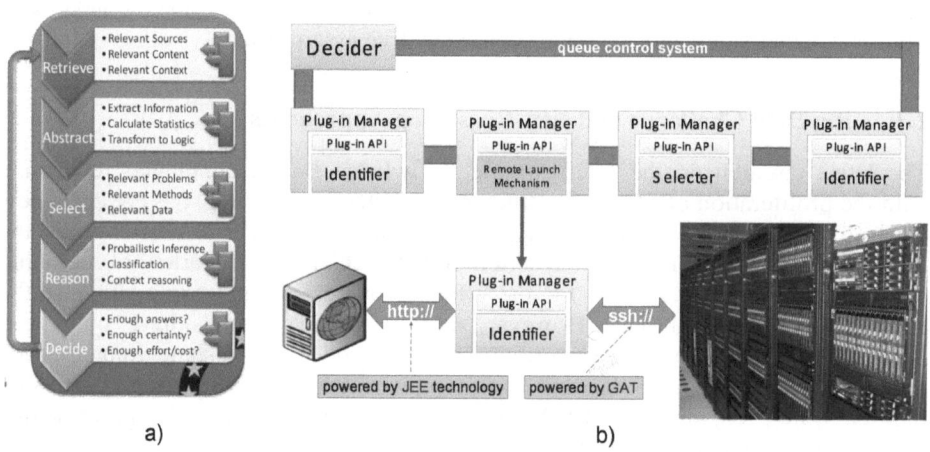

Fig. 1. Incomplete reasoning, the overall schema (a) and a service-oriented vision (b)

2.4   SOA Aspect in Semantic Reasoning

As we have discussed before, the standard reasoning methods are not valid in the existing configurations of the Semantic Web. Some approaches, such as incomplete reasoning, offer a promising vision how a reasoning application can overcome the "big data" limitation, e.g. by interleaving the selection with the reasoning in a single "workflow", as shown in Figure 1a. However the need of combining several techniques within a single application introduces new challenges, for example related to ensuring the proper collaboration of team of experts working on a concrete part of the workflow, either it is identification, selection, or reasoning. Another challenge might be the adoption of the already available solutions and reusing them in the newly developed applications, as for example applying selection to the JENA reasoner [2], whose original software design doesn't allow for such functionality. The SOA approach can help eliminate many of the drawbacks on the way towards creating new, service-based reasoning applications. Supposed that each of the construction blocks shown in Figure 1a is a service, with standard API that ensures easy interoperability with the other similar services, quite a complex application can be developed by a simple combination of those services in a common workflow (see Figure 1b).

Although the workflow concept is not new for the semantic reasoning [15], there was quite a big gap in realizing the single steps of the reasoning algorithms (Figure 1b) as a service. This was due to many reasons, among them complexity of the data dependency management, ensuring interoperability of the services, heterogeneity of the service's functionality. Realizing a system where a massive number of parties can expose and consume services via advanced Web technology was also a research highlight for Semantic Web. An example of very successful research on offering a part of the semantic reasoning logic as a service is the SOA4ALL[6] project, whose main goal was to study the service abilities of development platforms capable of offering semantic services. Several useful services wrapping such successful reasoning engines as IRIS and several others had been developed in the frame of this project. Nevertheless, the availability of such services is only an intermediate step towards offering reasoning as a service, as a lot of efforts were required to provide interoperability of those services in the context of a common application. Among others, a common platform is needed that would allow the user to seamlessly integrate the service by annotating their dependencies, manage the data dependencies intelligently, being able to specify parts of the execution that should be executed remotely, etc.

An outstanding effort to develop such a platform was performed in the LarKC (Large Knowledge Collider) [16] project. In the following sections, we discuss the main ideas, solutions, and outcomes of this project.

3   Large Knowledge Collider – Making the Semantic Reasoning More Service Oriented

3.1   Objectives and Concepts

In order to facilitate the technology for creation of trend-new applications for large-scale reasoning, several leading Semantic Web research organizations and technological companies have joined their efforts around the project of the Large Knowledge Collider (LarKC), supported by the European Commission. The mission of the project was to set up a distributed reasoning infrastructure for the Semantic Web community, which should enable application of reasoning far beyond the currently recognized scalability limitations [17], by implementing the interleaving reasoning approach. The current and future Web applications that deal with "big data" are in focus of LarKC.

---

[6] http://www.soa4all.eu/

To realize this mission, LarKC has created an infrastructure that allows construction of plug-in-based reasoning applications, following the interleaving approach, facilitated by incorporating interdisciplinary techniques such as inductive, deductive, incomplete reasoning, in combination with the methods from other knowledge representation domains such as information retrieval, machine learning, cognitive and social psychology. The core of the infrastructure is a platform – a software framework that facilitates design, testing, and exploitation of new reasoning techniques for development of large-scale applications. The platform does this by providing means for creating very lightweight, portable and unified services for data sharing, accessing, transformation, aggregation, and inferencing, as well as means for building Semantic Web applications on top of those services. The efficiency of the services is ensured by providing a transparent access to the underlying resource layer, served by the platform, involving high performance computing, storage, and cloud resources, and in the other way around, providing performance analysis and monitoring information back to the user. The platform is built in a distributed, modular, and open source fashion. Moreover, the platform offers means for building and running applications across those plug-ins, provide them a persistent data layer for storing data, facilitate parallel execution of large-scale data operations on distributed and high-performance resources [18].

The two main issues solved by LarKC are development of a reasoning application combining solutions and techniques coming from diverse domains of the Semantic Web and Computer Science disciplines (e.g. High Performance Computing), and ensuring the requested QoS requirements, in particular by targeting the modern e-Infrastructures such as grid and cloud environments.

Guided by the preliminary goal to facilitate incomplete reasoning, LarKC has evolved in a unique platform, which can be used for development of a wide range of semantic web applications, following the SOA paradigm. The sections below discuss the main functional properties and features of the LarKC platform.

3.2   Architecture Overview

The LarKC's design has been guided by the primarily goal to build a scalable platform for distributed high performance reasoning. Figure 2 shows a conceptual view of the LarKC platform's architecture and the proposed development life-cycle. The architecture was designed to holistically cover the needs of the three main categories of users – semantic service (plug-in) developers, application (workflow) designers, and end-users internet-wide. The platform's design ensures a trade-off between the flexibility and the performance of applications in order to achieve a good balance between the generality and the usability of the platform by each of the categories of users. In the following, we introduce some of the key concepts of the LarKC architecture and discuss the most important platform's services and tools for them.

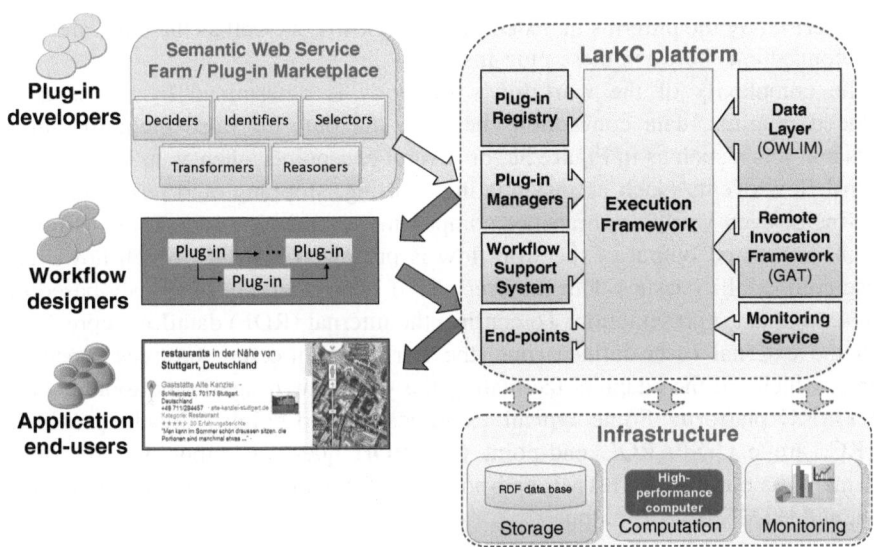

Fig. 2. Architecture of LarKC

1)   Plug-ins

Plug-ins are standalone services implementing some specific parts of the reasoning logic as discussed previously, whether it is selection, identification, transformation, or reasoning algorithm, see more at [17]. In fact, plug-ins can implement much broader functionality as foreseen by the incomplete reasoning schema (Figure 1), hence enabling the LarKC platform to target much wider Semantic Web user community as originally targeted, e.g. for machine learning or knowledge extraction. The services are referred as plug-ins because of their flexibility and ability to be easily integrated, i.e. plugged into a common workflow and hence constitute a reasoning application. To ensure the interoperability of the plug-ins in the workflows, each plug-in should implement a special plug-in API, based on the annotation language [19]. Most essentially, the API defines the RDF schema (set of statements in the RDF format) taken as input and produced as output by each of the plug-ins. The plug-in development is facilitated by a number of special wizards, such as Eclipse IDE wizard or Maven archetype for rapid plug-in prototyping. The ready-to-use plug-ins are uploaded and published on the marketplace – a special web-enabled service offering a centralized, web-enabled repository store for the plug-ins[7].

2)   Workflows

The workflow designers get access to the Marketplace in order to construct a workflow from the available plug-ins, combined to solve a certain task. In terms of LarKC, workflow is a reasoning application that is constructed of the (previously developed and uploaded on the Marketplace) plug-ins. The workflow's topology is

---

[7] Visit the LarKC Plug-in Marketplace at
http://www.larkc.eu/plug-in-marketplace/

characterised by the plug-ins included in the workflow as well as the data- and control flow connections between these plug-ins.

The complexity of the workflow's topology is determined by the number of included plug-ins, data connections between the plug-ins (also including multiple splits and joins, such as in Figure 3a, or several end-points, such as in Figure 3b), and control flow events (such as instantiating, starting, stopping, and terminating single plug-ins or even workflow branches comprising several plug-ins). Same as for plug-ins, the input and output of the workflow is presented in RDF, which however can cause compatibility issues with the user's GUI, which are not obviously based on an RDF-compliant representation. To confirm the internal (RDF) dataflow representation with the external (user-defined) one, the LarKC architecture foresees special end-points, which are the adapters facilitating the workflow usage in the tools outside of the LarKC platform. Some typical examples of end-points, already provided by LarKC, are e.g. SPARQL end-point (SPARQL query as input and set of RDF statements as output) and HTML end-point (HTTP request without any parameters as input and HTML page as output).

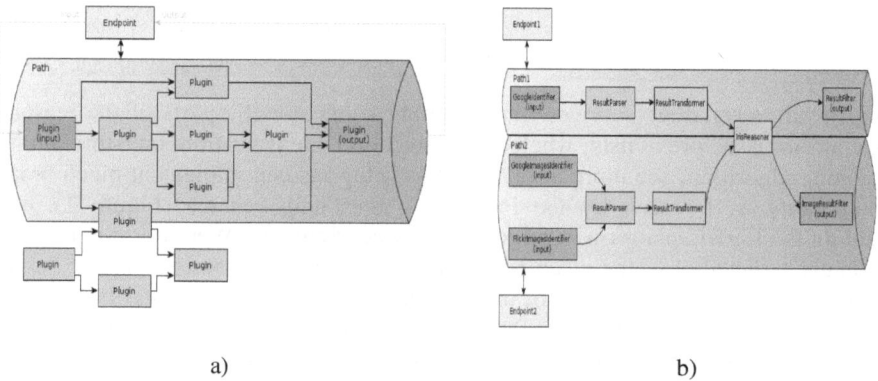

a)   b)

Fig. 3. Examples of LarKC application workflows: a) with non-trivial branched dataflow (containing multiple splits/joins), b) with multiple end-points

For the specification of the workflow configuration, a special RDF schema was elaborated for LarKC, aiming at simplification of the annotation efforts for the workflow designers. The schema also allows for specification of the additional features, such as remote plug-in execution, and can be used for tuning the front-end graphical interfaces of the applications to adapt them to the user needs. Listing 1 shows a simple example of the LarKC workflow annotation. The simple workflow consists of one plug-in (*plugin 1*), which is running on an external tomcat server (*host1*) and accessible over a SPARQL end-point (*ep*).

```
1
2 # Define plug-ins
3 _:plugin1 a <urn:eu.larkc.plugin.LLDReasoner> .
4 _:plugin1 a <urn:eu.larkc.FilteringPlugin.FilteringPlugin>
5 _:plugin1 larkc:runsOn _:host1 .
6
7    # Define hosts
8   _:host1 a <urn:eu.larkc.host.Tomcat> .
9   _:host1 larkc:hostType larkc:JEE .
0   _:host1 larkc:jeeUri <http://angelina.hlrs.de:8080> .
1
2 # Define a path to set the input and output of the workflow
3 _:path a larkc:Path .
4 _:path larkc:hasInput _:plugin1 .
5 _:path larkc:hasOutput _:plugin1 .
6
7 # Connect an endpoint to the path
8 _:ep a <urn:eu.larkc.endpoint.sparql.SparqlEndpoint> .
9 _:ep larkc:links _:path .
```

Listing 1. Example of an RDF workflow annotation

Creation of the workflow specification is pretty much simplified by using intuitive GUI's developed for LarKC, which enable constructing and executing workflows, such as Workflow Designer.

3)   Applications
Workflows are already standalone applications that can be submitted to the platform and executed by means of such tools as Workflow Designer discussed above. Nevertheless, workflows can also be wrapped into much more powerful user interfaces, adapted to the needs of the targeted end-user communities, e.g. Urban Computing, and using LarKC as a back-end engine. The SO approach makes possible hiding the complexity of the LarKC platform, by enabling its whole power to the end-users through such interfaces. We discuss some of the most successful examples of the LarKC applications in Section 4.

4)   Platform services
All above-described activities related to plug-in creation, workflow design, and application development are facilitated by an extensive set of the platform services, as shown in Figure 2.
Execution Framework is the "control centre" of the LarKC Platform. It is responsible for the services related to the plug-in (Plug-in Registry, Plug-in Managers), workflow (Workflow Support System, Workflow Designer), and application (End-points) support. It also provides a set of fundamental services indispensable for the organization of the data management (Data Layer), distributed execution (Remote Invocation Framework), and performance monitoring (Monitoring Services).
Plug-in Registry is a service that allows the platform to load the plug-ins as well as the external libraries needed by them to the internal plug-in knowledge base, where the plug-ins can be instantiated from when constructing and executing the workflow.
Plug-in Managers facilitate the integration of plug-ins within a workflow and the management of their execution. The managers allow a plug-in to be executed either locally or in a distributed fashion. The latter is facilitated by the remote invocation

framework that is based on Grid Access Toolkit (GAT) and support several categories of host, also in view of the Cloud paradigm.

Data Layer is a simplified realization of OWLIM – a high-performance RDF data base that supports the plug-ins and applications with respect to storage, retrieval (including streaming), and lightweight inference on top of large volumes of RDF data. In particular, Data Layer is used for storing the data passed between the plug-ins, so that only a reference is passed; this reveals the plug-ins from the need of handling the RDF data and hence make them applicable for large data volumes stored in the Data Layer.

5)   Infrastructure

With regard to the infrastructure layer, LarKC acts as a middleware that facilitates the successful application deployment and execution on the available resource base. The LarKC platform offers the plug-ins an abstraction layer, facilitated by the plug-in API, that allows applications based on those plug-ins to abstract from the specific resource layer properties, such as operating system, number of compute cores (for shared memory) or nodes (for distributed memory parallel systems), etc., hence making the deployment process as transparent as possible. This is facilitated by several know-how solutions for distributed execution, parallelization, and monitoring.

Distributed execution is the key feature of the LarKC execution model. It allows a plug-in to be executed on the resource that is remote with regard to the one where the platform is running. Standard cases where the applications can benefit from the distributed execution include but not restrict shipping the execution closer to the data being processed, running a part of the workflow on the resource that ensures better performance but forbids the full deployment of the LarKC platform, e.g. production high performance supercomputers, etc.

Monitoring is the essential feature of the LarKC platform that allows plug-ins to be (automatically) instrumented to produce some important metrics about their execution, e.g. execution time (performance), or size of the processed data (throughput). Those characteristics can be collected from different execution configurations and used for identifying possible bottlenecks or just collecting some interesting for the user statistics. The visualization tools are provided by the platform as well, so a very little efforts is needed to get the complete trace of the application run.

4   Success Stories and Application Examples

LarKC is the technology that not only enables the large-scale reasoning approach for the already existing applications, but also facilitates their rapid prototyping with low initial investments, leveraging the SOA approach through the solutions discussed in Section III. Furthermore, LarKC delivers a complete eco-system where the researches from very different domains can team up in order to develop new challenging mashup-applications, hence having a dramatic impact on a lot of problem domain. Below we describe some of the most prominent pilot applications developed with LarKC in 2010-2011.

4.1   BOTTARI

BOTTARI [20] is a location-based mobile application that leverages a place of interest recommendation system to support people who find themselves in the new place, which they are not familiar with. The application's front-end is implemented at Android tablets, whereas the back-end is served by LarKC. BOTTARI is collecting relevant information from social media networks such as Twitter and blog posts, elaborates it and provides contextualized suggestions. At the current stage, the application was implemented for one of the most popular touristic districts in Seoul, South Korea. The recommendations given by BOTTARI include places of interest nearby the current location of the user, reputation ranking of the suggested places according to the other users' feedback, identification of the most interesting place fitting well the user's profile. To the main innovations of BOTTARI can be referred offering a location-based service through a simple and intuitive interface, advanced semantic features, and hiding the complexity of reasoning from the end-user. BOTTARI become the winner of the International Semantic Web challenge 2011.

4.2   WebPIE

WebPIE (Web-scale Parallel Inference Engine) [21] is a MapReduce-based parallel distributed RDFS/OWL inference engine. Being implemented as a LarKC plug-in, WebPIE can be used for materialization of an RDF graph expressed in the OWL Horst semantics, which is required by a lot of semantic reasoning workflows. The workflows that use WebPIE can take advantages of the distributed and parallel reasoning, facilitated by the underlying MapReduce implementation with Hadoop. Thanks to the parallel implementation, WebPIE vastly outperforms all the existing inference engines when comparing supported language expressivity, maximum data size and inference speed. In LarKC, WebPIE can easily be integrated in any forward chaining reasoning workflow and thus improve its scalability. The distributed execution framework takes care of the execution of the WebPIE reasoner on a machine that can take full advantages of the parallel realization, e.g. a cluster of workstations or a parallel supercomputer. The WebPIE research won the first scalability prize at the IEEE Scale Challenge in 2010.

4.3   Genode-Wide Association Study

Genome-wide association study (GWAS) is a research domain aiming to identify common genetic factors that influence health and disease apparition. GWAS use bio-probes (gene markers) to look for higher levels of association between genes in a diseased subject as opposed to controls. The large numbers of markers mean that huge numbers of samples are needed to achieve sufficient statistical power. Semantic Web helps the GWAS researchers apply common statistical models to raw experimental data to find the relevance of each marker, and then rank them in order of relevance to the disease. Only the genes that are close to the top few markers are then studied in more depth by conventional techniques, to narrow the problem and achieve better results. This last bit is expensive, and improving rankings could improve both the efficiency and the economics of the technique. The WHO's cancer research unit, IARC, has chosen LarKC as the technology to combine prior knowledge about a gene with experimental data, thus improving statistical power [22]. The modular nature of LarKC plug-ins allowed for combination of those techniques with the modern advances of the Statistical Semantics as random indexing, term frequency inverse document frequency, or term expansion using UMLS. This allowed the researchers to scale knowledge discovery across the large amounts of biomedical knowledge now encoded in the data- and bibli-ome, and to apply it to the millions of data points in a typical GWAS.

5   Conclusion

LarKC is very promising platform for creation of new-generation semantic reasoning applications. The LarKC's main value is twofold. On the one hand, it enables a new approach for large-scale reasoning based on the technique for interleaving the identification, the selection, and the reasoning phases. On the other hand, through over the project's life time (2008-2011), LarKC has evolved in an outstanding, service-oriented platform for creating very flexible but extremely powerful applications, based on the plug-in's realization concept. The LarKC plug-in marketplace has already comprised several tens of freely available plug-ins, which implement new know-how solutions or wrap existing software components to offer their functionality to a much wider range of applications as even originally envisioned by their developers. Moreover, LarKC offers several additional features to improve the performance and scalability of the applications, facilitated through the parallelization, distributed execution, and monitoring platform. LarKC is an open source development, which encourages collaborative application development for Semantic Web. Despite being quite a young solution, LarKC has already established itself as a very promising technology in the Semantic Web world. Some evidence of its value was a series of Europe- and world-wide Semantic Web challenges won by the LarKC applications. It is important to note that the creation of LarKC applications, including the ones discussed in the paper, was also possible and without LarKC, but would have required much more (in order of magnitude) development efforts and financial investments.

We believe that the availability of such platform as LarKC will make a lot of developers to rethink their current approaches for semantic reasoning towards much wider adoption of the service-oriented paradigm. Another added value of LarKC is a number of very promising future researches that will be done as LarKC's spin-offs, including streaming data support, decision making in large systems, and many others. Among others, a lot of challenges are introduced by Smart Cities applications, which provide static data pools of Petabyte size range as well as deliver Terabytes of new dynamically-acquired data on the daily basis. We would be interested to apply LarKC to such challenging application scenarios and evaluate its ability to meet the real-time requirements of such large-scale systems.

Acknowledgments. We thank the consortium of the LarKC project (co-funded by the European Commission under the grant agreement ICT-FP7-215535) for materials provided.

References

[1] Sirin, E., Parsia, B., Cuenca Grau, B., Kalyanpur, A., Katz, Y.: Pellet: a practical owl-dl reasoner. Journal of Web Semantics,
http://www.mindswap.org/papers/PelletJWS.pdf

[2] McCarthy, P.: Introduction to Jena. IBM developerWorks,
http://www.ibm.com/developerworks/xml/library/j-jena/

[3] Fensel, D., van Harmelen, F.: Unifying Reasoning and Search to Web Scale. IEEE Internet Computing 11(2), 94–96 (2007)

[4] Broekstra, J., Klein, M., Decker, S., Fensel, D., van Harmelen, F., Horrocks, I.: Enabling knowledge representation on the Web by extending RDF schema. In: Proceedings of the 10th International Conference on World Wide Web (WWW 2001), pp. 467–478. ACM (2001)

[5] High Level Expert EU Group, Riding the wave - How Europe can gain from the rising tide of scientific data, Final report (October 2010), http://ec.europa.eu/information_society/newsroom/cf/document.cfm?action=display&doc_id=707

[6] Thompson, B., Personick, M.: Large-scale mashups using RDF and bigdata. In: Semantic Technology Conference (2009)

[7] Hustadt, U., Motik, B., Sattler, U.: Data Complexity of Reasoning in Very Expressive Description Logics. In: Proc. IJCAI 2005, Edinburgh, pp. 466–471 (2005)

[8] McKendrick, J.: Size of the data universe: 1.2 zettabytes and growing fast. ZDNet

[9] Della Valle, E., Ceri, S., van Harmelen, F., Fensel, D.: It's a streaming world! Rreasoning upon rapidly changing information. IEEE Intelligent Systems 24(6), 83–89 (2009)

[10] Fensel, D., van Harmelen, F.: Unifying Reasoning and Search to Web Scale. IEEE Internet Computing 11(2), 95–96 (2007)

[11] Cheptsov, A., Assel, M.: Towards High Performance Semantic Web – Experience of the LarKC Project. Inside - Journal of Innovatives Supercomputing in Deutschland 9(1) (Spring 2011)

[12] Huang, Z., van Harmelen, F., Teije, A.: Reasoning with inconsistent ontologies. In: Proceedings of the International Joint Conference on Artificial Intelligence, IJCAI 2005, pp. 454–459 (2005)

[13] Bozsak, E., Ehrig, M., Handschuh, S., Hotho, A., Maedche, A., Motik, B., Oberle, D., Schmitz, C., et al.: KAON - Towards a Large Scale Semantic Web. In: Bauknecht, K., Tjoa, A.M., Quirchmayr, G. (eds.) EC-Web 2002. LNCS, vol. 2455, pp. 304–313. Springer, Heidelberg (2002)

[14] Huang, Z.: Interleaving Reasoning and Selection with Semantic Data. In: Proceedings of the 4th International Workshop on Ontology Dynamics (IWOD 2010), ISWC 2010 Workshop (2010)

[15] Deelman, E., Gannon, D., Shields, M., Taylor, I.: Workflows and e-Science: An overview of workflow system features and capabilities. Future Generation Computer Systems 25(5) (2009)

[16] Fensel, D., van Harmelen, F., Andersson, B., Brennan, P., Cunningham, H., Della Valle, E., Fischer, F., Huang, Z., Kiryakov, A., Lee, T., Schooler, L., Tresp, V., Wesner, S., Witbrock, M., Zhong, N.: Towards LarKC: A Platform for Web-Scale Reasoning. In: Proceedings of the 2008 IEEE international Conference on Semantic Computing ICSC, pp. 524–529. IEEE Computer Society (2008)

[17] Assel, M., Cheptsov, A., Gallizo, G., Celino, I., Dell'Aglio, D., Bradeško, L., Witbrock, M., Della Valle, E.: Large knowledge collider: a service-oriented platform for large-scale semantic reasoning. In: Proceedings of the International Conference on Web Intelligence, Mining and Semantics, WIMS 2011 (2011)

[18] Assel, M., Cheptsov, A., Gallizo, G., Benkert, K., Tenschert, A.: Applying High Performance Computing Techniques for Advanced Semantic Reasoning. In: Cunningham, P., Cunningham, M. (eds.) eChallenges e-2010 Conference Proceedings. IIMC International Information Management Corporation (2010)

[19] Roman, D., Bishop, B., Toma, I., Gallizo, G., Fortuna, B.: LarKC Plug-in Annotation Language. In: Proceedings of The First International Conferences on Advanced Service Computing – Service Computation 2009 (2009)

[20] Celino, I., Dell'Aglio, D., Della Valle, E., Huang, Y., Lee, T., Kim, S., Tresp, V.: Towards BOTTARI: Using Stream Reasoning to Make Sense of Location-Based Micro-Posts. In: García-Castro, R., Fensel, D., Antoniou, G. (eds.) ESWC 2011. LNCS, vol. 7117, pp. 80–87. Springer, Heidelberg (2012)

[21] Urbani, J., Kotoulas, S., Maassen, J., van Harmelen, F., Bal, H.: OWL reasoning with WebPIE: calculating the closure of 100 billion triples. In: Aroyo, L., Antoniou, G., Hyvönen, E., ten Teije, A., Stuckenschmidt, H., Cabral, L., Tudorache, T. (eds.) ESWC 2010, Part I. LNCS, vol. 6088, pp. 213–227. Springer, Heidelberg (2010)

[22] Johansson, M., Li, Y., Wakefield, J., Greenwood, M.A., Heitz, T., Roberts, I., Cunningham, H., Brennan, P., Roberts, A., Mckay, J.: Using Prior Information Attained From The Literature To Improve Rankin. In: Genome-Wide Association Studies (2009)

Towards Integrating Emotion into Intelligent Context

Philip Moore[1,*], Cain Evans[1], and Hai V. Pham[2]

[1] School of Computing, Birmingham City University, Birmingham, UK
[2] Graduate School of Science and Engineering, Ritsumeikan University, Japan
ptmbcu@gmail.com

Abstract. Context-aware systems have traditionally employed a limited range of contextual data. While research is addressing an increasingly broad range of contextual data, the level of intelligence generated in context-aware systems is restricted by the failure to effectively implement emotional response. This paper considers emotion as it relates to context and the application of computational intelligence in context-aware systems. Following an introduction, personalization and the computational landscape is considered and context is introduced. Computational intelligence and the relationship to the Semantic Web is discussed with consideration of the nature of knowledge and a brief overview of knowledge engineering. Cognitive conceptual models and semiotics are introduced with a comparative analysis and approaches to implementation. Ongoing research with illustrative 'next generation' intelligent context-aware systems incorporating emotional responses are briefly considered. The paper concludes with a discussion where the challenges and opportunities are addressed; there are closing observations, consideration of future directions for research, and identification of open research questions.

Keywords: Intelligent Context, Emotion, Knowledge, Conceptual Models, Semiotics, *Kansei* Engineering, Computational Intelligence.

1   Background

Emotion represents an important element in an individual's response mechanism to a range of stimuli as emotional responses are fundamental to an individual's reaction to changing environments and social situations. Over time people develop an individual view of the world viewed through their personal dynamic perceptual filter created based on observation and experience; emotional response is the result of this individual view of the world. The factors that relate to emotional response form an important component in an individual's context which forms the basis upon which personalization and targeted service provision is achieved.

This paper presents a discussion around emotion and emotional response as it relates to the definition of context and its application in intelligent context-aware systems. Intelligence is a complex topic and can be viewed from two perspectives: (1) Human intelligence; in computational terms this relates to the Open World

---

* Corresponding author.

A. Haller et al. (Eds.): WISE 2011 and 2012 Combined Workshops, LNCS 7652, pp. 27–40, 2013.

Assumption (OWA), and (2) Computational intelligence (in computational terms this relates to the Closed World Assumption (CWA). While in an ideal world the OWA may be applied, in practice the CWA is the only currently realistic approach to achieve [albeit limited] intelligence in context-aware systems. The challenge lies in effective use of contextual information which includes emotional responses based on the OWA; the opportunities lie in the results that may be achieved if the OWA can be effectively implemented and computational intelligence which more closely approximates to human intelligence realized.

The paper is structured as follows: section 2 sets out an overview of personalization and the computational landscape with context and the nature of contextual information considered in section 3. Computational intelligence and the relationship to the Semantic Web is discussed in section 4. Section 5 presents a discussion on the nature of knowledge with consideration of explicit and tacit knowledge with a brief overview of knowledge engineering. In section 5 there is a discussion around conceptual models and Semiotics with a comparative analysis of the relationship between knowledge, conceptual models, and Semiotics. Implementation is addressed in section 6. Current research projects which represent our view of the 'next generation' of intelligent context-aware systems incorporating emotional responses are briefly considered in section 7. The paper concludes with a discussion where the challenges and opportunities are addressed; there are closing observations, consideration of future directions for research, and identification of open research questions.

2   Personalization and the Computational Landscape

Personalization on demand has gained traction driven by the demands of computer mediated P2P and B2B interactions. Concomitant with these developments is the revolution in the capability and ubiquity of mobile technologies (generally implemented in large scale distributed systems and ad-hoc wireless networks) and the growing use of Web 2.0 technologies in data intensive systems [20].

The traditional paradigm of centralized 'internet' (also termed an Intranet) based networks has been largely replaced by a new distributed 'Internet' based communications paradigm characterized by distributed ad-hoc wireless networks which are increasingly accessing geospatial, temporal, and cloud-based systems [13][23]. These developments have resulted in systems in which a user's context may be static or, in mobile systems highly dynamic. Mobile systems incorporate a diverse range of geographically diverse infrastructures with a potentially large user base and a broad range of fixed and wearable mobile devices [20].

The 'Internet' systems paradigm is frequently characterized by Large Scale Distributed Systems (LSDS) [13][14][23] in which interactions between infrastructure components and individual users (more accurately the users devices) are inherently complex; the complexity increasing exponentially as nodes are added to and removed from in the system dynamically. This complexity places great strain on communication systems with issues in the management of the interactions and the ability to enable personalization on demand which requires both user and network infrastructure knowledge to effectively target service provision.

LSDS are inherently context-aware, context performing an increasingly important role. The rule-based approach presented in this article is posited as an effective approach to enable: (1) targeted service provision in complex data intensive systems, (2) the processing of data from a diverse range of geographically and technologically divers sources in Wide Area Networks (WAN), Local Area Networks (LAN), and Personal Area Networks (PAN, (3) the capability to handle the inherent complexity of context, (4) the ability to manage CS, and (5) the capability to realize predictable decision- support under uncertainty [18][20][21].

As discussed in this paper emotion (more accurately emotional response) can be viewed in terms of contextual information if it can be codified, digitized, and implemented in an intelligent context-aware system. Emotion is characterized by cognitive response to changing states (contexts); such responses are characterized by an increasingly large range of contextual information in data intensive systems which may be both large and highly dynamic incorporating temporal, spatial, infrastructure, environmental, social, and personal data. There is a requirement for CS [20][21]; this, along with the volume and dynamic nature of the data calls for an approach capable of effectively handling both in geospatially diverse LSDS increasingly implemented in cloud-based solutions. The approach as discussed in the following sections and in [21] is designed to realize these aims.

3   Context

Context is central in realizing *Personalization*, context describing h/her prevailing dynamic *state*, as such it is inherently complex and domain specific [17][18][20][21]. A context is created using contextual information (context properties) that combine to describe an individual or entity, therefore a broad and diverse range of contextual information combines to form a context definition [20]. In actuality, almost any information available at the time of an individual's interaction with a context-aware system can be viewed as contextual information [20] including:

- The variable tasks demanded by users with their beliefs, desires, interests, preferences, and constraints.
- The diverse range of mobile devices and the associated service infrastructure(s) along with resource availability (connectivity, battery condition, display, network, and bandwidth etc), and nearby resources (accessible devices and hosts including I/O devices.
- The physical (environmental) situation (temperature, air quality, light, and noise level etc).
- The social situation (who you are with, people nearby - proximate information)
- Spatio-Temporal information (location, orientation, speed and acceleration, time of the day, date, and season of the year, etc).
- Physiological measurements (blood pressure, heart function - Electrocardiography (ECG or EKG from the German *Elektrokardiogramm*), cognitive functions related to brain activity (EEG from *Electroencephalography*), respiration, galvanic skin response, and motor functions including muscle activity).

- Cognitive and abstract contextual information such as an individual's emotional responses, intuition, feelings, and sensibilities.

The potential contextual information identified demonstrates the diverse nature and inherent complexity of context and context-aware systems. While the list includes cognitive properties, research is generally restricted to EEG and *Cognitive Behavioral Therapies* (CBT) [11][24]. Extending context to include the emotional factors is addressed in subsequent sections of this paper.

4   Computational Intelligence and the Semantic Web

Intelligence is an extremely complex topic and can be viewed from two perspectives: (1) Human intelligence (in computational terms this is analogous to the OWA), and (2) Computational intelligence (in computational terms this analogous to the CWA). The OWA is generally used where inference and reasoning is utilized; a function that is generally easy (albeit with frequent errors) for humans but is difficult for computer systems where sparse, brittle, and incomplete data results in the failure to reach a decision or conclusion - an essential function in context-aware systems where decision-support forms a central function [20].

In formal logic, the OWA functions on the basis that facts not explicitly defined and are not included in (or inferred from) knowledge recorded in a system are deemed to be unknown (rather than incorrect or false). The CWA is however predicated on the principle of *negation as failure*; i.e., *if it is not provably true, then conclude that it is false* [20]. For example, a database functions on the basis of the *unique names assumption* [7], this also holds true for ontologies [20][21]. The unique names assumption operates on the basis that a name is unique with no duplicate; this allows efficient and predictable searches of the database to be achieved. For ontology searching, merging, and matching synonyms have been used to identify similar concepts [20]; this however arguably reinforces the unique names assumption upon which ontologies function.

There is an ongoing debate in Semantic Web circles surrounding the OWA versus the CWA [25][32]. As observed, the CWA is predicated on negation-as-failure. While the Semantic Web and Semantic Web languages such as OWL are based on the OWA the CWA is also useful in certain applications [10]; a detailed exposition on the topic is beyond the scope of this paper however a discussion with extensive references can be found in [10][32]. In practical applications where context with predictable decision-support forms a central function and the Semantic Web technologies and OWL are employed, as is the case for OBCM [16], the CWA forms the basis for such applications.

5   The Nature of Knowledge

Having considered context and intelligence in computer systems we now turn to knowledge and its relationship context. Knowledge (in general and computational terms) falls into two general types: explicit and tacit knowledge. Conceptually, it is possible to distinguish between explicit and tacit knowledge however in actuality they are not independent but are interdependent where the creation of knowledge usable by individuals and computer systems is the aim.

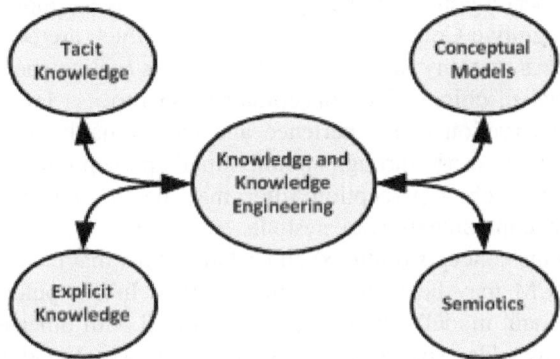

Fig. 1. Knowledge Concepts and the Knowledge Engineering Model

Explicit knowledge is: knowledge that can be clearly articulated and codified; as such it may be easily gathered and used in computer systems and applications. *Tacit* knowledge however represents a difficult challenge as it knowledge generated based on experience and observation in "real-world" situations (generally) in practicing a discipline or profession. Tacit knowledge is generally sub-conscious in nature and individuals may not be aware of the (tacit) knowledge they possess; as such an expert operates, makes judgments, and reaches conclusions without reference to explicit rules or principles [29].

Knowledge in the form of contextual information is the foundation upon which intelligent context-aware systems function; Knowledge Engineering (KE) [9] is the process which identifies and codifies the explicit and tacit knowledge (contextual information) which characterizes a domain of interest. KE is a function in software engineering and has been applied to the building, maintaining and development of applications and systems including: knowledge-based systems, expert systems, and decision support systems [9]. Additionally, KE has addressed cognitive science and socio-cognitive engineering where knowledge is structured according to the understanding of how human reasoning and logic functions [12][22]. KE describes the process of eliciting, and gathering knowledge from experts in a specific domain of interest. KE is a discipline that involves codifying and integrating knowledge into computer systems in order to solve complex problems normally requiring a high level of human expertise [9][20].

Identifying the contextual information however fails to address codification of tacit knowledge and its implementation in a context-aware system. An overview of the implementation of intelligent context is discussed in subsequent sections of this paper. For a detailed exposition on the implementation of intelligent context-aware systems with an evaluation and proof-of-concept see [20][21].

5.1   Conceptual Models and Semiotics

A model is (generally): (1) Physical Conceptual Models (PCM) which are representations of a process, state, or interaction with a physical object, device, or [for the
purpose of this paper] a computerized system (Figure 1 is a simple example of such
models), or (2) Cognitive Conceptual Models (CCM) which are cognitive conceptua-
lizations of a process or entity (as discussed later in this section such a model may for
example conceptualize color). The conceptualization process for a CCM manifests
itself based on observation and experience and can arguably form the basis upon
which humans view the world through an individual's perception filter. As we discuss
later in this paper, such a perception filter may be an important component in
inducing intelligence in context-aware systems.

A PCM represents concepts (entities) and relationships that exist between them; an
ontology and OBCM may be viewed in these terms. In computer science a PCM,
(also termed a domain model), should not be confused with other approaches to the
conceptual modeling addressing for example: data and logical modeling. Such models
may be created using for example the Unified Modeling Language (UML). While
these models are useful in the design and implementation process for computer systems the focus of this paper is on CCM's.

A CCM arguably has synergy with the concept of Semiotics. Semiotics (the
science of signs) has its genesis in the work the Swiss linguist Ferdinand de Saussure
(1857-1913) and the American Philosopher Charles Sanders Pierce (1839-1914) [3].
A discussion of their work is beyond the scope of this paper however a detailed exposition with extensive references can be found in [3]. In summary semiotics is defined
as the study of signs and sign processes (Semiosis). Semiosis is a process in which
currently experienced phenomena are interpreted as referring to other, experientially
absent, phenomena, thereby becoming meaningful entities, or signs. The reference of
a sign is made possible by memories of past interactions with the components of the
environment.

Generally applied to the media (film and text) [3] Semiotics is often divided into
three branches: (1) *Semantics*: Relation between signs and the things to which they
refer (their meaning which may differ between individuals based on experience and
observation), (2) *Syntactics*: Relations among signs in formal structures, and (3)
Pragmatics: Relation between signs and the effects they have on the people who use
them (again this may reflect individuals experience and observation)

Computational semiotics has addressed a diverse range of topics including: (1) logic, (2) mathematics, (3) theory and practice of computation, (4) formal and natural
language studies, (5) cognitive sciences generally, and (6) semiotics in a formal sense
with regard to cognition and signs. A common theme of this research is the adoption
of a "sign-theoretic" approach on issues related to artificial intelligence and knowledge representation [1]. Many applications of computational semiotics lie in research
addressing Human-Computer-Interaction (HCI) and the fundamental processes of
recognition [4]. For example, research in this field, termed 'algebraic semiotics',
combines aspects of algebraic specification and social semiotics [34]; this has been
applied to the design of user interfaces and to the representation of mathematical
proofs.

Emotional response to stimuli and events are influenced by CCM's and semiotic responses generated over time. Additionally, tacit knowledge is generally generated based on observation and experience over time. In considering CCM's, semiotics, and tacit knowledge as they apply to context (the focus of this paper); intuitively there is a synergy between these concepts and an individual's perceptual filter which as observed has a relationship with an individual's emotional response (emotion) to any given situation.

In considering CCM's, semiotic responses, and tacit knowledge: (1) they are generated over time based on experience and observation, and (2) effective description, documentation, and articulation of these concepts to another individual represents a challenging problem. For example, in computational terms, the color 'red' can be described in the RGB (the additive primary colors 'red' 'green' 'blue') scale as: 255-0-0 (or in Hexadecimal ff0000). This however fails to describe the color 'red' (or more accurately the specific shade of 'red' in the spectrum) to another person to enable the color to be recognized; additionally, every person will interpret a specific shade of 'red' differently.

In considering Semiosis and emotion an interesting phenomenon is Valence [6]. In everyday life, humans interact with and react to a range of stimuli; in such conditions (contexts) discrimination and categorization of "significant" stimuli forms a pivotal cognitive function. [6]. According to the widely accepted dimensional view of emotions [15], these "actions or action dispositions" are enabled using a valence categorization process (along the unpleasant/pleasant spectrum) in relation to the intensity (arousal) state that characterizes a situation. Based on this view, experimental data has pointed to the valence of the on-going stimulus being accounted for at a number of points in the information processing stream as indexed by the temporal aspects and the topography of event-related potentials (ERP) [5][30][33]. On the basis that humans react to emotional stimuli, the reactions being individual, valence may have a relevance and significance in context and the related issue of computational intelligence.

6   Implementation

The previous sections have considered Explicit and Tacit knowledge, knowledge engineering, Semiotics, and Conceptual Models; a synergistic relationship between tacit knowledge, semiotics, and CCM's has been drawn. This section addresses the implementation of these concepts in intelligent context-aware systems where CP with constraint satisfaction and preference compliance (CS) with decision-support form pivotal design goals. Following an overview of *Kansei* Engineering [20][28] intelligent context processing is addressed with an overview of the proposed approach to the implementation based on the context processing algorithm (CPA) [20][21].

6.1   *Kansei* Engineering

In semantic intelligent context-aware systems, contexts are dynamically influenced by user intuition, preferences, and emotions. An appropriate method termed *Kansei* Engineering has been developed as a methodology to deal with human feelings, demands, and impressions in context-aware applications.

Kansei is a Japanese term meaning sensibility, impression, and emotion [20]. *Kansei* words are given by adjectives describing human emotion, sensibility and impression; there is no equivalent term in English, the nearest applicable word is possibly intuition. *Kansei* evaluation is commonly used for evaluation methods to quantify impressions. For *Kansei* Evaluation, we have determined adjective pairs called *Kansei* words in pairs: (Synonym - Antonym) and (Synonym - Not Synonym). For instance, the pairs of adjectives (good - bad) and (successful - unsuccessful) are *Kansei* words.

6.2   Intelligent Context Processing

Central in the proposed approach to CP is the Context Processing Algorithm (CPA) [20] and the extended CPA [21] which employs context matching (CM) and provides a basis upon which contextual information can be processed in an intelligent context-aware system that enables CS with predictable decision support.

Prior to addressing the context-matching process it is necessary to briefly introduce the data structure which forms a fundamental component in the proposed approach. A detailed discussion on the topic can be found in [18][20] however in summary, the data structure is based on the *Semantic Context Modeling Ontology* (SCMO) created using the Web Ontology language (OWL) as discussed in [20].

The SCMO provides a generic, non-hierarchical, and readily extensible structure capable of adaptation to suit the domain specific nature of context with the capability to define the metadata, the context properties, and the literal values used in the context-matching process [18][20]. Additionally, while the approach presented in this paper does not currently use inference and reasoning (which generally applies subsumption and entailment) the CPA, as discussed in [21], is designed to accommodate this approach where required.

6.3   The Context-Matching Algorithm

The CPA approach is predicated on the processing of contextual information using the CM process [20][21]; CM (an extension of the data fusion concept) is designed to create the input context and access the output context(s) definitions to determine if the output (solution) context is an acceptable match with the input (problem) context. Essentially, the context-matching process is one of reaching a Boolean decision as to the suitability of a specific individual based on context [20][21]. Given that a perfect match is highly unlikely the CM algorithm must accommodate the PM issue along with a number of related issues as discussed in [20][21]. In CM the probability of a perfect match is remote therefore partial matching (PM) must be accommodated.

The CPA is predicated on the Event:Condition:Action (ECA) rules concept, the <condition> component employing the IF-THEN logic structure [20][21] which relates to the notion of <action> where the IF component evaluates the rule <condition> resulting in an <action>. The <action> in the proposed approach can be either: (1) a Boolean decision, or (2) the firing of another rule.

To address the PM issue the CPA applies the principles identified in fuzzy logic and fuzzy sets with a defined membership function which is predicated of the use of decision boundary(s) [2] (thresholds) as discussed in [21]. The membership function provides an effective basis upon which predictable decision support can be realized using both single and multiple thresholds to increase the granularity of the autonomous decision making process.

Conventional logic is generally characterized using notions based on a clear numerical bound (the crisp case); i.e., an element is (or alternatively is not) defined as a member of a set in binary terms according to a bivalent condition expressed as numerical parameters {1, 0} [11]. Fuzzy set theory enables a continuous measure of membership of a set based on normalized values in the range [0, 1]. These mapping assumptions are central to the CPA [20][21].

A system becomes a fuzzy system when its operations are: "entirely or partly governed by fuzzy logic or are based on fuzzy sets" and "once fuzziness is characterized at reasonable level, fuzzy systems can perform well within an expected precision range" [2]. Consider a use-case where a matched context mapped to a normalized value of, for example [0:80], has a defined degree of membership. This measure, while interesting, is not in itself useful when used in a decision-support system; in the CPA this is addressed using a distribution function (more generally referred to in the literature as a membership function) to implement the essential process of defuzzification as discussed in [2][21].

CM with PM imposes issues similar to those encountered in decision support under uncertainty, which is possibly the most important category of decision problem [31] and represents a fundamental issue for decision-support. For a detailed exploration of fuzzy sets and fuzzy logic see [12], a discussion around decision theory can be found in [31]. A comprehensive discussion on fuzzy system design principles can be found in [2] where a number of classes of decision problem are identified and discussed. In summary, Fuzzy Rule Based Systems have been shown to provide the ability to arrive at decisions under uncertainty with high levels of predictability [2][15][16]. A discussion on the CPA and rule strategies with the related conditional relationships for intelligent context-aware systems with example implementations and a dataset evaluation see [18][20][21].

7   Next Generation Intelligent Context-Aware Systems

It has been shown in [17][18][20][21] that context-aware systems are capable of realizing [albeit limited] intelligence in the processing of contextual information. This paper has considered the issues and challenges implicit in providing for improved levels of computational intelligence predicated on the integration of emotion (more accurately stated as emotional response) in the provision of personalized ser-vices. Our research into personalization, intelligent context processing, and the nature of computational intelligence has addressed the topic from a conceptual perspective and also as it relates to implementation in 'real-world' scenarios. This work has con-sidered use-cases in a range of domains including: the provision of tertiary education, the delivery of intelligent mobile marketing services, and importantly e-health monitoring.

In the case of tertiary education [35][19] the development of pedagogic systems us-ing a range of sensors to capture data (contextual information) relating to students' which is then intelligently processed to target resources and services has been investi-gated. Whilst many issues and challenges have been addressed the issue of measuring engagement in pedagogic systems, while partially solved (using principally atten-dance records and similar data), remains a significant challenge and an open research question. Emotional responses, if correctly measured and codified, may be used to more accurately assess levels of engagement to the benefit of both students in the learning experience and also to the university in improved outcomes.

We have identified in investigations around intelligent marketing solutions the po-tential benefits to be gained for business and individuals in the targeting of promo-tional advertisements in a mobile context predicated on a user's context. This research [8] has created and tested an intelligent mobile advertising system (iMAS). There is a large body of research which has considered the use of semiotics and cognitive con-ceptual models as they relate to the media (film and print); this work has included advertising and marketing. In considering targeted marketing empirical investigations have identified that, notwithstanding expressed preferences, individuals have differing responses to specific adverts based on life experiences which are unknown to the system. As with pedagogic systems, if emotional responses could be captured and utilized the benefits are potentially great for: (1) business with improved targeting of advertisements with an improved financial returns from the marketing and advertising budgets, and (2) for individuals where increased relevance in terms of *precision* and *recall* may result in reductions in irrelevant and therefore potentially annoying, demo-tivating, and poorly targeted messages.

Possibly the most important potential use of emotional response in the domain of e-health monitoring [36] in, for example, cognitive degenerative conditions on the Alzheimer spectrum where emotional response has important implications for both 'real-time' monitoring of patients and interventions which may be graduated as meas-ured against a patients current 'state'. Such 'states' currently include a broad range of contextual data [20] however there is a failure to capture emotional responses. Con-sider the potentially huge benefits for patients and carers in terms of quality of life and independent living for patients. Additionally, there are efficiency benefits to be derived from implementing intelligent context-aware assisted living solutions for healthcare professionals and the wider society where reductions in premature institu-tionalization offer the potential for huge financial savings on a global scale.

The challenges in realizing the integration of emotional response to a range of sti-muli is diverse domains are huge and are not underestimated. However, if the effec-tive use of emotion can be achieved the returns in both financial and personalization terms are very exciting.

8   Discussion

This paper has considered personalization to enable targeted service provision. Context has been addressed with a discussion around computational intelligence and the nature of knowledge including consideration of tacit and explicit knowledge, conceptual models, and semiotics. An overview of the approach to implementation of context in intelligent systems has been presented with examples of research where the integration of emotion offers the potential for the 'next generation' of intelligent context-aware systems. The focus of this paper is to consider how an increased range of contextual information utilizing tacit and explicit knowledge with conceptual models and semiotics can be used to improve computational intelligence in context-aware decision-support systems.

In considering emotion and emotional response, as it relates to context implemented in data intensive intelligent context-aware systems, we argue that: if the creation of computational intelligence which more closely approximates to human intelligence (which is characterized by sparse, brittle, and incomplete data) is to be realized, then the application of emotion using codified CCM, and semiotics in combination with tacit and explicit knowledge may provide a basis upon which this can be achieved. It is argued that in an ideal world the OWA is used however in practice the CWA is the only currently realistic approach to the realization of intelligence in context-aware systems where CS with predictable decision support is a central design requirement.

If computational intelligence which more closely resembles human intelligence can be realized there are exciting opportunities to exploit this capability in a range of domains including: e-Learning and e-Business applications. Perhaps the most interesting opportunity lies in increasing sophistication in levels of e-Health monitoring which is gaining traction in the field of assisted living in 'Smart Spaces". The challenges lie in effective use of knowledge in a context-aware system based on the OWA concept. We argue that addressing this challenge (at least in part) demands the codification of emotional response using CCM and semiotics. From an implementation perspective we postulate that the approach proposed in this paper using semantics and Kansei Engineering implemented using the CPA with OBCM provides a basis upon the codification of emotional response can be achieved.

In addressing pervasive computing [27] it has been observed that: all the basic component technologies exist today and in hardware, we have mobile systems and the related infrastructures, sensors, and smart appliances. Thomas *et al* [35][36] concur observing: components such as sensors, wireless mesh architectures, cloud services, and data brokerage/processing are all currently available and widely researched. It is argued in [20] that the challenges lie in the development of intelligent context middleware capable of processing the contextual information; this paper postulates that emotion and emotional responses form a part of an individual's response mechanism and as such can be viewed as contextual information. While the posited approach incorporates the ability to implement CP in LSDS the issues and challenges lie in the identification of the data points (contextual data), data capture and representation, addressing these questions represents the basis for future ongoing research.

While the research discussed in this paper has begun to resolve a number of issues relating to the processing of contextual information, including potentially the data that relates to emotional response, a number of challenges identified remain as open research questions. Such questions relate to: (1) the identification of the data (knowledge) that identifies emotional responses, (2) the development of a [non-invasive] approach to data capture of such information, (3) the development of a suitable Semiotic grammar and valence measurement system for emotion, and (4) representing the knowledge (data) in a suitable data structure; the OBCM currently utilized is recognized as an effective but sub-optimal solution. For a detailed discussion on the issues and challenges identified in the research with consideration of potential solutions including issues with alternative approaches to context processing see [11][18][20][21][24][37].

References

1. Andersen, P.B.: A Theory of Computer Semiotics. Cambridge University Press (1991)
2. Berkan, C., Trubatch, S.L.: Fuzzy Systems Design principles: Building Fuzzy IF-THEN Rule Bases. IEEE Press, Piscataway (1997)
3. Chandler, D.: Semiotics The Basics. Routledge, NY (2002)
4. de Souza, C.S.: The Semiotic Engineering of Human-Computer Interaction. MIT Press, Cambridge (2005)
5. Delplanque, S., Lavoie, M.E., Hot, P., Silvert, L., Sequeira, H.: Modulation of cognitive processing by emotional valence studied through event-related potentials in humans. Neurosci. Lett. 356, 1–4 (2004)
6. Delplanque, S., Silvert, L., Hot, P., Rigoulot, S., Sequeira, H.: Arousal and valence effects on event-related P3a and P3b during emotional categorization. International Journal of Psychophysiology 60(2006), 315–322 (2005)
7. Elmasari, R., Navathi, S.B.: Fundamentals of Database design, 2nd edn. Benjamin/Cummins, Redwood City (1994)
8. Evans, C., Moore, P., Thomas, A.M.: An Intelligent Mobile Advertising System (iMAS): Location-Based Advertising to Individuals and Business. In: Second International Workshop on Intelligent Context-Aware Systems (ICAS 2012), Proc. of The 6th International Conference on Complex, Intelligent, and Software Intensive Systems (CISIS 2012), Palermo, Italy, July 4-6, pp. 959–964 (2012)
9. Gonzalez, A.J., Dankel, D.D.: The engineering of knowledge-based systems theory and practice. Prentice-Hall, New Jersey (1993)
10. Horrocks, I., Patel-Schneider, P.F., Boley, H., Tabet, S., Grosof, B., Dean, M.: SWRL: A Semantic Web rule language combining OWL and RuleML (2004), http://www.w3.org/Submission/SWRL/
11. Hu, B., Majoe, D., Ratcliffe, M., Qi, Y., Zhao, Q., Peng, H., Fan, D., Jackson, M., Moore, P.: EEG-Based Cognitive Interfaces for Ubiquitous Applications: Developments and Challenges. In: IEEE Intelligent Systems - Brain Informatics. IEEE, USA (September / October 2011)
12. Klir, G.K., Yuan, B.: Fuzzy sets and fuzzy logic: theory and applications. Prentice Hall, NJ (1995)
13. Kermarrec, A.M., Massoulie, L., Ganesh, A.J.: Probabilistic Reliable Dissemination in Large-Scale Systems. IEEE Transactions on Parallel and Distributed Systems 14(2) (February 2003)

14. Lan, Z., Zheng, Z., Li, Y.: Toward Automated Anomaly Identification in Large-Scale Systems. IEEE Transactions on Parallel and Distributed Systems 21(2) (February 2010)
15. Lang, P.J., Greenwald, M.K., Bradley, M.M., Hamm, A.O.: Looking at pictures: affective, facial, visceral, and behavioral reactions. Psychophysiology 30, 261–273 (1993)
16. Moore, P., Hu, B., Zhu, X., Campbell, W., Ratcliffe, M.: A Survey of Context Modeling for Pervasive Cooperative Learning. In: International Symposium on Information Technologies and Applications in Education (ISITAE 2007), Kunming, Yuan, P.R, China, November 23-25, pp K5-1–K5-6. IEEE, USA (2007)
17. Moore, P., Hu, B., Wan, J.: 'Intelligent Context' for Personalised Mobile Learning. In: Caballe, S., Xhafa, F., Daradoumis, T., Juan, A.A. (eds.) Architectures for Distributed and Complex M-Learning Systems: Applying Intelligent Technologies. IGI Global, Hershey (2009)
18. Moore, P., Hu, B., Jackson, M.: Fuzzy ECA Rules for Pervasive Decision-Centric Personalised Mobile Learning. In: Xhafa, F., Caballé, S., Abraham, A., Daradoumis, T., Juan Perez, A.A. (eds.) Computational Intelligence for Technology Enhanced Learning. SCI, vol. 273, pp. 25–58. Springer, Heidelberg (2010),
http://dx.doi.org/10.1007/978-3-642-11224-9
19. Moore, P.: Anytime-Anywhere Personalised Time-Management in Networking for e-Learning. eLC Research Paper Series 3, 48–59 (2011)
20. Moore, P., Pham, H.V.: Rule Strategies and Ontology-Based Context Modelling in Human-Centric Data Intensive Intelligent Context-Aware Systems. In: Kołodziej, J., González-Vélez, H. (eds.) Intelligent Computing in Large-Scale Environments. The Knowledge Engineering Review (KER). Cambridge University Press, Cambridge (2013) ISSN: 0269-8889
21. Moore, P., Pham, H.V.: Intelligent Context with Decision Support under Uncertainty. In: Second International Workshop on Intelligent Context-Aware Systems (ICAS 2012), Proc. of The 6th International Conference on Complex, Intelligent, and Software Intensive Systems (CISIS 2012), Palermo, Italy, July 4-6, pp. 977–982 (2012)
22. Negnevitsky, M.: Artificial Intelligence: A Guide to Intelligent Systems. Addison Wesley (2005) ISBN 0-321-20466-2
23. Nurmi, D., Wolsk, R., Grzegorczyk, C., Obertelli, G., Soman, S., Youseff, L., Zagorodnov, D.: The Eucalyptus Open-source Cloud-computing System. In: 9th IEEE/ACM International Symposium on Cluster Computing and the Grid, Shanghai, China, May 18-21, pp. 124–131 (2009)
24. Peng, H., Hu, B., Liu, Q., Dong, Q., Zhao, Q., Moore, P.: User-centered Depression Prevention: An EEG Approach to pervasive healthcare. In: Proc. of the 5th International Conference on Pervasive Computing Technologies for Healthcare (Pervasive Health), Dublin, Ireland, May 23-26, pp. 325–330 (2011)
25. Russell, S.: Readings about the question: It is said that reasoning on the Semantic Web must be monotonic. Why is this so, when human reasoning, which seems to have served us well, is nonmonotonic? (2003), http://robustai.net/papers/Monotonic_Reasoning_on_the_Semantic_Web.html (retrieved)
26. Russell, S., Norvig, P.: Artificial intelligence: A modern approach. Prentice-Hall, NJ (1995)
27. Saha, D., Mukherjee, A.: Pervasive Computing: A Paradigm for the 21st Century. IEEE Computer 36(3), 25–31 (2003)
28. Salem, B., Nakatsu, R., Rauterberg, M.: Kansei experience: aesthetic, emotions and inner balance. Int. J. of Cognitive Informatics and Natural Intelligence 3, 54–64 (2009)

29. Schmidt, F.L., Hunter, J.E.: Tacit knowledge, practical intelligence, general mental ability, and job knowledge. Current Directions in Psychological Science 2, 8–9 (1993)
30. Schupp, H.T., Jungho fer, M., Weike, A.I., Hamm, A.O.: Attention and emotion: an ERP analysis of facilitated emotional stimulus processing. NeuroReport 14, 1–5 (2003)
31. Shackle, G.L.S.: Decision, Order and Time in Human Affairs. Cambridge University Press, Cambridge (1961)
32. Sheth, A., Ramakrishnan, C., Thomas, C.: Semantics for the Semantic Web: The Implicit, the Formal and the Powerful. International Journal on Semantic Web & Information Systems 1(1), 1–18 (2005)
33. Smith, N.K., Cacioppo, J.T., Larsen, J.T., Chartrand, T.L.: May I have your attention, please: electrocortical responses to positive and negative stimuli. Neuropsychological 41, 171–183 (2003)
34. Tanaka-Ishii, K.: Semiotics of Programming. Cambridge University Press (2010)
35. Thomas, A.M., Shah, H., Moore, P., Rayson, P., Wilcox, A.J., Osman, K., Evans, C., Chapman, C., Athwal, C., While, D., Pham, H.V., Mount, S.: E-education 3.0 Challenges and Opportunities for the Future of iCampuses. In: Proc. The 6th International Conference on Complex, Intelligent, and Software Intensive Systems (CISIS 2012), Palermo, Italy, July 4-6, pp. 953–958 (2012)
36. Thomas, A.M., Moore, P., Shah, H., Evans, C., Sharma, M., Xhafa, F., Mount, S., Pham, H.V., Wilcox, A.J., Patel, A., Chapman, C., Chima, P.: Smart Care Spaces: Needs for Intelligent At-Home Care. Int. J. Space-Based and Situated Computing 3(1) (2013)
37. Zhang, X., Hu, B., Moore, P., Chen, J., Zhou, L.: Emotiono: An Ontology with Rule-Based Reasoning for Emotion Recognition. In: Lu, B.-L., Zhang, L., Kwok, J. (eds.) ICONIP 2011, Part II. LNCS, vol. 7063, pp. 89–98. Springer, Heidelberg (2011), doi:10.1007/978-3-642-

Contextual Modelling in Context-Aware Recommender Systems: A Generic Approach

Christos Mettouris and George A. Papadopoulos

University of Cyprus
Department of Computer Science, University of Cyprus, Nicosia, Cyprus
1 University Avenue, P.O.Box 20537, CY-2109
{mettour,george}@cs.ucy.ac.cy

Abstract. Context-aware recommender systems (CARS) use context data to enhance their recommendation outcomes by providing more personalized recommendations. Context modelling is a basic procedure towards this direction since it models the contextual parameters to be used during the recommendation process. Most literature works however build domain specific contextual models that only represent information of a particular domain, excluding the possibility of model sharing and reuse among other CARS. In this paper we focus on this issue and study whether a more generic modelling approach can be applied for CARS. We discuss a possible solution and show through literature review on relevant systems that the proposed solution has not yet been applied. Next, we present a novel generic contextual modelling framework for CARS, discuss its advantages and evaluate it.

Keywords: Context Modelling, Context Modelling Framework, Context-Aware Recommender systems, Context-Awareness, MDA.

1 Introduction

According to Adomavicius and colleagues [1, 2] important research issues related to Context-Aware Recommender Systems (CARS) have to do with contextual modelling, more important of which are how to model the context in order for a CARS to be able to use contextual information directly in (or prior/after) the recommendation process, and develop appropriate methods - or extend existing 2 dimensional ones (2D) to multidimensional (MD) - in order to include more dimensions than the user and item (i.e. the context) in the recommendation process. All aforementioned, as well as more challenges require that the context has been appropriately modeled.

Another critical contextual modelling issue has to do with the most common practice followed in recommender systems that model the context: developing domain specific models that only represent information of the particular application domain (e.g. music, restaurants nearby). Domain specific models cannot be applied in other domains, while most of them are also application specific, meaning that they cannot be applied to other recommenders even of the same application domain. By

A. Haller et al. (Eds.): WISE 2011 and 2012 Combined Workshops, LNCS 7652, pp. 41–52, 2013.

constructing application specific contextual models, many different and very specific models are produced with no reuse and sharing capabilities. Moreover, developers and researchers struggle to design their own models as they think appropriate and according to their own knowledge and skills, with no reference model to use, no guidance and strictly based on the application at hand, often resulting in the production of overspecialized, inefficient and incomplete contextual models.

We argue that the above contextual modelling problems can be addressed to some extent by defining a generic, abstracted contextual modelling framework for CARS: a model template in essence that will be able to uniformly model the most important contextual parameters for these systems and provide a good, solid reference to developers who will be able to use the framework and extend it both at model level and code level in order to build their own application driven models. Developers will be guided by the modelling framework, and through its objects, properties and relations, will be directed towards a more efficient, effective and correct selection and usage of context properties for their own application model. Moreover, the framework will introduce developers to modern concepts derived by CARS research that might not be familiar with, such as the "context dependent rating data" [5], the "supposed context" [3], the "static/dynamic context", the "context weights", etc., as well as the role such concepts can play in a context model and a recommendation process.

Researchers will benefit as well from such a contextual modelling framework for CARS. The framework will provide a spherical view of CARS research and its concepts, assisting new researchers in understanding these concepts, as well as research problems, issues and challenges. The modelling framework will inevitably project any inconsistencies that might occur after an addition of a new concept, and therefore model corrections after additions will be made easier. The most important advantage though of using a generic contextual modelling framework is that following research works can be based on this abstracted framework, enhance and extend it in order to solve important contextual modelling problems in CARS research, while avoiding the risk of being over-focused on a particular domain.

This work, through the bibliographic review of section 2, shows that a generic contextual model, model template or modelling framework for CARS does not exist in the bibliography, and proceeds with our first attempt towards this goal. In particular, section 3 provides our first attempt to design and build a generic contextual modelling framework for CARS. In section 4 we theoretically evaluate the proposed framework by using existing works from the CARS literature, showing how the framework can be used to model application specific CARS and novel research modelling methods. Section 5 completes the paper with conclusions and future work.

2   Related Work

In order to infer whether any attempts towards a generic contextual model for CARS exist in the bibliography, we have reviewed recommender systems that use contextual and conceptual models. Towards this direction, a number of CARS have been reviewed, as well as semantic recommender systems, since many semantic recommenders use semantics to model information, including the context. The review was focused on whether the contextual or conceptual model used was application specific or generic. To better define the terms "application specific" and "generic", we use the definition of Peis [18]. Peis and colleagues classify semantic recommenders as *generic recommender systems* those that do not focus extensively on a particular domain and as *domain specific recommenders* those that do. Examples of each class of recommenders can be found at [18]. Peis's categorization into generic and domain specific can be applied to all semantic and context aware recommenders: such systems either apply to some generic application area (generic systems), such as recommendation of products, web services, etc. or apply to a particular domain (domain specific systems) such as recommendation of movies, books, etc. In this work, we categorize systems as generic and domain specific based on Peis's categorization, with one additional condition: we also categorize systems that attempt to facilitate *any* application specific domain as generic recommenders.

Our review revealed that most CARS and semantic recommenders in the literature are domain specific, which confirms our initial statement [4, 6, 7, 8, 9, 11, 19, 20, 21, 23]. Unfortunately, the models derived from such works cannot be applied for use in other domains. A number of generic recommenders also exists [10, 13, 16, 22] that either apply to some generic application area, or can be applied to more than one domain by linking domain specific ontologies to their own data and knowledge pool in order to gain domain-aware knowledge and provide domain-aware functionality. One of the most representative examples of a generic recommender system is the one of Loizou and Dasmahapatra [15], who propose a generic, abstracted model in the form of an ontology which could be used by many different types of recommender systems and which ontologically models, not only data and context, but also the recommendation process.

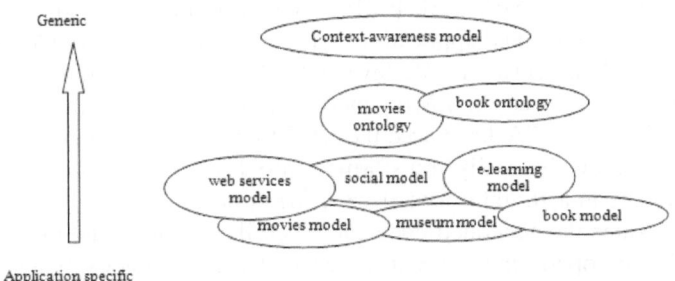

Fig. 1. How conceptual models for recommender systems move from application specific to generic

Although some of the semantic and contextual models try to be more generic, the majority represent information that either concern particular application domains (e.g. tourism, movies, museums), or more abstracted domains (e.g. products in general, web services, e-learning, etc.). Moreover, a common practice is to use general purpose ontologies for facilitating sharing and reuse among semantic recommenders. The aforementioned are depicted in Fig. 1. Lower are shown conceptual models of recommender systems that are application specific. A step higher, but still application specific are conceptual models that can be adopted by a number of recommenders such as web service recommendations, e-learning recommendations, etc. Examples of generic conceptual models are the general purpose ontologies, which truly facilitate sharing and reuse among many recommenders.

3   A Generic Contextual Modelling Framework for Cars

To the best of our knowledge no attempts have been made towards developing a truly generic contextual model for CARS that will define the basic contextual entities of such systems, their properties and associations so that CARS will be able to *extend* this model to construct application specific models for the needs of the application at hand. A generic contextual model would *simplify* the process of contextual modelling in CARS and enable context *uniformity*, *share* and *reuse*. Moreover, it would introduce developers and new researchers to important concepts of CARS research in order to assist them in building more effective context-aware recommenders, while researchers will be aided by using the model to apply their solutions to research problems relevant to CARS context modelling. Fig. 1 displays the generic nature of the proposed contextual model ("Context-awareness model"). The model has to be generic enough to be able to describe any contextual definition related to CARS. In this work we propose such a model in the form of a modelling framework.

3.1   The Modelling Framework

The modelling framework is essentially a model template, which itself is an abstracted model designed and built as a UML class diagram by using the Eclipse Modeling Framework (EMF) [12]. EMF is a Java framework and code generation facility for building tools and applications based on a model. It provides the means to transform a model into customizable Java code. After designing our framework as an EMF model, we have used the EMF generator to create a corresponding set of Java implementation classes. We have used the EMF tool for three main reasons: (i) so that our framework could be easily transformed into Java code in order to be used by CARS developers in a straightforward way, (ii) in order to be highly extendable and customizable, since the generated code can be easily extended and modified and (iii) EMF provides the opportunity to edit the generated code classes by adding methods and variables and still be able to regenerate code from the modelling framework, as all additions will be preserved during the regeneration. In this way, developers are able not only to extend the code generated from the framework, but also to extend the framework as well, regardless of any code extensions that might occur before the framework extension.

Fig. 2 depicts the contextual modelling framework for CARS. The boxes represent context entities (or classes). There exist two types of associations: the solid arrow from entity "a" to entity "b" (e.g. from "Context" to "itemContext") depicts that "b" is a context property of "a". The other type of arrow from "a" to "b" (e.g. from "itemContext" to "Item") depicts that "a" is a subclass of, and therefore inherits class "b".

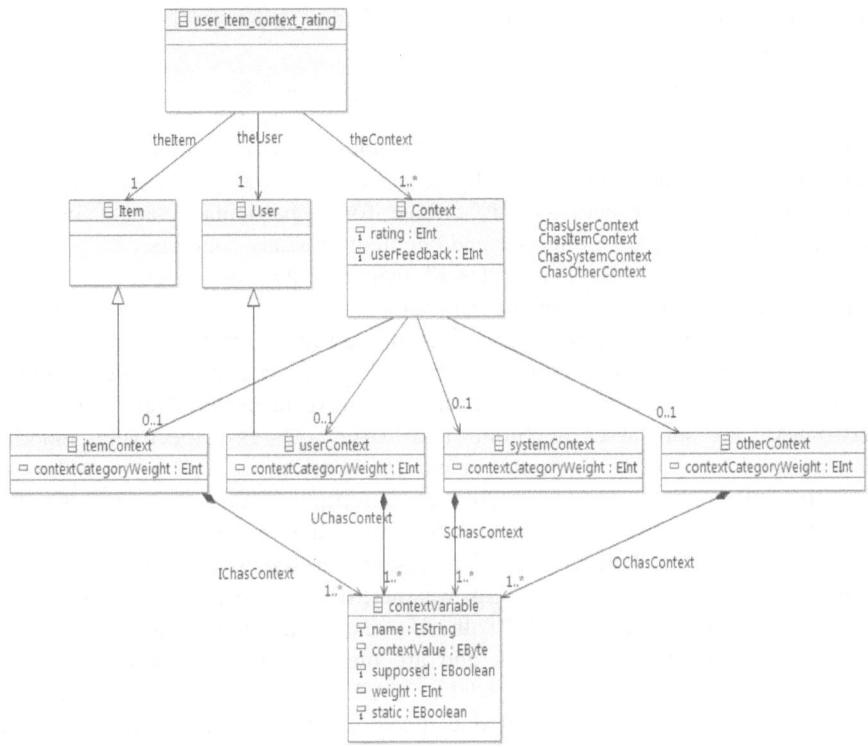

Fig. 2. The proposed contextual modelling framework for CARS

From top-to-bottom, the level of abstraction decreases. The top entity "user_item_context_rating" reflects on the fundamental concept of CARS research: to include the context in the recommendation process in order to result from the 2D un-contextual recommenders: Users × Items → Ratings to the multidimensional context-aware recommenders: Users × Items × Context → Ratings [2]. The "user_item_context_rating" entity represents a single complete recommendation process. For each recommendation attempt, a recommender must examine whether an item is suitable for a user in a certain context. This can be depicted through the question: "what is the rating a particular user would assign to a particular item under a certain context"? This rating score is what a recommender must calculate. Therefore, the "user_item_context_rating" entity has exactly one "Item", exactly one "User" but one or more "Context" entities, each including a property "rating" to ensure that in each context a user is able to rate the same item with a different rating score. In CARS literature this is known as "Context dependent ratings" [5]. Such user ratings are assigned on items in particular contexts: a user may rate multiple times an item, each rating taking place in a different context. Baltrunas and colleagues [5] note that user preferences (and hence ratings) on items depend on the context and therefore context dependent rating data on items should be available. With the aforementioned configuration, our modelling framework provides the means for including "Context dependent ratings".

The "Context" entity in Fig. 2 represents the *context instance*: the set of context variables with their corresponding values that constitute the context for a single recommendation. The "Context" entity also includes the "userFeedback" property that represents user feedback information for a particular recommendation. As with the "rating" property, "userFeedback" is unique for a particular context but many "userFeedback" can exist for one user and one item: in many contexts.

Regarding the entities "Item" and "User", from a context-awareness point of view, we are only interested in the context information related to the item and the user at hand. This context information is represented in the framework by "itemContext" and "userContext" entities: these entities represent any item and user related context information that participates in a particular context instance. "itemContext" and "userContext" are subclasses of "Item" and "User" classes respectively, inheriting their characteristics to use them as context information, as well as extending and overwriting information and functionality as appropriate. Similarly, "systemContext" represents any system related context information that participates in a particular context instance, while "otherContext" represents any contextual information other than item, user and system context (e.g. weather, time, temperature). "itemContext", "userContext", "systemContext" and "otherContext" constitute the four *main context classes* in our modelling framework and are meant to be perceived as the main context entities for any contextual model of CARS; any context information of any CARS should be able to be represented as a context property of one (or more) of the main context classes, as an entity of the type "contextVariable". The entity "contextVariable" can be a context property of any one of the four main context classes (or more than one in the case where the main context classes share a contextual information).

The "Context" entity may include zero or one of any of the four main context classes (via the corresponding associations "ChasUserContext", "ChasItemContext", "ChasSystemContext" and "ChasOtherContext" appeared to the side of "Context" in Fig. 2 for readability reasons); we have provided the zero possibility in the case where a CARS does not need to use the particular context class for any reason.

Each of the four main context classes mentioned above includes zero or one weight property so that developers are able to denote the importance of each class, and by this provide a hierarchy about which context class(es) is(are) more important. This importance hierarchy is necessary to be included in the recommendation process.

The "contextVariable" entity is positioned at the lowest level of the framework representing the least abstracted entity: the context variable that contains the actual contextual information. Each main context class has to include at least one "contextVariable" entity. Each "contextVariable" has a name and a value, while a weight property is also included in case the CARS developer would like to denote a particular importance for a certain variable. The property "static" refers to whether the context is static or dynamic (static=true/false). Static context cannot change dynamically (e.g. user profile information), as opposed to dynamic that can (e.g.

weather). Hinze and Buchanan [14] propose that context-awareness can help in reducing the amount of data to be accessed real time by pre-retrieving any relevant pre-known data, e.g. the static context. This increases efficiency. By using the "static" attribute, a CARS developer may assign to context variables whether they are a part of static or dynamic context and by that, specify a different functionality for them.

Asoh and colleagues investigate in [3] whether the answers of users during questionnaires about their preferences on items differ when they are in a given context, as opposed to not actually being in that context but only imagine being in it (supposed context). Their results suggest that the ratings of users in supposed contexts may be different than the ratings of the same users in real contexts. Their findings are very important since much information about user preferences often results from user questionnaires on supposed contexts, meaning that these results could be proved misleading, even false. Back to our modelling framework, the purpose of the property "supposed" is to denote whether the particular context variable is a part of the supposed context. Since supposed context usage may negatively influence the rating of a user, it is very important that this type of context can be represented in any context model. In the case where a recommender includes many "supposed" context information, the recommendation results could be misleading. Thus, a context model has to be able to denote whether the context it represents can be fully trusted or not; we use the "supposed" property of each "contextVariable" entity to denote whether the particular context variable can be fully trusted (supposed=false) or needs caution (supposed=true).

3.2   Code Generation

The contextual modelling framework for CARS presented above can be used to automatically generate the Java code by using the EMF. For each of the 9 classes in the diagram of Fig. 2, a Java interface is created, as well as a Java implementation file containing the implementation code. All aforementioned files are generated automatically by the Eclipse tool. For example, for the class "contextVariable" a file contextVariable.java is created that contains the "contextVariable" interface, and another one named contextVariableImpl.java that contains the interface implementation code. contextVariable.java includes the abstracted methods (setter/getter), as well as the variables of the class. The contextVariableImpl.java includes the implementations of the abstracted interface methods. The automatically generated Java code can be freely extended and modified by developers in order to become tailored to a specific application domain. Developers and researchers are free to implement the abstracted methods of the interface as they think appropriate, modify them, as well as extend the code by adding new methods and variables. Moreover, the EMF provides the ability to change the modelling framework of Fig. 2 and regenerate the code multiple times. In the case a method was manually changed by the developer prior to a code regeneration, the EMF can be prevented from overwriting the particular method (the developer must remove the tag "@generated" from the particular method).

4   Theoretic Evaluation

Proper evaluation of the modelling framework would require application of the framework in real scenarios, i.e. making the modelling framework publicly available and invite CARS developers to use it for their applications. The aforementioned would provide us with good feedback regarding the framework's strong and weak points, as well as any shortfalls. Due to lack of time, we leave this as future work. In order to theoretically evaluate our framework, we have used three research works from the context-aware recommendations literature: an application specific context-aware recommender system and two research oriented works. The aim of the theoretic evaluation is, on the one hand, to examine whether our generic contextual modelling framework for CARS is able to successfully model the context used by these systems and how this can be done, and on the other hand to observe whether the framework can be used for realizing novel research-oriented context modelling methods.

4.1   Modelling an Application Specific CARS

We have chosen the media recommendation system of Yu and colleagues [23] because the authors consider four different types of context, while most works consider significantly less. The four types are: content context, operating context, user context and terminal context. We will focus on context modelling by examining whether our contextual modelling framework for CARS could be used to successfully model the context in the particular application. According to the authors, content context is the context of an item, operating or situational context is the user's current location, time and activity, user context consists of user preferences and terminal context is relevant to terminal capabilities.

We aim to observe whether our modelling framework for CARS can be used to represent the four different types of context. Starting with content context, our framework provides the context entities "Item" and "itemContext" which can be used for modelling the items as follows: generic item characteristics can be assigned to class "Item" while strictly contextual information can be assigned to the class "itemContext". Note that a context instance includes only one "Item", one "itemContext" but many "contextVariable" entities representing the many pieces of context information related to that item. A contextual information that is assigned as context property to the class "itemContext" is in essence a "contextVariable" assigned to the class "itemContext" via the association "IChasContext" (Item Context hasContext). For example, to assign the contextual information "actor" to "itemContext", the developer will create a "contextVariable" by the name of "actor" and will assign it to "itemContext" via the association "IChasContext". The benefit of our approach is that any context entity represented as context variable can be assigned as property to any of the four main context classes of our framework, even to more than one. This provides developers with the freedom to assign context variables to main context classes as they think appropriate and according to the application at hand. In the case where a context variable is assigned to many context classes, then the developer can specify a different functionality for the particular variable according to the context class currently used (e.g. context variable "user's activity" can be treated differently when perceived as part of "userContext" and in another way when perceived as part of "systemContext").

The next context type is the operating or situational context that includes the user's current location, time and activity. Such context type does not exist as a single entity in our system, since situational context can vary a great deal among different applications and domains. Instead, we chose to contextually model only entities that are well defined and not controversial among different domains. Hence, we model such context by denoting it as non static context in "contextVariable" entity. By this approach, *any* context information of any type can be denoted as dynamic, which is a developer's decision. Regarding the work under examination, situational context is modeled by our framework as follows: "user's current location" and "user's activity" are context properties under "userContext" and possibly under "systemContext" and "otherContext" (in case user's current location and activity also affect system functionality and other events), while "time" is an "otherContext" property. Finally, "User preferences" are assigned under "userContext" and "terminal context" is assigned under "systemContext".

Regarding "contextVariable" properties "supposed", "weight" and "static", we assign "supposed=false", "weight=0...10" depending on the perceived importance for each one of the context variables by the developer and specify "static=true" for static context: any contextual information regarding the item ("itemContext", e.g. actor, genre, language), a part of "userContext" that is static ("User preferences") and the "systemContext" which is mainly static. "static=false" is specified for dynamic context such as the operational/situational context (see paragraph above).

From the above discussion we state that the proposed generic contextual modelling framework for CARS is able to model the context as specified by the system of Yu and colleagues, and result in an application specific model for media recommendations, which we name as "media model". The advantage of using the contextual modelling framework for CARS instead of the model proposed by Yu and colleagues is that the resulting "media model" allows for sharing and reuse among various applications, and can easily be further extended and modified to suit the developer's demands. Moreover, both the generic contextual modelling framework for CARS and the application specific "media model" can be used as reference and guidance by developers to implement their own application specific models for media recommenders.

4.2   Modelling Research Oriented Works

For each user u and context k, Panniello and Gorgoglione [17] define the user profile in context k, i.e. the contextual profile Prof(u, k). For example, if the contextual variable "Season" has two values (e.g., "Winter" and "Summer"), then the authors assign two contextual profiles for each user, one for the winter and the other for the summer and use the appropriate one according to the context. Similarly,

Baltrunas and Amatriain [4] propose using micro-profiles of the user, which are snapshots of the user profile in certain time periods (e.g. morning, noon, night), instead of using the whole user profile (they use the time as context). By using the reduced time-based micro-profile of the user instead of the whole profile they manage to reduce the input dataset of the recommendation algorithm and thus improve accuracy. Both approaches are based on the same idea: using specific user profiles that are defined based on a particular context instance instead of using the whole user profile to provide recommendations.

After studying the above contextual modelling approaches, we examined whether they could be successfully implemented using our generic contextual modelling framework for CARS. In the framework we provide researchers the ability to define in their models a user profile instance for a particular context instance by using the "userContext" entity which is directly associated with the "Context" entity (see Fig. 2). An instance of the context is composed of all context variables associated with it having a particular value. Schematically, we can say that an instance of the context is a "Context" entity composed of all the "contextVariable" entities associated to it through the four main entities "itemContext", "userContext", "systemContext" and "otherContext" (Fig. 2). On the other hand, the contextVariable: name="Time", value="morning" participates in a number of context instances, each of which is valid when the time is morning. These context instances define the context: Time="morning". The same applies for each "contextVariable" in the modelling framework. In the case where a researcher needs to use time-based user micro-profiles, our framework provides by itself such functionality as follows (suppose time is divided to 3 distinct time slots: morning, noon and night): define three context variables, one for each time slot: contextVariable: name="Time" value="morning", contextVariable: name="Time" value="noon", and contextVariable: name="Time" value="night". Each of the three "Time" contextVariable entities corresponds to a different context in respect to time: morning, noon and night. For each of the three "Time" contextVariable entities, a number of context instances are created which are valid for the particular contextVariable's value. These context instances however, also include a "userContext" entity that contains the user context information that is valid for the particular context instance, and consequently for the particular time. In this way, user context-aware time-based micro-profiles are automatically constructed through the modelling framework. Similarly, by selecting a different context variable than time, e.g. season, we can automatically produce context-aware season-based micro-profiles of the users (or any other context entity).

The advantage of using our modelling framework is that, automatically and by default, the framework's context instances define all valid contextual information around a fact or event (in the example above around a specific time slot). Hence, by using the framework, a researcher is given the opportunity to explore more easily and straightforward the benefits of context-awareness, as in the example above where time-based user micro-profiles are automatically created through the framework.

5   Conclusions and Future Work

After confirming that no existing work attempts to define a generic contextual model for CARS, in this paper we have proposed such a model in the form of a contextual modelling framework and theoretically evaluated it with positive results. As future work we will conduct practical evaluation of the modelling framework by applying it in real scenarios. To test whether our framework is indeed capable of facilitating any CARS, we aim to make the framework publicly available and invite CARS developers to use it for their own applications. This will provide us with valuable feedback about the framework's strong, weak points and shortfalls. Moreover, we will extend our modelling framework by conceptually modelling functionality (i.e. recommendation algorithms) in addition to the context. The goal is to research whether by including conceptual sub-models of the various recommendation algorithms in the framework the implementation of more efficient CARS can become easier for developers and researchers. This can possibly lead to a fully model based CARS development, which is a very important concept to be studied.

References

1. Adomavicius, G., Sankaranarayanan, R., Sen, S., Tuzhilin, A.: Incorporating contextual information in recommender systems using a multidimensional approach. ACM Transactions on Information Systems (TOIS) 23, 103–145 (2005)
2. Adomavicius, G., Tuzhilin, A.: Context-aware recommender systems. In: Proceedings of the 2008 ACM Conference on Recommender Systems, pp. 335–336 (2008)
3. Asoh, H., Motomura, Y., Ono, C.: An Analysis of Differences between Preferences in Real and Supposed Contexts. In: Proceedings of the 2nd Workshop on Context-Aware Recommender Systems (2010)
4. Baltrunas, L., Amatriain, X.: Towards Time-Dependant Recommendation based on Implicit Feedback. In: Workshop on Context-Aware Recommender Systems, CARS 2009 ACM Recsys, vol. 2009, pp. 1–5 (2009)
5. Baltrunas, L., Kaminskas, M., Ricci, F., Rokach, L., Shapira, B., Luke, K.: Best Usage Context Prediction for Music Tracks. In: Proceedings of the 2nd Workshop on Context-Aware Recommender Systems (2010)
6. Blanco-Fernández, Y., Pazos-Arias, J.J., Gil-Solla, A., Ramos-Cabrer, M., Barragáns-Martínez, B., López-Nores, M., García-Duque, J., Fernández-Vilas, A., Díaz-Redondo, R.P.: AVATAR: An Advanced Multi-Agent Recommender System of Personalized TV Contents by Semantic Reasoning. In: Zhou, X., Su, S., Papazoglou, M.P., Orlowska, M.E., Jeffery, K. (eds.) WISE 2004. LNCS, vol. 3306, pp. 415–421. Springer, Heidelberg (2004)
7. Bogers, T.: Movie Recommendation using Random Walks over the Contextual Graph. In: Proceedings of the 2nd Workshop on Context-Aware Recommender Systems (2010)
8. Bu, J., Tan, S., Chen, C., Wang, C., Wu, H., Zhang, L., He, X.: Music recommendation by unified hypergraph. In: MM 2010 Proceedings of the International Conference on Multimedia, p. 391 (2010)
9. Cantador, I., Castells, P.: Semantic Contextualisation in a News Recommender System. In: Workshop on Context-Aware Recommender Systems CARS 2009 in ACM Recsys, vol. 2009 (2009)

10. Costa, A., Guizzardi, R., Guizzardi, G., Filho, J.: CoReS: Context-aware, Ontology-based Recommender system for Service recommendation. In: UMICS 2007, 19th International Conference on Advanced Information Systems Engineering, CAISE 2007 (2007)
11. Drumond, L., Girardi, R., Leite, A.: Architectural design of a multi-agent recommender system for the legal domain. In: Proceedings of the 11th International Conference on Artificial Intelligence and Law, ICAIL 2007, p. 183 (2007)
12. Eclipse Modeling Framework Project (EMF),
    http://www.eclipse.org/modeling/emf/
13. Emrich, A., Chapko, A., Werth, D.: Context-Aware Recommendations on Mobile Services: The m:Ciudad Approach. In: Barnaghi, P., Moessner, K., Presser, M., Meissner, S. (eds.) EuroSSC 2009. LNCS, vol. 5741, pp. 107–120. Springer, Heidelberg (2009)
14. Hinze, A., Buchanan, G.: Context-awareness in mobile tourist information systems: challenges for user interaction. In: International Workshop on Context in mobile HCI at the Conference for 7th International Conference on Human Computer Interaction with Mobile Devices and Services, Salzburg, Austria (September 2005)
15. Loizou, A., Dasmahapatra, S.: Recommender Systems for the Semantic Web. In: ECAI 2006 Recommender Systems Workshop, Trento, Italy, August 28- September 11 (2006)
16. Moscato, V., Picariello, A., Rinaldi, A.M.: A recommendation strategy based on user behavior in digital ecosystems. In: Proceedings of the International Conference on Management of Emergent Digital EcoSystems, MEDES 2010, p. 25 (2010)
17. Panniello, U., Gorgoglione, M.: A Contextual Modeling Approach to Context-Aware Recommender Systems. In: Proceedings of the 3rd Workshop on Context-Aware Recommender Systems (2011)
18. Peis, E., Morales-del-Castillo, J.M., Delgado-López, J.A.: Semantic Recommender Systems. Analysis of the State of the Topic [en linea]. Hipertext.net (6) (2008),
    http://www.hipertext.net
19. Santos, O.C., Boticario, J.G.: Modeling recommendations for the educational domain. Procedia Computer Science 1(2), 2793–2800 (2010)
20. Sielis, G.A., Mettouris, C., Papadopoulos, G.A., Tzanavari, A., Dols, R.M.G., Siebers, Q.: A Context Aware Recommender System for Creativity Support Tools. Journal of Universal Computer Science 17(12), 1743–1763 (2011)
21. Sielis, G.A., Mettouris, C., Tzanavari, A., Papadopoulos, G.A.: Context-Aware Recommendations using Topic Maps Technology for the Enhancement of the Creativity Process. In: Educational Recommender Systems and Technologies. IGI Global (2011)
22. Uzun, A., Räck, C., Steinert, F.: Targeting more relevant, contextual recommendations by exploiting domain knowledge. In: HetRec 2010 Proceedings of the 1st International Workshop on Information Heterogeneity and Fusion in Recommender Systems, pp. 57–62 (2010)
23. Yu, Z., Zhou, X., Zhang, D., Chin, C., Wang, X., Men, J.: Supporting Context-Aware Media Recommendations for Smart Phones. IEEE Pervasive Computing 5(3), 68–75 (2006)

Introduction to the Proceedings of the Workshop on Cloud-enabled Business Process Management (CeBPM) 2012

Dana Petcu[1] and Vlado Stankovski[2]

[1] West University of Timisoara, Romania
petcu@info.uvt.ro
[2] University of Ljubljana, Slovenia
vlado.stankovski@fgg.uni-lj.si

Introduction

The First Workshop on Cloud-enabled Business Process Management (CeBPM) in 2012 was held in conjunction with the WISE 2012 conference in Paphos, Cyprus. The Workshop focused on an emerging area aiming to address the gap between the automatisation and optimization of business operations on one side and the offering of software service utilities needed to support such business operations on the other. A new set of technologies are needed that would support the increasingly resource demanding daily business operations of enterprises. Some expected benefits of CeBPM approaches are higher availability of business processes on demand, scalable and elastic provision of needed computational resources, lower startup costs for new enterprises and possibilities to chose and optimise the service utilities based on non-functional requirements, such as reliability, securty, cost and so on.

Along this line, CeBPM 2012 is a forum for researchers and practitioners allowing them to identify the latest progress in the field and future directions that need intensive research and technology development. The six full papers were selected after a thorough peer-review by the Workshop Program Committee Members and represent a range of relevant topics. Following is a brief overview of the contributions.

The paper "Cloud Storage of Artifact Annotations to Support Case Managers in Knowledge-Intensive Business Processes" by the authors Marian Benner-Wickner, Matthias Book, Tobias Brückmann and Volker Gruhn proposes a Cloud-based architecture for annotating artifacts with document- and content-level metadata to support case managers' cognitive effort of organizing, relating and evaluating artifacts involved in each case, and describe a tool that provides a consolidated environment for working with artifacts from heterogeneous sources.

The paper entitled "Collaborative Process Design in Cloud Environment" by Jiří Kolar and Tomáš Pitner presents an approach for end-to-end BPM adoption in an organization, with emphasis on collaboration with the process participants. Their work is also focused on the use of Cloud-based environments for process design.

The paper "Challenges for Migrating to the Service Cloud Paradigm: An Agile Perspective" by Stavros Stavru, Iva Krasteva and Sylvia Ilieva deals with the chalenges of migrating agile methods and techniques to the Cloud. A thorough literature review and an expert judgment is presented on different agile techniques, taken from Scrum and Extreme Programming (XP), that could address the identified challenges. A ranked list of applicable agile techniques is presented and suggestions for their adoption.

The paper "Towards a Trust-manager Service for Hybrid Clouds" by Fatma Ghachem, Nadia Bennani, Chirine Ghedira and Parisa Ghoddous proposes an approach to help Private Clouds in the selection of trustworthy Public Cloud services. The solution takes the form of a service called trust manager that analyses Private Cloud service needs and bases the decision-making on the Private Cloud past invocation analysis.

The paper "Designing an SLA Protocol with Renegotiation to Maximize Revenues for the CMAC Platform" by the authors Adriano Galati, Karim Djemame, Martyn Fletcher, Mark Jessop, Michael Weeks, Simon Hickinbotham and John McAvoy focuses on new models to negotiate and manage Service Level Agreements (SLAs). It analyses the possibility of integrating an SLA approach for Cloud services based on the Condition Monitoring on A Cloud (CMAC) platform, which offers services to detect events on assets as well as data storage services.

The paper entitled "Levi - A Workflow Engine using BPMN 2.0" by Eranda Sooriyabandara, Ishan Jayawardena, Keheliya Gallaba, Umashanthi Pavalanathan and Vishaka Nanayakkara focuses on a cloud-ready BPMN 2.0 execution engine called Levi, which executes BPMN 2.0 processes natively.

We sincerely thank the Program Committee Members of the CeBPM 2012 Workshop for their time and support throughout the reviewing period.

Dana Petcu,
Vlado Stankovski

CeBPM 2012 Workshop Chairs

Collaborative Process Design
in Cloud Environment

Jiří Kolář and Tomáš Pitner

Masaryk University, Faculty of Informatics
Botanická 68a, 602 00 Brno, Czech Republic
{kolar,tomp}@fi.muni.cz
http://fi.muni.cz/~xkolar2/

Abstract. This paper presents an approach to adoption of BPM in an
organization, with emphasis on collaboration with process participants.
We present subset of our methodology for end-to-end BPM adoption,
aimed to describe collaborative processes mapping, iterative process design and further process improvement. Such technique preserves organization's flexibility as it helps to obtain realistic processes easily adaptable
to changing business requirements. We further explain how to foster collaboration by use of a cloud-based environment for process design and
define some more general requirements on such environment. We approach general obstacles of BPM adoption observed by practitioners and
scientist and explain how the methodology can help to deal with some
of those obstacles by involving process participants to collaboration on
process design and improvement.

1 Introduction

Business Process Management (BPM) is often considered as the quite rigid approach to managing organizations. As many recent successful adoptions of BPM
were implemented in large enterprises, BPM is very often recognized as mostly
suitable for large organizations. Thus many Small and Medium Eenterprises
(SME) stay away from this management approach as they consider it clumsy
and threatening to hinder their main competitive advantage [6]. We see elim-
ination of the rigid flavour of BPM as a challenge and work on end-to-end
methodology suitable for agile adoptions of BPM in SME sized organizations.
[16] The methodology puts emphasis on agility and collaboration during adoption process, which should result in establishment of realistic processes, foster
interactions among process participants and provide hospitable environment for
continous process improvement. In this paper we present a subset of the methodology focused mainly on collaboration between process experts responsible for
process modeling and process participants – subject matter experts performing
the actual work within particular processes. Further we set requirements on sup-
portive Cloud-based Process Collaboration Environment (PCE) which supports
the collaboration namely during process mapping, design, further improvements
and provide space for rich feedback and discussion.

A. Haller et al. (Eds.): WISE 2011 and 2012 Combined Workshops, LNCS 7652, pp. 55–69, 2013.
© Springer-Verlag Berlin Heidelberg 2013

Firstly we briefly introduce modern BPM and state of the art in the area of Cloud-enabled BPM. Then we focus on recent research related to common obstacles of BPM adoptions, choose several of them relevant to our context and leading to problem definition. In the following section we evaluate some existing research approaches applied to solve the problem and highlight the main points that served as an inspiration for our approach. In the last section we present subset of our methodology and set requirements for the Cloud-based PCE supporting principles defined in the methodology. In the last part we conclude our findings and outline directions of further research.

1.1 Shift of Focus in Modern BPM

The contemporary understanding of Business Process Management can be seen from two different perspectives. From the Management perspective, BPM is a dynamic management approach where operations of an organization are described by processes. A process is defined as a repeatable sequence of activities, linked to organizational business goals. Execution of the processes contributes to fulfillment of those goals [5] [4]. On the other hand, BPM can be seen from the technical perspective which embodies design of Enterprise Information Systems (EIS) and the way of thinking about system's behavior. Such EIS design often incorporates use of Business Process Management Systems (BPMS) for process design and execution. However BPM can be still adopted without engine-based process automation [7]. Modern approach to BPM is often called "holistic BPM" encompasses both perspectives, addresses strategy, people, business processes and technology and puts emphasis on continuous process improvement after initial adoption phase [1] [2] [11]. In last few years quite some of BPM technologies have reached acceptable maturity level and the focus of many practitioners and researchers has consequently shifted from the technological perspective of BPM to the adoption process itself and the organizational changes towards the process-oriented principles [7] [4]. Such adoption process often involves significant changes in target organization, such as flattening the organization structure, definition of processes and adoption of role-based model. [7] Large organizations usually have performed their flattening in a natural way due to their size, as they use some form role-based model and often they have some kinds of workflow definitions. Therefore the BPM adoption does not mean complete change of mindset. A bit different situation is observed in SME's, as they stick very often with functional hierarchical organizational models, their tasks are often tight with concrete persons and a lot of their work is organized at hoc. In this case BPM adoption means a big step forward and significant organizational changes [14] [15].

1.2 BPM in Cloud Context

Cloud based BPMS is a wish of many BPM experts since the times of introduction of Software as a Service (SaaS) cloud model. As BPM technologies are complex and hard to deploy and maintain, SaaS sounds like perfect solution, where the entire complex of BPMS could be provided as service [3]. However in practice most of BPMSes are far from being provided as SaaS due to several reasons. Even the most complete BPMSes are still being tailored and modified for most of deployments, which in not so easy to do if they would be provided as SaaS. Secondly, BPMS is often used to integrate services running local intranets whereas integration of such local services with remote Cloud environment is still seen as an issue in the integration context [21]. Although we can see some efforts to provide entire BPMS functionality in SaaS, these services have still not matured at the moment. Something a bit closer to the state of maturity are Cloud based environments for process design (PCEs). Such environments can be provided in cloud quite easily. They operate similarly to other popular SaaS collaborative applications, and can be very helpful during process design and foster efficient collaboration, one of critical factors of well designed processes.

2   Problem Definition

Recent research, case studies and reports from practice [24] [22] [23] identify several obstacles of adoptions of BPM in organizations. In this section we are going to describe some of them considered as important in context of our research focused on SME sized adoptions, discuss how they are related to each other and formulate the problem we are trying to solve.

As BPM in its holistic form is quite modern approach, we still lack methodologies and best practices for end-to-end BPM adoptions. [7] [13] [10] We can find several useful techniques for initial phases of adoption, initial business analysis and organization assessment such as Business Motivation Model [19] and later implementation of a BPM solution, usable in later technical phases of adoption [25] [26],but we lack end-to-end methodologies guiding from the early phases of adoption, such as gathering the information for process modeling, process design, mapping business goals to processes and linking business KPI's to process metrics. [1] [13] [27] According to mentioned sources, lack of methodologies covering end-to-end adoption process with respect to principles of holistic BPM is a valid problem to be solved.

Second important obstacle is related to external subjects conducting BPM adoptions. BPM projects are often being conducted by team of BPM specialists , an external subject or eventually internal team operating independently from the rest target organization. [8] Such conducting subjects often acquire only simplified external view on organization's business and they do not involve target organization's process participants as much as they should. They often perform contracted part of the job, results are handed to target organization, necessary changes are executed and eventual SW solution deployed. Usually adoption process organized according to such waterfall model ends at this point, there is no space for feedback from process participants, correction of faulty or inefficiently modeled processes. Such waterfall adoption often does not bring core

BPM adoption benefits, as BPM adoption should be iterative process which continue after initial adoption further by continuous process improvement and maintenance [28].

One of very important outcome of BPM is building systematic knowledge-base in target organization. Mapping and defining processes do not only help to codify the existing know how, but also bring opportunity to share knowledge across process participants. [28] [8] It is an opportunity to review how the work is currently done, evaluate efficiency, provide space for fresh ideas how to make the work better and help to maintaining and extend shared knowledge base. [8] [7] [10]. Well expressed by [30]: "Process can serve as transitional object for mental models". In other words, involvement of process participants into process design phase can be seen as a social activity, which lead to extension of their knowledge. At the same time more space for efficient collaboration across all process participants is being created by observing such "mental models" of the others. [28] [8]. Thus according to those findings, systematic involvement of process participants and organization stakeholders in process design has positive effect on quality of modeled processes, make participants more accountable for their tasks within their process and initiate collaboration among process participants. According to mentioned obstacles, we believe there is a need for complex methodology which provide guidelines for performing end-to-end BPM adoption, performed in agile manner in short iterations, with strong involvement of target organization's participants in adoption process. Such methodology should also guide organization trough the later phases following initial adoption and describe how to further maintain defined processes, update them to reflect changes in business and perform continuous process improvement.

3   Literature Review

3.1   Collaborative Process Methods

One of the important sources of inspiration is "G-MoBSA framework" [8]. This research heavily focused on socio-cognitive perspective of process modeling provide some significant new ideas how to extend the concept of knowledge creation and sharing during adoption process and propose complex methodology for group model building and complementary argumentation schema. The framework also proposes a BPM experimentation module, which serve as a discrete simulation environment. Despite the fact this framework bring several highly innovative ideas about collaboration on process design, it put a lot of requirements on process participant's knowledge of the proposed framework, which can lead to waste of participant's time dedicated to process design. Because in most cases the time process participants can spend on process design collaboration is limited, the tool they use should be simple and intuitive and their activity should be very straightforward, to capture maximum of their "subject matter knowledge". Also the argumentation schema seems to be a bit overkill for most of cases, as several iteration and concluding discussion can provide same results as long and complicated argumentation according to the schema, which also do not reflect the fact that arguments of people with stronger position on the organization have usually higher importance and taken as authoritative.

We can find more simple and intuitive approaches to the same problem such as [29]. This research effort is focused primary on design of simple collaboration tool based on MediaWiki software with semantic extension. Use of such intuitive technology seems to be very close to our approach, however this paper also introduce simple process language defined inside Semantic Wiki used for process description. Such approach can be useable for very simple processes, nevertheless as there is BPMN 2.0 L1 subset, intuitive enough to be even understandable by people without knowledge of BPMN notation, but still extendable to complete process model, use a non-standardized process description language does not make much sense today.

One of the most complete methodologies for end-to-end BPM adoption, which can serve in many ways as inspiration for development for mode light-weighted methodologies is CBM-BPM-SOMA developed in IBM. It is a merge of three separate methods linked to each other. This triplet cover technique used for organization assessment and business analysis (Component Business Modeling - CBM), the core method focused on process analysis (BPM) and technically-oriented Service Oriented Modeling and Architecture (SOMA) mainly focused on efficient identification, definition and composition of services [9]. However this methodology is designed for adoption of large scale full featured BPM solution, which includes automation by usage of one IBM BPM products and integration of various services and systems. Such solutions fit well complex BPM solutions of large enterprises, but they seem not suitable for agile small scale BPM adoptions. There are some other approaches which outline the whole adoption lifecycle like [27], but those does not seem to be detailed enough to be successful.

3.2 Process Collaboration Environment (PCE)

We mentioned before an idea of PCE an environment for collaborative process design is not completely new and many important BPM vendors also visible in Gartner's magic quadrant such as IBM, Signavio, Intalio, Pega [12] make extensive efforts to develop server-side environments for collaborative process design. However most of them allow only local installations on private servers, which get them closer to "private cloud" concept. Some existing public cloud services rely on open-source technologies. Probably the most popular tool of that type is Oryx editor, developed as open source project [20]. Oryx is being tailored by several BPMS vendors for example Signavio and Alfresco. Oryx itself is only visual modeling tool, and for full blown PCE it have to be extended for some advanced features such as mechanism for providing and managing feedbacks, real-time multi-user collaboration and change tracking.

4   Results

4.1   Agile Methodology for Collaborative Approach to Process Design

In this section we will present subset of our methodology focused on small-scale BPM adoptions. This subset is focused primarily on collaboration of initial process design and also on further collaborative improvement of processes. We will put emphasis on involvement of process participants, as they play key role in gathering of requirements in initial process design as well as consequent iterations focused on process improvement. Early draft of the methodology was applied in practice so far in two case studies. First case study was performed in commercial environment, SME software company: IT Logica s.r.o [18] ,focused on Web-Application development. Second case study was performed in ICT department of Masaryk University in Brno and was focused primarily on ICT services provided at University [17]. In both cases agility and need for more iterative approach to process design and need for further process maintenance was identified as a drawback of our methodology, so we did recently several changes towards more iterative agile principles. Planning the BPM adoption Adoption consists of several phases. At the end of each phase results should be reviewed and the plan for forthcoming phases should be detailed. In general estimation of effort for each phase is not easy at the beginning and many details about next phase are uncovered at the end of preceding phase. We should also keep in mind that BPM adoption often means changes in both organizational structure and used ICT technologies. This means that changes should be committed iteratively and all new systems should run in parallel and migration should be very careful. Obvious seems to be usage of conventional project management tools which help project manager to deal with planning complexity and make the plan systematic and understandable.

4.2   Adoption Participants

BPM adoption should start with identification of participants. Key participants should be chosen very carefully as their contribution can significantly influence the whole adoption. We have to make sure all participants are properly informed about the adoption process, they understand the adoption goals and they should be convinced about potential benefits of adoption process.

We are going to describe following participant roles:

- Sponsor
- Organization's management
- Adoption coordinator
- Process analyst
- EIS designers and developer
- Process participant
- Process maintainer

Sponsor. This role usually belongs to organization owner or CEO. Sponsor provides resources for adoption process such as funding and allocates internal human resources. His commitment is absolutely necessary for success of adoption and he has to clearly understand potential benefits, risks and overall impact on organization. Organization's management Each manager has to be fully familiar at least with impact of adoption on his area of responsibility and also understand the big picture of the adoption. On the side of lower management we face often fear of loss of responsibility and importance. This is very important to be solved, managers play important role in the adoption and we have to carefully explain all benefits adoption can bring to them and make sure all their fears are dispelled. Adoption coordinator Usually member of external "BPM team". He usually acts as Project manager of the adoption and he is the core person responsible for entire adoption process. He has to plan the adoption process carefully, execute it and periodically monitor the progress. He should be familiar with organization's business context, cooperate closely with Sponsor and Organization's management. He should be experienced process analyst familiar with issues of process modeling and manage team of process analysts.

Process Analyst. Usually also member of external "BPM team", responsible for interviewing process participants, modeling and documenting organization's processes. Good communication skills are a must. He has to have strong knowledge of process modeling techniques and he should have at least basic knowledge of organization's business domain as well.

EIS Designers and Developer. Internal or external person responsible for design of EIS in target organization. He should have at least basic knowledge of BPMS technologies if a BPMS is used and understand at least basic BPM concepts. He should be aware of desired impact of adoption on organization's EIS.

Process Participant. Internal organization's worker performing activities of modeled processes. He usually has a key knowledge about how the process works in details and he should serve as main sources of information about modeled processes. Similarly to organization's managers, participants are often afraid of negative impact of BPM adoption on his work. Thus we have to carefully explain all benefits adoption can bring to him and make sure he is willing to collaborate.

Process Maintainer. Internal person made responsible for further maintenance and improvement of processes after adoption. He should work closely with adoption coordinator and team of process analysts and learn as much as possible. He should learn how to model and modify processes, synchronies changes between organization's business goals & objectives and processes, how to set measures on processes and transform measured data into KPIs. In short, he should be able to perform those steps periodically after end of initial adoption on his own and further develop the organization's processes.

4.3   Setting Preceding the Adoption

There are several activities, which should be done shortly after kickoff the adoption process.

Introductionary Meeting. There should be a meeting which introduce the plan of adoption and create common understanding across all involved subjects.
  Such meeting should be attended at least by:

- Sponsor and part of organization's management directly involved in adoption process
- Adoption coordinator, eventually some process analysts
- As much as possible process participants
- Process maintainer

On such meeting we should present most important facts about the adoption and provide space for discussion Presentation should cover:

- Basic facts about the adoption, such as purpose, goals and expected outcomes
- Highlight the importance collaboration across all the involved subjects
- Outline the whole adoption plan and rough time schedule
- Brief introduction of process used process modeling technique
- Introduction of used PCE
- Rough structure of process interviews

PCE Setting. We have to make sure all users of our PCE are able to access it and know how to use it. We should also provide a person supporting PCE users to achieve maximum contribution. There should be some example processes as well as feedbacks, so users can use it as a template.

4.4   Adoption Phases

Adoption consists of several phases performed in a recommended order. However in some cases the sequence of these phases has to be tailored to the situation. For example when the business goals and objectives of the organization are relatively simple, but the business of the organization itself is built on critical mass of EIS components and ICT services, the analysis of those systems turns to be more important and it can be performed earlier. However this leads to the bottom-up approach to BPM adoption, which is not really in scope of the researched methodology.
  We are going to describe following phases:

- Organization assessment phase
- Initial process mapping phase
- Iterative process improvement

Organization Assessment Phase. In this phase we gather context information about organization and its business, collect business related information and use it as an input for process analysis and design. These activities are done by Adoption coordinator by performing interviews with organization's management and root stakeholders.

Roles involved: Sponsor, Organization's management, Organization's management, Adoption coordinator

Phase inputs:

- Previous efforts of organization assessment
- Business plan
- Any documents describing organization structure
- Definitions of metrics and previous business data
- ICT services documentation

Phase activities:

1. Review and refine business plan & vision
2. Review and refine goals and objectives (G&O)
3. Review and specify business metrics and KPIs mapped to objectives
4. Describe in detail organizational structure, including roles and responsibilities
5. Describe business components (organization units)
6. Describe ICT services both consumed and provided internally and externally
7. Create priority list of business activities
8. Create complete list of relevant processes mapped to business activities

We first collect the AS-IS state, discuss it with the management and define initial TO-BE state. Nevertheless TO-BE state should not involve much reengineering at this stage. It can involve:

- Business plan re-engineering
- KPI's and metrics definition and re-engineering
- Estimation of quality and costs of ICT services
- proper mapping of G&O to processes
- clear definition of roles

For more formal description of organization business plan&vision and Goals&Objectives we can use some more formal techniques such as Business Motivation Model (OMG 2008). However BMM is quite complex technique and can fit only for organizations with more complex business planning. Phase outputs:

- Refine business plan, vision,
- G&O and related KPI definitions
- Description of organizational structure with subordinations, roles and responsibilities
- Prioritized list of business activities mapped to existing processes

Initial Process Mapping Phase. To obtain realistic processes that correspond to reality, the involvement of each process participant to the process definition in "design time" is crucial. Otherwise we can easily end up with idealistic process definitions dreamed by management that have nothing to do with reality. The more intuitive technology we use for sharing the modeled processes with process participants, the more efficient collaboration we achieve.

Phase inputs:

- Prioritized list of business activities mapped to existing processes (from previous phase)
- Any documents describing activities involved in modeled processes
- KPI definitions (from previous phase)

Phase activities:

1. Complete prioritized process list (existing and new) with process owners assigned
2. Interview process participants and define initial processes
3. Create Detailed BPMN 2.0 models of chosen processes and write complementary descriptions
4. Define roles within processes and map them to organization's roles
5. Identify and refine process metrics linked to KPI's
6. Set up PCE and publish processes there.

Phase outputs:

- Prepared PCE
- Complete list of prioritized processes with assigned owners and roles
- Initial version of process BPMN 2.0 models and descriptions published in PCE
- Clear definitions of process metrics and mapping to KPI's
- Initial feedbacks about processes from participants stored in PCE

The main responsibility of good process design of the modeled processes lies on Adoption coordinator. It is generally assumed that the processes should be modeled by Process Analysts who are dedicated to this activity, but they do not usually understand each process in detail. Thus they have to cooperate with process participants who are involved in the activities performed within the process. Initial set of defined processes should be also approved by organization's management and sponsor of the BPM solution. Steps of the initial process mapping phase are described in Figure 1. Here the adoption coordinator captures the scope of the organization and creates list of processes. Then he models and describes the selected processes and publishes the draft to the PCE. At this step the process participants and organization's management should provide rich feedback and comments, they have to identify parts of the process which are faulty, unclear or too general. Such feedback is stored in the PCE. After the predefined period of time, Adoption coordinator collects the provided feedback and closes the initial phase.

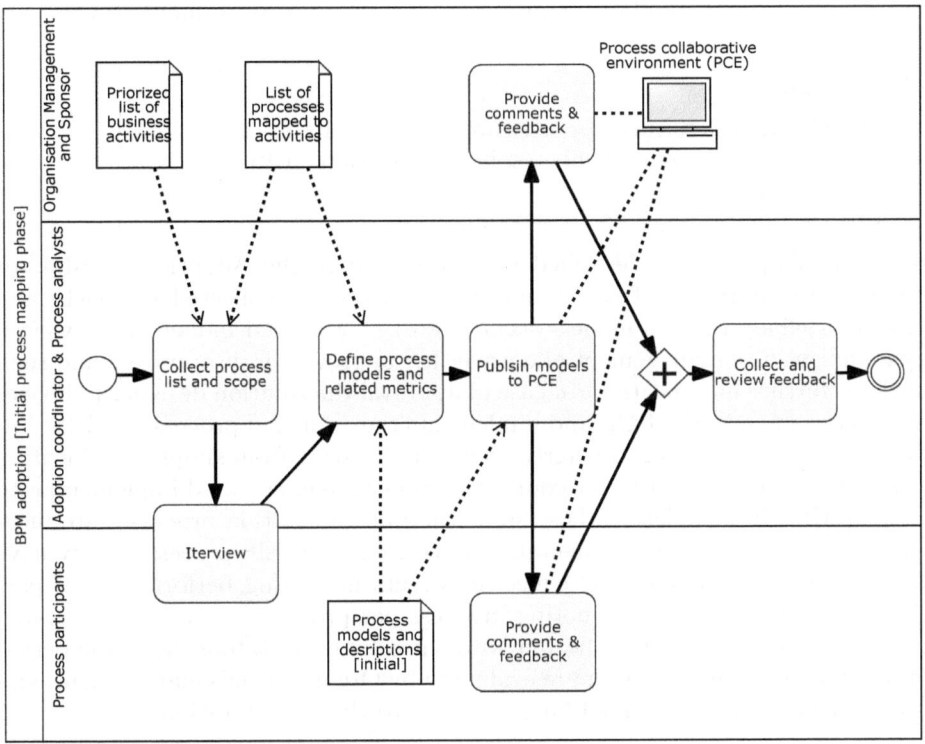

Fig. 1. (Initial process mapping phase)

Iterative Process Improvement. This phase should be performed in short iteration cycles (I would recommend 1-6 months), the anticipated changes should be also of reasonable size, corresponding to the available human resources. Phase inputs:

- Feedbacks about processes from participants and management stored in PCE
- Process update requests (2+ iteration)
- Process data (2+ iteration)

Phase activities:

1. Modify process models and descriptions according to feedbacks and change requirements
2. Discuss changes and get approval with Organization's management and Sponsor
3. Publish updated processes to PCE and open for discussion
4. Implement changes in processes in EIS
5. Measure process execution automatically or manually
6. Collect process data
7. Let the Organization's management and Sponsor to evaluate measured data

8. Collect Process update requests from Organization's management and Sponsor

Phase outputs:

- Modeled and described processes published to PCE
- Updated processes implemented in organization's EIS
- Process data
- Process update requests for next iteration

Steps of this phase are described in Figure 2. Here the Adoption coordinator initiates first iteration of improvement phase, reviews collected feedbacks and modifies defined process models according to it. Modified models are reviewed by organization's management and are either approved or disapproved and send back for further modification. In case of approval the solution designer publishes modified version to the PCE and implements the approved processes in EIS. Implementation depends on the agreed level, it can start from simple modification of existing activities in EIS for completing process-engine based implementation in a BPMS. By completing these steps the implementation processes are measured. In case of basic implementation of conventional EIS processes, they have to be measured manually, by collecting events indicating performance of particular activities or even by noting progress per process. In case of automated monitoring tools, data are collected automatically by such tool. After the period of measurement, process data are evaluated by Organization's management, and process changes are requested for processing to the next iteration.

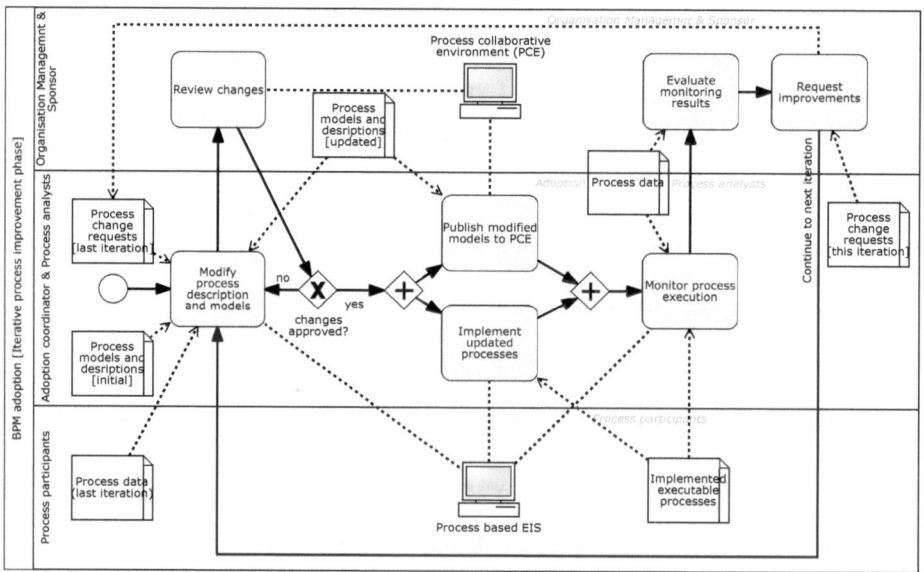

Fig. 2. Iterative process improvement phase

4.5   Cloud Environment Requirements

BPM adoptions including any form of BPMS-based implementation are usually done context of a chosen technology. Choice of the technology often depends on factors such as desired level of automation, amount of human tasks in processes, ICT technologies already present in target organization,budget for BPMS software and many others. Because we want to keep our methodology versatile and technology independent we set just very general requirements on PCE and let the subject performing the adoption to choose the most suitable tool which will fit project size, complexity and general adoption context. PCE, a software application used to support the methodology providing an environment for broad collaboration on designed processes should fulfill following general requirements:

- Provide simple interface or integration with modeling tool for publishing process models and their text descriptions
- Track revisions and changes done by particular user in published processes back in history at the very beginning
- Display categorized list of published processes to a user
- Display diagram of each process with complementary description and chosen comments of other users
- Allow users to comment on particular process or it's part
- Allow users to mark faulty or inefficient part of process model or description (optional)
- Allow user to propose process changes and improvements
- Allow administrator to manage visibility of comments among users

In most simple cases any kind of WIKI, or web Content Management System which allow users to write comments to content can be considered as the tool for PCE.. Ad-hoc solution built on organization's Document Management System can serve for this purpose as well. Some process authoring environments or modeling tools also provide PCE functionalities and allow users to collaboratively model processes. However this can be too complicated for process participants, as they usually have poor knowledge of process modeling.

There are several potential advantages of moving PCE into cloud environment:

- Efficient sharing and real-time collaboration
- PCE is easily accessible from any environment and OS, it does not require any local installations of the dedicated tools
- Centralized storage allows proper versioning, tracking of changes and history

5   Conclusions and Future Research

According to amount of existing research around collaborative process design we believe that we are working on valid research problem. Extending the knowledge in collaboration can help to perform BPM adoptions better and leverage full potential of BPM even in SME context. The presented subset of methodology should contribute to solution of defined problem and requirements on collaboration tool should help to choose or develop the right technology that fit for particular situations. We believe in further improvement of our methodology according to feedback from practice and we want to end up with comprehensive set of guidelines of the whole adoption process. Our aim is to keep the methodology versatile and technologically independent, provide guidelines for as much aspects of the adoption as possible, but also provide space for ad-hoc customizations, so the methodology still can be tailored to particular situation. Modern Cloud-enabled collaborative tools help to foster better collaboration in many cases and we believe that process modeling can be one of them. However the human factor remains still most crucial influence in BPM adoption context and the will to collaborate across all subjects have to be initiated by wise decisions made by people and a PCE still remain just a tool that can support such collaboration.

Acknowledgements. "This work was supported by the European Union's territorial cooperation program between Austria and the Czech Republic of the under the EFRE grant M00171, project "iCom" (Constructive International Communication in the Context of ICT)."

References

1. Ackoff, R.L.: Re-creating the Corporation: A Design of Organizations for the 21st Century. Oxford University Press, New York (1999)
2. Jackson, M.C.: Systems Thinking: Creative Holism for Managers. Wiley, Chichester (2003)
3. Jiang, J., Le, J., Wang, Y., Sun, J., He, F.: The BPM Architecture Based on Cloud Computing. In: KAM 2011: Proceedings of the 2011 Fourth International Symposium on Knowledge Acquisition and Modeling. IEEE Computer Society (October 2011)
4. Smith, H., Fingar, P.: Business Process Management: The Third Wave. Meghan-Kiffer Press, Tampa (2003)
5. Harmon, P.: Business Process Change: A Manager's Guide to Improving, Redesigning, and Automating Processes. Morgan Kaufmann, San Francisco (2003)
6. Raymond, L., Bergeron, F., Rivard, S.: Determinants of business process reengineering success in small and large enterprises: an empirical study in the Canadian context. Journal of Small Business Management 36(1), 72–85 (1998)
7. Jeston, J., Nelis, J.: Business Process Management: Practical Guidelines to Successful Implementations. Butterworth-Heinemann (2008)
8. Adamides, E., Karacapilidis, N.: 'A knowledge centred framework for collaborative business process modelling'. International Journal of Business Process Management 12(5), 557–575 (2006)
9. Fiammante, J.: Dynamic SOA and BPM: Best Practices for Business Process Management and IBM Press (2009)
10. Singer, R., Zinser, E.: Business process management — S-BPM a new paradigm for competitive advantage? In: Buchwald, H., Fleischmann, A., Seese, D., Stary, C. (eds.) S-BPM ONE 2009. Communications in Computer and Information Science, vol. 85, pp. 48–70. Springer, Heidelberg (2010)
11. Gartner Research: Gartner position on Business Process Management, Gartner Research note, ID: G00136533 (2006)

12. Gartner Business Process Management Summit, Gartner Research note (April 2011)
13. Indulska, M., Chong, S., Bandara, W., Sadiq, S., Rosemann, M.: Major issues in business process management: a vendor perspective. In: Proceedings of the Pacific Asia Conference on Information Systems, PACIS 2007 (2007)
14. Chapman, R., Sloan, T.: Large firms versus small firms - do they implement CI the same way? The TQM Magazine 11(2), 105–110 (1999)
15. Spanos, Y., Practacos, G., Papadakis, V.: Greek firms and EMU: contrasting SMEs and large-sized enterprises. European Management Journal 19(6), 638–648 (2001)
16. Kolar, J.: A framework for Business Process Management in small and medium enterprises. Unpublished Dissertation Proposal. Masaryk University (2011)
17. Kolář, J.: Procesní analýza pro Centrum Informačních Technologií Filozofké fakulty MU. Unpublished commercial case study (September 2011b)
18. Kolář, J.: Procesní analýza pro společnost IT Logica s.r.o., Unpublished commercial case study (July 2011c)
19. OMG. Business Motivation Model (BMM) Specification, v1.0. (2008), http://www.omg.org/cgi-bin/doc?formal/08-11-02.pdf (retrieved June 2012)
20. Oryx website from, http://bpt.hpi.uni-potsdam.de/Oryx/ (retrieved June 2012)
21. Dillon, T., Wu, C., Chang, E.: Cloud Computing: Issues and Challenges. In: 2010 24th IEEE International Conference on Advanced Information Networking and Applications (2010)
22. Mertens, W., Van den Bergh, J., Viaene, S., Schroder-Pander, F.: How bpm impacts jobs: An exploratory field study. IEEE Computer Society (2011)
23. Chong: Business process management for SMEs: an exploratory study of implementation factors for the Australian wine industry. Journal of Information Systems and Small Business 1(1-2), 41–58 (2007)
24. Imanipour, N., Talebi, K., Rezazadeh, S.: Obstacles in Business Process Management (BPM) Implementation and Adoption in SMEs (January 23, 2012), SSRN: http://ssrn.com/abstract=1990609
25. Aksu, F., Vanhoof, K., De Munck, L.: Evaluation and Comparison of Business Process Modeling Methodologies for Small and Midsized Enterprises Intelligent Systems and Knowledge Engineering, ISKE (2010)
26. Roser, S., Bauer, B.: A Categorization of Collaborative Business Process Modeling Techniques. In: 7th IEEE International Conference on E-Commerce Technology Workshops (CEC 2005 Workshops), Munchen, Germany, July 19, pp. 43–54. IEEE Computer Society (2005)
27. Benjamin, P., Erraguntla, M., Mayer, R., Painter, M., Marshall, C.: A Framework for Adaptive Process Modeling and Execution (FAME), wetice. In: Seventh International Workshop on Enabling Technologies: Infrastructure for Collaborative Enterprises, p. 3 (1998)
28. Bruno, G., Dengler, F., Jennings, B., Khalaf, R., Nurcan, S., Prilla, M., Sarini, M., Schmidt, R., Silva, R.: Key challenges for enabling agile BPM with social software. Journal of Software Maintenance and Evolution: Research and Practice 23(4), 297–326 (2011)
29. Dengler, F., Lamparter, S., Hefke, M., Abecker, A.: Collaborative Process Development using Semantic MediaWiki. In: Proc. of 5th Conf. of Prof. Knowl. Management (2009)
30. Morecroft, J.: Mental models and learning in system dynamics practice. In: Pidd, M. (ed.) Systems Modelling: Theory and Practice, pp. 101–126. Wiley, Chichester (2004)

Towards a Trust-Manager Service for Hybrid Clouds

Fatma Ghachem[1], Nadia Bennani[1], Chirine Ghedira[2], and Parisa Ghoddous[1]

[1] Université de Lyon, CNRS, LIRIS, Lyon, France
firstname.lastname@liris.cnrs.fr
[2] Université de Lyon, MAGELLAN, MODEM, Lyon, France
Chirine.ghedira-guegan@univ-lyon3.fr

Abstract. Cloud computing changed recently business view regarding their Information System through an on-demand provisioning of computing resources. Recent discussions about data security requirements in cloud computing environment tend to conflict with other requirement including usability and economic. In hybrid clouds that combine private and public clouds usage, private clouds are able both to externalize resources and invoke services from public cloud when needed. However, in such specific inter-cloud environment risks arise. Indeed, private clouds aren't sufficiently assured about how credible is the data computed by these resources they entrusted. This is due to clouds autonomy preservation, difference in control policy definitions and lack of transparency in clouds. In this position paper, we tend to propose an approach to help private cloud selecting a trustworthy public cloud service. The idea consists in a trust manager as a service that bases the decision-making on the private cloud past invocation analysis.

Keywords-component: Cloud computing, data credibility, trust management, hybrid cloud, SLA.

1   Introduction

Cloud computing changed recently business view regarding their Information System through an on-demand provisioning of computing resources. Recent discussions about data security requirements in cloud computing environment tend to conflict with other requirement including usability and economic. In our opinion, these discussions are too undifferentiated and neglect the real-world security reasoning and concerns. In the context of cloud, security problems increase tenfold. Indeed, the invoking schema is both horizontal as many cloud services use other cloud services not necessarily deployed on the same cloud, and vertical, as most cloud offering are deployed over several layers (IaaS, PaaS and SaaS) with different security requirements.

This observation is accentuated in the context of hybrid cloud computing where data security problems can be classified into two categories: security problems due to data externalization towards services deployed on the public cloud to ensure systems responsiveness; and security problems coming from data issued by public cloud services and that require to be trusted by the invoking private cloud. In the first case, the problems encountered are mainly privacy and integrity problems due to data

A. Haller et al. (Eds.): WISE 2011 and 2012 Combined Workshops, LNCS 7652, pp. 70–76, 2013.

externalization. Many solutions have been proposed such as [10]. In the second case, private cloud services often interact with cloud providers for services provisioning such as infrastructure, platforms and software. However, they aren't sufficiently assured about how trustable and credible the data computed by these resources are. The lack of trust in a data produced by a service deployed on a public cloud, can impact private cloud services security. Indeed, a data misuse by a private cloud service (which is generally kept inside due to its sensitivity) can lead to produce bad data or to make inappropriate decisions. In this position paper we concentrate on the last case. The case of hybrid cloud is slightly more complex as the private cloud service has to base its decision on a set of aspects such as the trust level of the invoked service, the public cloud service execution environment (the reliability of the physical or the virtual machine (if machine virtualization stands) and so on). Solutions to the trust challenge are hence to be crucial ingredient to a more wide spread deployment of hybrid cloud computing.

Solving this problem is not an obvious task due to clouds autonomy preservation, differences in control policy definitions, and lack of transparency in clouds.

One way of coping with the problem is to identify how to determine data credibility through evaluation of service trustworthiness in hybrid cloud computing environments? In other words, proposing a methodology that enables private cloud to determine the trustworthiness of public cloud entities based on past service invocation information inference. Selection of a particular service could be motivated by information such as services provider's reputation in terms of performance, security and quality of service.

The rest of the paper is organized as follows. Section 2 briefly mentions recent work on trusting public cloud entities. Section 3 presents our current research. At the end, we summarize conclusions and challenges for future work.

2   Trusting Public Cloud Entities: Recent Work

In the cloud context, trust solutions can be classified based on the following criteria:

- Fine grained vs. coarse grained: fine-grained solutions are aimed at gathering information about service deployment environment characteristics (e.g. information about Virtual machines and/or Physical machine run on). While coarse grained are more concerned about having a global overview of trust (e.g. whether it meets trust requirements or not).
- Proactive vs. Reactive: there are proactive approaches using preventive controls or reactive approaches using detective controls. The preventive controls are defined as procedures used to alleviate an issue to never take place. The detective controls are used when we request using a service or after an issue happens to check for causes.

In literature assessment, there are a lot of solutions treating trust issue in the cloud. We find out three categories of trust in cloud environment: trust based on accountability, trust based on third party and trust based on reputation feedback and social networking. Accountability has been widely used as a solution to detect faulty entities in distributed systems [1]. In [2, 3], this approach has been applied to cloud computing in order to find and expose faulty nodes and establish trust. TrustCloud [2] allows to answer, inside a cloud, questions like when, where and how an intruder threatened confidentiality or integrity of data inside a file. It is a file-centric solution that considers five levels of granularities in monitored data, from the data containers (e.g. files, databases), data provenance collection, creation and evolution and finally the data workflow. Similarly, DataProve [3] gathers data provenance from logs across system and application layers in a cloud. It redefines granularity of provenance by making it data-centric instead of file-centric. Data gathered concerns provenance of applications, virtual machines, physical machine and cloud location. This category does not treat the problem in a inter-cloud context. In *third party based trust* solutions [16, 17], a third entity establishes trust between two parties interested in interacting with each other. The trust management, in this category, is coarse grained, preventive and located on a third party which is often an online web interface. They help services providers to select reliable cloud providers based on a rating in form of a questionnaire filled in by current cloud consumers. This category focuses on helping users (instead of services) choosing the right cloud. Moreover, these solutions are not fine-grained. Finally in the third solutions category, trust is reputation based. In those works, inter-cloud trust is of concern. In [4], Abawajy et al. propose a distributed framework that enables interested actors determining the trust rate of federated cloud computing entities. It helps final consumers in associating a rate to exiting feedbacks estimating a prospective services provider. In [5], authors propose a solution where users are supported in choosing a trustworthy cloud provider by defining a set of evaluation attributes. In contrast, in [6], the reputation is gathered from peering entities, which represent each cloud. The reputation calculation is based on two factors: cloud's own experience and reputation rating. The cloud own experience is a value given by a cloud to another to rate how satisfactory the interactions were. While the reputation rating represents a value of confidence a cloud had about another cloud's behavior. The cloud own experience rates are published to other cloud peers. For this last category, the trust reasoning is coarse-grain and not especially adapted to the hybrid cloud context.

Considering those works, there is a real lack of trust solutions targeting public entities in the hybrid cloud context. Besides, there is no solution that considers fine grain features in this context. With this position paper, we so aim to close this gap. Besides, we think also that pointing public cloud data monitoring heterogeneity constitutes a notable argument to manage hybrid cloud security.

Thus, we believe that a trust manager service targeting public cloud entities should be an interesting expectation. The proposed service should gather data about previous service invocations and store them at the private cloud side and analyze them to meet the needs (in trust) of the invoking service towards a public cloud entity. This service should consider contextual constraints such as different service delivery and different service deployments models. In the following we discuss this idea.

3   A Premise of a Trust Management Service for Hybrid Cloud

Considering the aforementioned challenges, the service has to incorporate the following features:

Invoking Scenario Diversity. As we can see on top of figure 1, the possible service invoking scenarios are numerous and have to be taken into account in the proposed trust service. Indeed, the private cloud invoking service could be of SaaS, PaaS or IaaS type. It can invoke a public service of the same layer or lower. To illustrate the feature, a PaaS service invokes either a PaaS or an IaaS service in the public cloud. In the first case, the public entities to be trusted are for instance the physical and virtual machine where in the second one only physical machine is implied. Thus, in the first case the associated fine grain information that allow to make a reasoning about the trust level are a subset compared to the one considered in the second case.

Heterogeneity in Information Supplied by Public Cloud. By nature provided public cloud are able to supply heterogeneous data about their entities. Furthermore, data provided by public cloud provider depend on their governance leading to a form of heterogeneity. Hence a trust service has to base its reasoning on heterogeneous and probably incomplete set of information.

By analyzing the previous features, we come out with the two basic principles of the targeted trust manager service:

Data Set Layering. As mentioned before, the entities considered in the trust decision-making are more numerous when the invoked service is a SaaS and decreases for PaaS and IaaS services. Figure 1 illustrates a possible required data sets corresponding to each kind of service invocation.

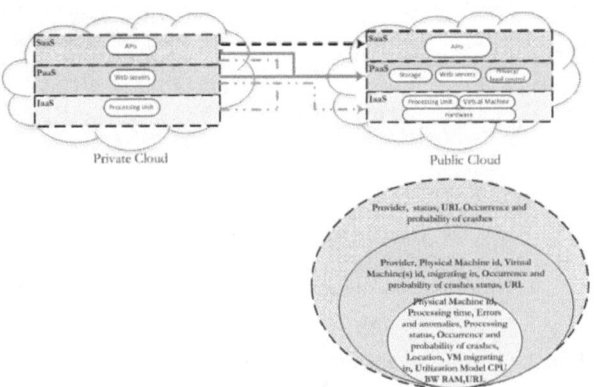

Fig. 1. Information gathered by service type

Reasoning Based on Incomplete Information to deal with the second feature. Using Bayesian network constitute an original and interesting faculty. Indeed, Bayesian learning takes a probability-based approach for reasoning and inferring results. Each training example that is encountered can change the probability that a hypothesis is correct. Moreover, rather than using knowledge from the current data set or training examples only, prior knowledge can be combined with the observed and incomplete data to arrive at more meaningful results. In our case, the decision of invoking (or not) a service should depend on its trust level. To do so, we plan to classify the list of services proposed by the public cloud to be invoked in two categories: trustable or untrustable services. Four variables may be considered: service reputation (SR), service provider reputation (SPR), physical (PMR) and virtual (VMR) (if virtualization stands on) machine reliabilities. Both service and service provider parameters should be instantiated in the training example according to their punctual reputation level. These values are taken from the local private cloud or from external (social network sources). Based on the training model, thanks to the Bayesian network, we can obtain a probability of service invocation trust on which stands a set of trusted services selection. We think about two selection algorithms. The first one located at the public cloud side tends to propose a subset of deployed and ready-to-use services that correspond to the invoking service needs. This algorithm should filter the set of services according to a given SLA extension by the private cloud trust requirements. The second selection algorithm is to be localized at the private cloud side and will help in choosing the best service among those proposed by the public cloud service taking into account information coming from the reasoning mechanism described below.

Attempting the previous original approach, we thought about a two sides' architecture illustrated by figure 2 and that should be composed of five main independent components:

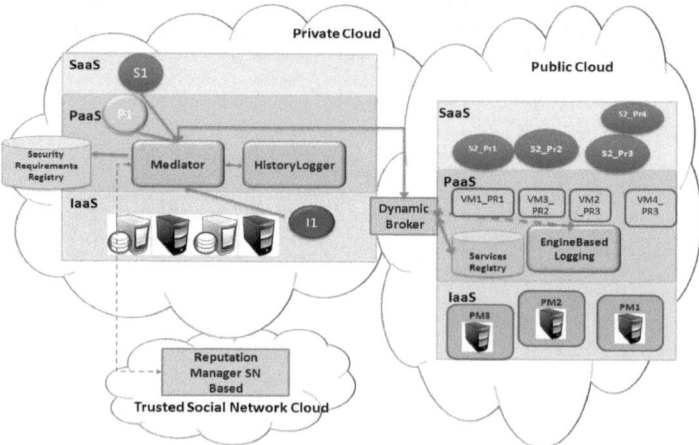

Fig. 2. Our trust management architecture for hybrid clouds

First, the *trust manager* -which should be a PaaS service located within the private cloud - runs the reasoning process and the private cloud algorithm in order to retrieve the appropriate service from the selection made by the public cloud. It should tightly collaborate with a *History Logger* also located at the private cloud to acquire past experiences obtained by already invoked services. At the public cloud side, we find the *dynamic Broker* whose main role in our case is to discover and select the appropriate service (SaaS, PaaS or IaaS) according to security restriction attributes inspired from the approach stated in [7]. Each new service, once published, can hence be easily reached, discovered and then selected for use. The public cloud algorithm will be run in this component. When the first selection achieved based on function criteria, an *engine Based Logging* service should filter more keeping only services that enforce the private clouds security constraints and expectations.

4   The Roadmap

In this position paper, we have first emphasized the need to be confident with information coming from public cloud service invoking in order to keep secure, more sensitive services located in the private cloud. This should be achieved taking into account public cloud entities trust level. We have then tried to define the features and the requirements of such trust service invoking and have given a preliminary architecture in term of PaaS services implementing this functionality both at the public and the private side. In addition, we would like to take into account public cloud heterogeneity in logging and disseminating information about their services towards their consumers.

The road ahead will be first to identify the sweet spot i.e the adequate scenario that will prove the importance of our proposal. We are thinking about an environmental scenario where information can be obtained from a set of cloud services associated with a corresponding sensor network. Second, we have to identify which fine-grain information from the public cloud, are interesting to measure the trust level of each public entity implied in a service invoking. This with the objective to determine which information is necessary to enforce service trustworthiness. Finally, we intend to investigate how to be able to give guarantees with respect to the effectiveness of such security/trust service. We will consider how to assess efforts on implementing the reasoning process thanks to a proposed Bayesian network. We will try to validate it using a simulation platform like the Weka platform [8]. Furthermore, we will assess the service feasibility using a cloud simulator like CloudSim [9].

References

1. Andreas, H., Peter, D., Petr, K.: PeerReview: Practical Accountability for Distributed Systems. In: Proceedings of Twenty-first ACM SIGOPS Symposium on Operating Systems Principles, vol. 41, pp. 175–188 (2007)
2. Ko, R.K., Jagadpramana, P., Mowbray, M., Pearson, S., Kirchberg, M., Liang, Q., Lee, B.S.: TrustCloud: A Framework for Accountability and Trust in Cloud Computing. In: 2011 IEEE World Congress on Services (SERVICES), pp. 584–588 (2011)
3. Olive, Z., Qing, K., Markus, K., Ryan, K.L., Bu Sung, L.: How To Track Your Data: The Case for Cloud Computing Provenance
4. Abawajy, J.: Determining Service Trustworthiness in Intercloud Com-puting Environments. Presented at the 10th International Symposium on Pervasive Systems, Algorithms, and Networks (ISPAN), Kaoh-siung, pp. 784 – 788 (2009)

5. Sheikh Mahbub, H., Sebastian, R., Max, M.: Cloud Computing Land-scape and Research Challenges regarding Trust and Reputation. In: Presented at the UIC-ATC 2010 Proceedings of the 2010 Symposia and Workshops on Ubiquitous, Autonomic and Trusted Computing, pp. 410–415 (2010)
6. Khan, K.M., Malluhi, Q.: Establishing Trust in Cloud Computing. IT Professional 12(5), 20–27 (2010)
7. Andrzej, G., Michael, B.: Toward dynamic and attribute based publication, discovery and selection for cloud computing. Future Generation Computer Systems 26(7), 947–970 (2010)
8. Weka, http://www.cs.waikato.ac.nz/ml/weka/
9. Cloudsim, http://www.cloudbus.org/cloudsim/
10. Rajak, S., Verma, A.: Secure Data Storage in the Cloud using Digital Signature Mechanism. International Journal of Advanced Research in Computer Engineering & Technology 1(4) (June 2012)

Challenges for Migrating to the Service Cloud Paradigm: An Agile Perspective

Stavros Stavru, Iva Krasteva, and Sylvia Ilieva

Sofia University "St. Kliment Ohridski" 5, James Bouchier Str.,
P.B. 48 1164 Sofia, Bulgaria
{stavross,ivak,sylvia}@fmi.uni-sofia.bg

Abstract. Migrating to the Service Cloud Paradigm implies the migration of legacy software systems to a service-oriented architecture with deployment in the cloud. Although this specific software modernization paradigm promises numerous strategic and operational advantages, it poses also many complex organizational and technical challenges, among which is the lack of mature processes, methods and techniques. This paper examines the questions of whether agile methods and techniques could be scaled to fit the migration to the Service Cloud Paradigm and how they could help overcoming the challenges of software modernization in this specific context. The research methodology presented here first extracts the challenges of the migration to Service Cloud Paradigm through a systematic literature review and then, using expert judgment, evaluates how different agile techniques, taken from Scrum and Extreme Programming (XP), could address the identified challenges. As a result, a ranked list of applicable agile techniques is presented and suggestions for their adoption in software modernization projects are drawn.

Keywords: Agile Software Development, Software Migration, Software Modernization, Cloud Computing, Service-Oriented Architecture.

1 Introduction

Building a modernized system from scratch often requires a huge investment in time and efforts, making this modernization approach almost unfeasible in real industrial settings. But software organizations, continuously pressured by their dynamic business and IT environment, have to modernize. Thus new and more undemanding modernization approaches are needed. With the increasing popularity of Service Oriented Architecture (SOA), the reuse of legacy system by exposing its functionality through services is identified as a feasible and very promising modernization approach. One particular modernization paradigm is the Service Cloud Paradigm, which stands for combination of Cloud Computing and SOA for development of Software as a Service (SaaS) systems and their deployment on the Cloud. Although this paradigm promises numerous strategic and operational advantages, it poses also many complex organizational and technical challenges, including but not limited to extensive business context analysis and complementary strategic efforts, steep

A. Haller et al. (Eds.): WISE 2011 and 2012 Combined Workshops, LNCS 7652, pp. 77–91, 2013.

learning curve, external dependencies and lock-ins, complex governance, interoperability, etc. Being in its early adoption stage, the migration to the Service Cloud Paradigm still lacks mature process models, methods and techniques, so needed for providing systematic guidelines for approaching software modernization in this specific context and overcoming its challenges. Agile methods and techniques, on the other hand, has been widely adopted in the industry and successfully applied in various contexts, different from traditional software development. Thus the question of whether agile methods and techniques could be scaled to the migration to Service Cloud Paradigm becomes a question of present interest.

The study presented in the paper proposes a systematic approach for reviewing the challenges of SOA and Cloud Computing relevant to Service Cloud Paradigm. Further it evaluates, through expert judgment, various agile techniques in terms of their potential to address these challenges. These agile techniques are taken from Scrum and XP software development methods since they are the most widely adopted methods through the agile community in the recent years (in more than two thirds of the projects surveyed by VersionOne[1]). The results of the evaluation step are used for making suggestions on the applicability of the techniques and their inclusion in a particular agile method for migration to Service Cloud Paradigm. The design of the agile method is out of the scope of this paper since it involves an extensive research, which cannot be presented within the limits of this paper.

The present study has been carried as part of FP7 European research project for reuse and migration of legacy applications to interoperable cloud services (REMICS[2]). The main objective of the REMICS project is to specify, develop and evaluate a tool-supported model-driven methodology for migration to the Service Cloud Paradigm. The migration process consists of several steps: (1) understanding the legacy system in terms of its architecture and functions; (2) designing a new SOA application that provides the same or better functionality; and (3) verifying and implementing the new application in the cloud.

The reminder of the paper is organized as follows: Section 2 describes the methodology used for conducting literature review and evaluating agile techniques; Section 3 presents the challenges from SOA and Cloud Computing fields, extracted by the review process and relevant for the migration to the Service Cloud Paradigm; Section 4 discusses the results of the evaluation of agile techniques and their potential to address the identified challenges; and Section 5 concludes the paper and outlines directions for future research.

2   Methodology

The research methodology used in the present study contains two basic steps. During the first step a systematic literature review on challenges of SOA and Cloud Computing, which are relevant to Service Cloud Paradigm, is performed. The second step involves a particular evaluation technique to study whether agile techniques can be used to address the challenges identified on the first step. The steps are presented in details in the subsections below.

---

[1] http://www.versionone.com
[2] http://www.remics.eu/

2.1   Review

Articles were eligible for inclusion in the review based on their relevance to the review objectives, which are: (1) they describe the current state of research and practice in either SOA and/or Cloud Computing; and (2) they identify and discuss different challenges these areas poses to both academia and industry. The relevance was evaluated by reviewing the abstracts of the articles and grading them as either relevant or irrelevant. The inclusion was also restricted by the type of the study, including only review articles and excluding theoretical (conceptual) or empirical studies. No restrictions were made in regard to the publication year of the articles thus covering all the years available in the included electronic database at the time of the review (1 January, 2012). Other exclusion criteria used were: (1) the article does not have abstract or the abstract is not available from the included electronic database; (2) the access to the full text of the article is restricted; and (3) the full text of the article is not available in English.

The search strategy included both journals and conference papers, and was limited to the Scopus electronic database. Scopus is the largest abstract and citation database of research literature and quality web sources, which ensured the coverage of nearly 18,000 titles from more than 5,000 publishers. The titles of both journals and conference papers were searched using the following search terms:

(1)   Service-Oriented Architecture - ("Service-Oriented" AND (Challenges OR Review OR Landscape OR Roadmap OR "State of"));
(2)   Cloud Computing - ("Cloud Computing" AND (Challenges OR Review OR Landscape OR Roadmap OR "State of").

Applying the search strategy resulted in an initial pool of 121 articles, 65 articles for SOA and 56 for Cloud Computing. Some additional articles, not covered by the search strategy, were also included as being recommended by the research community. Thus, by using the inclusion and exclusion criteria the initial pool of articles was limited to 58 articles. Their full texts were thoroughly examined in order to extract the challenges of SOA and Cloud Computing, presented in the subsequent sections.

2.2   Evaluation

Agile techniques were evaluated using expert's opinions through Delphi technique. Delphi is a technique frequently used for eliciting consensus from within a group of experts and has many advantages over other methods of using panel decision making [1]. Various researchers [1-3] all found that one of the major advantages of using it as a group response is that consensus will emerge with one representative opinion from the experts. Other advantages include its simplicity, anonymity, controlled feedback from the interaction and others [4]. Some limitations include that judgments are derived from the subjective opinions of experts and may not be representative, it requires adequate time and participant commitment, its validity extremely depends on the expertise and experience of the panelists, etc. [4] However, Linstone [2] recommends the Delphi technique when the examined issue does not allow the use of analytical techniques but can benefit from the subjective judgments on a collective basis, which we believe is our case.

The process followed was the one proposed by Pfeiffer [5] and included the following three steps:

(1) Experts' recommendations - A questionnaire was sent to a panel of four experts (with an average of 9 years of both academic and industrial experience in agile software development), asking them to review the list of challenges extracted by the review process and make subjective judgment and recommendations on which agile techniques (from Scrum and XP) could be used to address these challenges. From each expert a list of agile techniques was obtained.

(2) Experts' ratings - From the individual recommendations, a consolidated list of agile techniques was created. The list was then sent to each expert to further evaluate the relevance (on a five-point scale) of all techniques in regard to all challenges.

(3) Experts' consensus - The consolidated list, together with experts' ratings was sent again in order to discuss big differences in ratings. At the end consensus was gained, resulting in a final list of agile techniques and addressed challenges.

3   Challenges in Migrating to the Service Cloud Paradigm

The challenges identified by the review process were sorted into two categories:

(1) Organizational challenges, including challenges from all levels of the organization as strategic challenges (e.g. process reengineering, external dependencies, etc.), managerial challenges (e.g. governance, competence acquisition) and operational challenges (e.g. lack of methodologies, tools, etc.), and which were mostly process and people oriented;

(2) Technical challenges, which included design, implementation, verification challenges and deployment challenges, and were product and technology oriented.

As the focus of this study was on the migration to the Service Cloud Paradigm, we expected that not all of the challenges discussed by the reviewed articles would be relevant. For that reason, we limited the extraction of challenges to: (1) for Service-Oriented Architecture we included challenges relevant for both service consumers and providers, as software migration could involve both developing of new services and using external ones; (2) for Cloud-Computing we included only challenges relevant for the cloud consumers, excluding challenges covering the development of cloud infrastructure, how security should be achieved within this infrastructure and other challenges more relevant for the cloud provider.

3.1   Challenges of Service-Oriented Architecture

Total of 31 articles were thoroughly reviewed in order to extract the challenges relevant to SOA. As many challenges were found, they were further consolidated into total of 17 challenges, 10 of which were organizational and 7 were technical. A summary of these challenges, together with their references, is presented in Table 1.

Table 1. Challenges of Service-Oriented Architecture

#	Challenge	Description	Ref.
Organizational challenges			
O1	Identification and availability of service consumers	In some cases services are first developed and then offered to real service consumers (mostly due to customer unavailability). Therefore the service provider defines and prioritizes requirements based on its assumptions of what "potential consumers" would need or how "potential usage patterns" would look like.	[6-7]
O2	Identification and reengineering of business processes / tasks	Business process modeling and analysis might be an intensive activity and a prerequisite for the specification, design and implementation of services (if top-down or domain composition approach is incorporated)	[7-14]
O3	Lack of software development methods	Due to its early adoption stage, there is still scarce availability of software development methods and techniques for guiding the migration to and the development of software systems based on Service-Oriented Architecture.	[15-17]
O4	Unclear system ownership	Service-Oriented Architecture tends to blur the boundaries of software systems ownership, so who owns what (in terms of services) might become an issue.	[8-9, 11-12, 18-19]
O5	Complex governance	Service-Oriented Architecture raises unique challenges and can be derailed unless an effective governance framework is established to clearly identify roles and responsibilities.	[9-10, 12-13, 16-22]
O6	External dependencies	By developing software systems using external services, the organization is exposed to higher risk due to third party dependencies. Thus risk mitigation becomes important issue.	[11, 15]
O7	Acquisition of competencies and expertise	For an organization adopting or migrating to Service-Oriented Architecture, thorough understanding of the underlying technologies remains highly critical. At the same time too many technologies and standards could be involved and/or not enough expertise could be available in the market.	[6, 8, 17-18, 21, 23]
O8	Addressing increased complexity	The plethora and diversity of parties involved (service consumers, service developers, integrators, infrastructure developers, etc.), technologies incorporated, diversity of types of services available as run time components, the middleware and the infrastructure required to make these components interoperable and deployable, the technical skills and expertise required etc. increase significantly the complexity of the developed or migrated software system (System of systems).	[17-19, 24]

Table 1. (*continued*)

O9	Immature tools and integrated development environments	Due to its early adoption stage, there are few and still under development tools and integrated development environments to assist software engineers in their migration or development efforts.	[8-9, 12, 16-19]
O 10	Evolving standards	While the core web services standards (i.e. XML, SOAP, WSDL, and UDDI) are relatively mature and stable, many of the additional standards that address important issues such security and reliability (e.g. WS-Coordination, WS-Atomic Transaction, WSDM, WS-Reliability, etc.) are still under development.	[8-9, 12, 16-19, 25]

Technical challenges

T1	Addressing service design	Service design needs to determine what constitutes a service and its operations (or interface), and make decisions about the level of service aggregation (or service granularity). There is no straightforward approach (neither design methodology) and often the design process becomes challenging and cumbersome.	[6, 9, 11-12, 14-19, 21-23, 25-28]
T2	Securing reconfiguration and composition	As services could be used by different consumers to perform different business tasks (covering various business scenarios), they should be reconfigurable and facilitate composition and orchestration.	[7, 9, 12, 16, 18, 22, 28-31]
T3	Securing compatibility and standards compliance	Service interfaces cannot be changed very often due to issues such as backward and forward compatibility and compliance with standards.	[11, 16, 18, 22, 26, 32]
T4	Addressing security, interoperability and other quality aspects	There could many challenges related to security (due to publicized interfaces, distributed and no trusted network and channel, unintended orchestration, the use of open standards, etc.), interoperability (due to different standards, middleware and service infrastructure, service semantics, etc.), conducting performance and reliability analysis, ensuring correctness and reliability of test cases, provenance, etc.	[6, 8, 18-19, 23, 28, 30-35]
T5	Simulation of deployment environment	Test instances of the services and simulated deployment environment could be required during implementation as real services could be unavailable or too critical to be executed (e.g. ordering purchase).	[8-9, 11-12, 16]
T6	Testing services	Testing in Service-Oriented Architecture might require complex logistics, test instances of services, more levels of testing (due to diversity of parties involved), new techniques for issues localizations and troubleshooting, greater and more diverse exception handling, more complex route cause and impact analysis, etc.	[6, 9-11, 15-16, 18, 23, 26, 33, 36]
T7	Securing SLA and QoS contracts	Fulfillment of Service Level Agreement (SLA) and Quality of Service (QoS) contracts during services usage could be challenging due to network connectivity issues, unpredicted load, etc.	[9-12, 18, 22-23, 30]

As seen from Table 1, most of the reviewed literature (84% of the reviewed articles) was concerned with technical challenges. This was expected as SOA has to achieve some degree of technical excellence before it is widely accepted by the industry, and with the extensive use of SOA in real industrial settings the majority of organizational challenges would emerge.

The challenges in Table 1 have various implications on the incorporation of agile methods and techniques into the Service Cloud Paradigm, especially when it comes to organizational challenges. For example, working with potential service consumers and potential usage patterns (O1) contradicts with agile methods and techniques, which are mostly customer centric and require intensive interaction and collaboration with the customer. Not involving real customers (or at least somebody who could represent them) in the early phase could result in customer feedback received late in the development lifecycle, customer negotiation and relationship issues when real customers interests are conflicting (e.g. in terms of service functionality, interface, etc.); complex impact analysis and lack of flexibility and agility because of unknown customers (especially when the services are publicly available), etc. The prerequisite to define business processes and tasks in advance (O2), before the actual design, implementation and verification of services (if top-down or domain composition approach is incorporated) also poses some limitations on the use of agile methods and techniques, which emphasize on working software and thus applies very short increments (from 2 to 4 weeks) with potentially shippable products. This might also hinder achieving organizational agility and responding to change, because it limits the possible introduction of changes in the predefined business models late in the development life cycle. Challenges as unclear system ownership (O4, e.g. who should test external services, included in the current iteration; who should be responsible for the infrastructure required; etc.) and the need for complex service governance (O5, e.g. requiring adherence to codes, policies and standards in order to secure process and services compliance) set the focus on defining formal processes (incl. roles, responsibilities, activities, artifacts, etc.) and extensive usage of tools (incl. integrated development environments), while agile emphasize on individuals and interactions. In terms of technical challenges, there were also many implications. Coding and testing (mostly unit and integration) are not the only major activities in SOA. Considerable attention is paid also to service design, composition and orchestration (T1, T2), which in some cases could even replace traditional coding (e.g. when the system is built from external services only). In terms of testing, system testing (T5, T6) becomes crucial as there could be many parties involved (e.g. service developers, integrators, consumers, infrastructure developers, etc.). Software quality (T3, T4, T7) is much more emphasized because security, interoperability, reliability and compatibility, etc. are real life concerns in the SOA. There are also other implications to be considered, including increased complexity (O8, O10), evolving standards, etc.

3.2   Challenges of Cloud Computing

A total of 27 articles in the area of Cloud Computing were reviewed. The extracted challenges were further consolidated into 12 challenges, including 8 organizational challenges and 4 technical. They are shown in Table 2.

Table 2. Challenges of Cloud Computing

#	Challenge	Description	Ref.
Organizational challenges			
O1	Trust in the cloud provider	Trust becomes an issue when the data and computation are decentralized and distributed beyond the boundaries of the organization.	[37-47]
O2	External dependencies and vendor lock-ins	Exit strategies and lock-in risks (ability to switch cloud providers) are important concerns for organizations exploiting Cloud Computing, as well as vendor dependencies (e.g. only using services the provider is willing to offer).	[37-42, 46, 48-52]
O3	Security and privacy	The success or failure of Cloud Computing is highly dependable on how confident consumers feel that the services of the cloud provider are reliable and available, secure and safe, and that privacy is protected.	[37-38, 40-49, 51-60]
O4	Delegating data governance	By moving data into the Cloud, organizations might lose some capabilities to govern their own data, including data creation and receipt, distribution, use, maintenance and disposition.	[37-38, 43, 46-47, 49, 51-54, 57, 60-61]
O5	Introduction of new roles and responsibilities	The responsibilities for integrating in-house IT with the cloud infrastructure, adapting the in-house software systems with the cloud platform, etc. should be clearly defined and appropriate roles assigned (e.g. cloud infrastructure integrators, cloud platform integrators).	[61]
O6	Acquisition of competencies and expertise	Cloud Computing involves new technologies and standards, requiring the acquisition of new competences and expertise.	[61]
O7	In-house integration	Integration of in-house IT to the Cloud infrastructure could be hard and time consuming undertaking.	[37, 42]
O8	Early adoption stage	Due to its early adoption stage, Cloud Computing still lacks enough major cloud service providers, lacks variety of mature tools (incl. integrated development environments) and standards are continuously evolving (and even competing).	[49-50, 61-62]
Technical challenges			
T1	Addressing architectural and technical constraints	Cloud Computing poses some architecture constraints on the way software systems are build, incl. decomposition, decoupling, componentization, etc. and some technical constraints as what should be the type of database, etc.	[52]
T2	Maintenance and troubleshooting difficulties	Defects detection, localization and troubleshooting could be problematic as there might be no access to or sufficient control over the cloud infrastructure.	[51, 54, 63]
T3	Lack of cloud provider support	There might be lack of provider support or it might be provided as a paid service.	[54]
T4	Cloud interoperability	Cloud providers are not using any kind of common open standards, which makes cloud interoperability a serious concern.	[50, 61, 63]

The challenges mostly studied by the research community are related to security and privacy (81% of the reviewed articles), and trust (60%). This is expected since organizations tend to be extremely sensitive when crucial business assets (as business data and computation) are delegated to external parties. In cloud environment there might be no control (and even visibility in some cases) over these assets and no guarantees (even if there are strong service level agreement, strict standards, etc.) that the cloud provider would act in accordance to the interests of the organization and no external parties (or events) would significantly interfere in their relations (e.g. through business acquisition). Thus mistrust and lack of security and privacy might become the greatest barrier for the adoption of cloud solutions.

Cloud Computing and its challenges further affect the way agile methods and techniques could be incorporated into the Service Cloud Paradigm. Trust (O1), security and privacy (O3) of data and computation are as much important for the organization as for its customers, so the organization-customer collaboration, central to agile software development, might need to be extended to include the cloud provider (in order to increase transparency, visibility, responsiveness, etc., so needed for the building trust and confidence). Vendor lock-ins (O2) might further hinder organizational flexibility and agility (e.g. as one could not change its cloud provider effortless and in a timely manner), while external dependencies (O2) could decrease the business value delivered to customers (e.g. due to new requirements coming from the cloud infrastructure or the organization is pressured to use specific and expensive software licenses coming from the could provider, etc.) and could further decrease organizational responsiveness (e.g. due contracting). In terms of technical challenges, the maintenance and troubleshooting difficulties (T1), together with the lack of cloud support (T3), might require more involvement from upper management (in order to assure the commitment of the cloud provider). Also, the architectural and technical constraints (T2) and the need to consider quality aspects (T2, T4, such as interoperability, security, performance, etc.) might require considerable efforts to be made for architecture and design, except for coding and testing.

4   Results

The present section discusses the results of the evaluation of agile techniques based on the Delphi technique. The results are shown in Table 3, where the techniques are sorted by the total number of challenges they are expected to address (shown in brackets next to the technique's name).

Table 3. Evaluation of Agile techniques based on the challenges they are expected to address

Agile Technique	SOA Challenges	CC Challenges
Extreme Programming (XP)		
Small Releases (20)	O1, O6, O7, O8, O10, T1, T2, T3, T4, T5, T6, T7	O1, O2, O3, O4, O6, O8, T2, T3
Planning Game (14)	O1, O2, O4, O6, T1, T2, T3	O1, O2, O3, O4, O5, O7, T3

Table 3. (*continued*)

Pair Programming (13)	O5, O7, O8, O9, T1, T2, T3, T4, T5, T5	O6, T1, T2
Whole Team (13)	O2, O4, T1, T2, T3, T4, T5, T6	O5, O6, T1, T2, T3
Continuous Integration (12)	O5, O8, O8, T2, T3, T4, T5, T6	O3, O4, T1, T2
Test-Driven Development (11)	O5, O8, T1, T2, T3, T4, T5, T6, T7	T1, T2
System Metaphor (10)	O5, O8, T1, T2, T3, T5, T6	O6, O7, T1
Collective Code Ownership (7)	O4, O5, O7, T5	O6, O7, T2
Refactoring (6)	O5, O8, T5, T6	T1, T2
Simple Design (5)	O8, T1, T4, T5	T2
Coding Standards (4)	O5, O8, T3	T2
Scrum		
Sprint (20)	O1, O6, O7, O8, O10, T1, T2, T3, T4, T5, T6, T7	O1, O2, O3, O4, O6, O8, T2, T3
Sprint Planning Meetings (18)	O1, O2, O4, O5, O6, O7, O8, T1, T2, T3, T7	O1, O2, O3, O4, O5, O7, T3
Cross-Functional Teams (17)	O4, O5, O6, O7, O8, T2, T3, T4, T5, T6, T7	O5, O6, O8, T1, T2, T3
Product Backlog (15)	O1, O2, O4, O8, T1, T2, T3, T4, T7	O3, O4, O6, O8, T1, T2
Spring Backlog (15)	O1, O2, O4, O8, T1, T2, T3, T4, T7	O3, O4, O6, O8, T1, T2
Daily Scrum (12)	O4, O5, O6, O7, T2, T3, T4	O2, O6, T1, T2, T3
Scrum of Scrums (9)	O4, O5, O6, O8	O6, O7, T1, T2, T3
Scrum Master (9)	O4, O5, O6, T2, T3, T7	O2, O5, T3
Product Owner (9)	O2, O3, T1, T2, T3, T7	O2, O3, O4
Sprint Review Meeting (5)	O1, O5, T4, T3	T3
Sprint Retrospective (5)	O6, O7, O8	O6, T3
Sprint Burn Down Chart (4)	O5, O8	T2, T3

As seen from Table 3, the agile techniques recommended the most by the experts was Small Releases (from XP) and Sprints (from Scrum). Their arguments include receiving feedback quickly (e.g. identifying security, interoperability, privacy, etc. issues earlier in the development lifecycle), increasing responsiveness to change (e.g. limiting the impact of evolving standards on software delivery and cost/time overruns, overcoming vendor lock-ins by deploying increments on different cloud infrastructures, etc.) and issues (e.g. broken SLAs or QoS contracts), building trust and confidence (e.g. through frequent communication, increased visibility and traceability, etc.), effective competency acquisition (through learning by doing), early delivery of business value, etc. Next, Planning Poker (from XP) and Sprint Planning

Meeting (from Scrum) were rated second. The motivation for that was the active involvement of customers or customers' representatives in the development process and enhanced collaboration (e.g. resulting in more effective business process modeling and analysis by taking into consideration also the technical aspects, whether architectural or technical, of the migrated software system, better identification and prioritization of customer requirements based on customer value and collective estimations, effective identification of possible composition/choreography workflows, etc.), early detection of risks due to external dependencies and their timely mitigation (e.g. gaining support and commitment by customers and upper management to remove impediments coming from external parties), clarification of team responsiveness and service ownership (e.g. by identifying external services to be used by third parties, who will integrate them, who will test them, etc.), early escalation of quality concerns (e.g. compatibility, standards compliance, etc.), etc. The third most rated agile techniques were Pair Programming (from XP) and Cross-Functional Teams (from Scrum). Among the reasons for recommending these techniques were more effective acquisition of competencies and expertise (e.g. through daily knowledge transfer between programming pairs, homogeneous distribution of knowledge and expertise within the cross-functional team, etc.), managing increased complexity (e.g. through highly collaborative and knowledgeable workers), increased team responsiveness and support (e.g. team support for issues troubleshooting and resolution, improvement of used tools and procedures, identifications of impediments and external dependencies, etc.), secured quality in terms of security, interoperability, performance, etc. Other agile techniques, highly rated were Whole Team and Continuous Integration (from XP) and Product Backlog, Spring Backlog and Daily Scrum (from Scrum).

Based on the presented results and following the Pareto principle (80% of the effects come from 20% of the causes) [64], we would recommend that an organization, migrating to the Service Cloud Paradigm and interested in Agile Software Development, should start with small releases (or sprints), incorporate planning meetings similar to either Planning Game or Sprint Planning Meeting, and encourage Cross-Functional Teams. Afterwards, it might continue with Product / Sprint Backlogs, Daily Scrums, Continuous Integration and On-Site Customer in order to further address the challenges of the migration process. Finally, as all examined agile techniques have the potential to contribute to the Service Cloud Paradigm, an organization might also consider full implementation of either XP or Scrum (or a hybrid), as this will ensure cohesiveness and will allow the organization to take full advantage of these methods.

5 Conclusions

Agile methods and techniques have the potential to address many of the challenges possessed by the Service Cloud Paradigm. Thus an agile service- and cloud-oriented modernization method could be reasonable not only because the target organization might had already adopted the agile values and principles, but also because it promises to overcome the challenges of migrating to the Service Cloud Paradigm. The future work is the identification of challenges from other related areas as Model-Driven Development and Software Modernization, and the development of a complete agile method to support organizations in the adaptation of their legacy systems to the Service Cloud Paradigm.

Acknowledgments. The research presented in this paper was partially supported by FP7 project REMICS, contract No. 257793 and by National Scientific Fund, Bulgarian Ministry of Education and Science, Research Project agreement n. DO-02-182.

References

1. Helmer, O., Helmer-Hirschberg, O.: Looking forward: a guide to futures research. Sage Publications (1983)
2. Linstone, H.A., Turoff, M.: The Delphi method: techniques and applications. Addison-Wesley Pub. Co., Advanced Book Program (1975)
3. Dalkey, N.C., Corporation, R.: Delphi: Rand (1967)
4. Yousuf, M.I.: Using Experts' Opinions through Delphi Technique. Practical Assessment Research & Evaluation 12 (2007)
5. Pfeiffer, J.: New look at education: systems analysis in our schools and colleges. Odyssey Press (1968)
6. Tilley, S., et al.: Migrating to SOA: approaches, challenges, and lessons learned. Presented at the Proceedings of the 2010 Conference of the Center for Advanced Studies on Collaborative Research, Toronto, Ontario, Canada (2010)
7. Bano, M., Ikram, N.: Issues and Challenges of Requirement Engineering in Service Oriented Software Development. In: 2010 Fifth International Conference on Software Engineering Advances (ICSEA), pp. 64–69 (2010)
8. Mahmood, Z.: Service oriented architecture: potential benefits and challenges. Presented at the Proceedings of the 11th WSEAS International Conference on Computers, Agios Nikolaos, Crete Island, Greece (2007)
9. Lewis, G.A., et al.: Effects of service-oriented architecture on software development lifecycle activities. Software Process: Improvement and Practice 13, 135–144 (2008)
10. Kontogiannis, K., et al.: A research agenda for service-oriented architecture. Presented at the Proceedings of the 2nd International Workshop on Systems Development in SOA Environments, Leipzig, Germany (2008)
11. Kontogiannis, K., et al.: The Landscape of Service-Oriented Systems: A Research Perspective. In: International Workshop on Systems Development in SOA Environments, SDSOA 2007, ICSE Workshops, p. 1 (2007)
12. Lewis, G.A., et al.: Common Misconceptions about Service-Oriented Architecture. In: Sixth International IEEE Conference on Commercial-off-the-Shelf (COTS)-Based Software Systems, ICCBSS 2007, pp. 123–130 (2007)
13. Zheng, L., et al.: Facing Service-Oriented System Engineering challenges: An organizational perspective. In: 2010 IEEE International Conference on Service-Oriented Computing and Applications (SOCA), pp. 1–4 (2010)

14. Hutchinson, J., et al.: Evolving Existing Systems to Service-Oriented Architectures: Perspective and Challenges. In: IEEE International Conference on Web Services, ICWS 2007, pp. 896–903 (2007)
15. Nasr, K.A., et al.: Realizing service migration in industry—lessons learned. Journal of Software Maintenance and Evolution: Research and Practice, n/a–n/a (2011)
16. Papazoglou, M., et al.: Service-Oriented Computing: A Research Roadmap. International Journal of Cooperative Information Systems 17, 223 (2008)
17. Kokko, T., et al.: Adopting SOA: Experiences from Nine Finnish Organizations. In: 13th European Conference on Software Maintenance and Reengineering, CSMR 2009, pp. 129–138 (2009)
18. Mahmood, Z.: The Promise and Limitations of Service Oriented Architecture. International Journal of Computers 1, 74–78 (2007)
19. Becker, A., et al.: Value Potentials and Challenges of Service-Oriented Architectures. Business & Information Systems Engineering 3, 199–210 (2011)
20. Maurizio, A., et al.: Service Oriented Architecture: Challenges for Business and Academia. In: Proceedings of the 41st Annual Hawaii International Conference on System Sciences, p. 315 (2008)
21. Bhallamudi, P., Tilley, S.: SOA migration case studies and lessons learned. In: 2011 IEEE International Systems Conference (SysCon), pp. 123–128 (2011)
22. Papazoglou, M.P., et al.: Service-Oriented Computing: State of the Art and Research Challenges. Computer 40, 38–45 (2007)
23. Tilley, S.: Report from the 5th and 6th international workshops on adoption-centric software engineering: Migrating to SOA. In: 2011 IEEE International Systems Conference (SysCon), pp. 135–139 (2011)
24. Nigul, L., et al.: The SOA programming model: challenges in a services oriented world. Presented at the Proceedings of the 2009 Conference of the Center for Advanced Studies on Collaborative Research, Ontario, Canada (2009)
25. Feuerlicht, G.: Enterprise SOA: What are the benefits and challenges? In: Systems Integration, pp. 36–43 (2006)
26. Lewis, G.A.: SMART: The Service-oriented Migration and Reuse Technique: Carnegie Mellon University, Software Engineering Institute (2005)
27. Kulkarni, N., Dwivedi, V.: The Role of Service Granularity in a Successful SOA Realization A Case Study. Presented at the Proceedings of the 2008 IEEE Congress on Services - Part I (2008)
28. Issarny, V., et al.: Service-oriented middleware for the Future Internet: state of the art and research directions. Journal of Internet Services and Applications 2, 23–45 (2011)
29. Brown, P.C.: Succeeding with SOA: realizing business value through total architecture. Addison-Wesley (2007)
30. Choudhury, P., et al.: Deployment of Service Oriented architecture in MANET: A research roadmap. In: 2011 9th IEEE International Conference on Industrial Informatics (INDIN), pp. 666–670 (2011)
31. Yi, W., Blake, M.B.: Service-Oriented Computing and Cloud Computing: Challenges and Opportunities. IEEE Internet Computing 14, 72–75 (2010)
32. Balasubramaniam, S., et al.: Challenges for assuring quality of service in a service-oriented environment. In: ICSE Workshop on Principles of Engineering Service Oriented Systems, PESOS 2009, pp. 103–106 (2009)
33. Simanta, S., et al.: Information assurance challenges and strategies for securing SOA environments and web services. In: 2009 3rd Annual IEEE Systems Conference, pp. 173–178 (2009)

34. Venters, C.C., et al.: Provenance: Current directions and future challenges for service oriented computing. In: 2011 IEEE 6th International Symposium on Service Oriented System Engineering (SOSE), pp. 262–267 (2011)
35. Phan, C.: Service Oriented Architecture (SOA) - Security Challenges and Mitigation Strategies. In: Military Communications Conference, MILCOM 2007, pp. 1–7. IEEE (2007)
36. Canfora, G., Di Penta, M.: Testing services and service-centric systems: challenges and opportunities. IT Professional 8, 10–17 (2006)
37. Verma, A., Kaushal, S.: Cloud Computing Security Issues and Challenges: A Survey. In: Abraham, A., Mauri, J.L., Buford, J.F., Suzuki, J., Thampi, S.M. (eds.) ACC 2011, Part IV. CCIS, vol. 193, pp. 445–454. Springer, Heidelberg (2011)
38. Xiu-ping, Z.: Study on the opportunities and challenges of the cloud computing for Chinese medium-sized and small enterprises. In: 2011 International Conference on E - Business and E -Government (ICEE), pp. 1–3 (2011)
39. Habib, S.M., et al.: Cloud Computing Landscape and Research Challenges Regarding Trust and Reputation. In: 2010 7th International Conference on Autonomic & Trusted Computing (UIC/ATC), pp. 410–415 (2010)
40. Dillon, T., et al.: Cloud Computing: Issues and Challenges. In: 2010 24th IEEE International Conference on Advanced Information Networking and Applications (AINA), pp. 27–33 (2010)
41. Al-Qirim, N.: A Roadmap for success in the clouds. In: 2011 International Conference on Innovations in Information Technology (IIT), pp. 271–275 (2011)
42. Kim, W., et al.: Adoption issues for cloud computing. Presented at the Proceedings of the 11th International Conference on Information Integration and Web-based Applications & Services, Kuala Lumpur, Malaysia (2009)
43. Oh, T.H., Lim, S., Choi, Y.B., Park, K.-R., Lee, H., Choi, H.: State of the Art of Network Security Perspectives in Cloud Computing. In: Kim, T.-h., Stoica, A., Chang, R.-S., et al. (eds.) SUComS 2010. CCIS, vol. 78, pp. 629–637. Springer, Heidelberg (2010)
44. Roberts II, J.C., Al-Hamdani, W.: Who can you trust in the cloud?: a review of security issues within cloud computing. Presented at the Proceedings of the 2011 Information Security Curriculum Development Conference, Kennesaw, Georgia (2011)
45. Hay, B., et al.: Storm Clouds Rising: Security Challenges for IaaS Cloud Computing. In: 2011 44th Hawaii International Conference on System Sciences (HICSS), pp. 1–7 (2011)
46. Timmermans, J., et al.: The Ethics of Cloud Computing: A Conceptual Review. In: 2010 IEEE Second International Conference on Cloud Computing Technology and Science (CloudCom), pp. 614–620 (2010)
47. Takabi, H., et al.: Security and Privacy Challenges in Cloud Computing Environments. IEEE Security & Privacy 8, 24–31 (2010)
48. Ovadia, S.: Navigating the Challenges of the Cloud. Behavioral & Social Sciences Librarian 29, 233–236 (2010)
49. Mathisen, E.: Security challenges and solutions in cloud computing. In: 2011 Proceedings of the 5th IEEE International Conference on Digital Ecosystems and Technologies Conference (DEST), pp. 208–212 (2011)
50. Petcu, D.: Portability and interoperability between clouds: Challenges and case study. In: Abramowicz, W., Llorente, I.M., Surridge, M., Zisman, A., Vayssière, J. (eds.) ServiceWave 2011. LNCS, vol. 6994, pp. 62–74. Springer, Heidelberg (2011)
51. Mathur, P., Nishchal, N.: Cloud computing: New challenge to the entire computer industry. In: 2010 1st International Conference on Parallel Distributed and Grid Computing (PDGC), pp. 223–228 (2010)

52. Chang, H., Choi, E.: Challenges and security in cloud computing. In: Kim, T.-H., Vasilakos, T., Sakurai, K., Xiao, Y., Zhao, G., Ślęzak, D. (eds.) FGCN 2010. CCIS, vol. 120, pp. 214–217. Springer, Heidelberg (2010)

53. Kossmann, D., Kraska, T.: Data Management in the Cloud: Promises, State-of-the-art, and Open Questions. Datenbank-Spektrum 10, 121–129 (2010)

54. Joshi, K.R., et al.: Dependability in the cloud: Challenges and opportunities. In: DSN, pp. 103–104 (2009)

55. Morrell, R., Chandrashekar, A.: Cloud computing: new challenges and opportunities. Network Security 2011, 18–19 (2011)

56. Al-Azzoni, I., et al.: Abstract only: performance evaluation for software migration. SIGSOFT Softw. Eng. Notes 36, 42 (2011)

57. Roberts, J.C., Al-Hamdani, W.: Who can you trust in the cloud?: a review of security issues within cloud computing. Presented at the Proceedings of the 2011 Information Security Curriculum Development Conference, Kennesaw, Georgia (2011)

58. Lar, S.U., et al.: Cloud computing privacy & security global issues, challenges, & mechanisms. In: 2011 6th International ICST Conference on Communications and Networking in China (CHINACOM), pp. 1240–1245 (2011)

59. Zhao, W.: An Initial Review of Cloud Computing Services Research Development. In: 2010 International Conference on Multimedia Information Networking and Security (MINES), pp. 324–328 (2010)

60. Zhang, Q., et al.: Cloud computing: state-of-the-art and research challenges. Journal of Internet Services and Applications 1, 7–18 (2010)

61. Dudin, E., Smetanin, Y.: A review of cloud computing. Scientific and Technical Information Processing 38, 280–284 (2011)

62. Loutas, N., et al.: Cloud Computing Interoperability: The State of Play. In: 2011 IEEE Third International Conference on Cloud Computing Technology and Science (CloudCom), pp. 752–757 (2011)

63. Grobauer, B., Schreck, T.: Towards incident handling in the cloud: challenges and approaches. Presented at the Proceedings of the 2010 ACM workshop on Cloud Computing Security Workshop, Chicago, Illinois, USA (2010)

64. Pareto, V.: Manual of political economy: Scholars Book Shelf (1971)

Cloud Storage of Artifact Annotations to Support Case Managers in Knowledge-Intensive Business Processes

Marian Benner-Wickner[1], Matthias Book[2], Tobias Brückmann[1], and Volker Gruhn[1]

[1] Paluno – The Ruhr Institute for Software Technology
University of Duisburg-Essen
[2] Dept. of Computer Science, Chemnitz University of Technology
{marian.benner-wickner,tobias.brueckmann,
volker.gruhn}@paluno.uni-due.de
matthias.book@informatik.tu-chemnitz.de

Abstract. In many domains, there are tasks for which no strict process can be prescribed, but which require the expertise of case managers who work with information from a broad set of sources. To support case managers' highly individual work, we present an approach that enables them to structure their work along a flexible agenda and a diverse collection of artifacts. We propose a cloud-based architecture for annotating artifacts with document- and content-level metadata to support case managers' cognitive effort of organizing, relating and evaluating the multitude of artifacts involved in each case, and describe a tool that provides a consolidated environment for working with artifacts from heterogeneous sources.

Keywords: case management, agendas, artifacts, annotations, cloud storage.

1 Introduction

Business process management and workflow management ensure that process instances follow the definitions and guidelines described in the form of process models. This works nicely for all kinds of processes that are well-understood and highly structured. Such business processes tend towards automation (as a consequence of ongoing industrialization) [1]. Usually, however, certain parts of business processes cannot be automated; very often, they cannot even be described in detail. They are less structured and depend much more on the expertise of a responsible person (called a case manager in the following). Very often, such a case manager does not follow a predefined process model, but a more or less coarse agenda that he arranges flexibly as the case demands. Here we leave the traditional field of business process management and arrive at so called case handling [2]. In case handling, we observe that case managers employ other cognitive approaches for planning and structuring their work than they would need to "blindly" follow predefined processes. Besides the agenda, the artifacts that the case manager works with are in the focus of the attention, and a lot of cognitive effort is spent on finding them, relating them, evaluating them, basing decisions on them, etc. To support these activities, case managers form a mental model of

A. Haller et al. (Eds.): WISE 2011 and 2012 Combined Workshops, LNCS 7652, pp. 92–104, 2013.

the information landscape and the relevance and relations of all information fragments (artifacts) that takes a lot of effort and focus to keep in one's head without suitable tool support.

In this paper, we introduce a cloud-based toolset for flexible agenda and artifact management. The toolset has been developed in the context of our research agenda concerning the support of case manager. Based on the observation that some business activities cannot be expressed by process models, but are highly dependent on searching, exploring, and extracting knowledge from a broad body of information stemming from various (enterprise-internal as well as publicly available) data sources of different format and nature, we provide a workspace for artifact-oriented working in business processes.

In summary, the contribution of this paper is as follows:

- We introduce the notion of knowledge-intensive business processes to indicate that business processes may contain parts cannot be exhaustively described *a priori*.
- We introduce the notion of flexible agendas, which let case managers organize their process work in different styles.
- We introduce the notion of a workspace, which supports knowledge work that is completely artifact-oriented.
- We introduce a mechanism for annotating artifacts that is independent of their file format and source location, by storing all annotation in a central cloud service.

In the following section, we introduce the core concepts of case management. Section 3 then discusses an architecture for annotating artifacts with metadata in a unified way, and storing those annotations in a central cloud service. Sect. 4 gives an overview of a toolset supporting case managers. We conclude with a discussion of related work in Sect. 5 and an outlook on further research in Sect. 6.

2   Agenda- and Artifact-Driven Case Management

Even though the tasks of a case manager do not have a clearly defined internal structure (and therefore cannot be supported by traditional workflow management systems), we believe that solving these tasks can be tool-supported. Our mechanisms focus on supporting different working styles of case managers in their knowledge work, where they shall be free to plan and structure their work according to the cognitive approach that seems most useful to them. To accomplish this, we introduce the concept of agendas (Sect. 2.1) and the notion of a workspace containing all relevant artifacts (Sect. 2.2). Artifacts are documents, links, e-mails and all the other data sources needed by the case manager during completion of the task. (At this time, we do not explicitly consider collaboration with other stakeholders and experts. However, artifacts can also link to shared repositories like Google drive. These shared sources constitute another important class of data sources, and context comes into play as a third class of data source when the case manager is not office-based, but works on location with a mobile device [3]. The explicit integration of collaborative and mobile aspects into the approach is subject of our ongoing work.)

We will introduce the concepts of agendas, templates and artifacts using an example from the rehabilitation management domain in the context of disability insurance. Here, for certain kinds of injuries, rehabilitation managers are responsible for driving the medication and therapy process in order to avoid unnecessary costs and to ensure quick recovery of the injured person. While some of the activities performed by a rehabilitation manager can be explicity defined and supported by structured dialogs and algorithms, the details of certain other parts (such as selecting the most appropriate care provider) are rather undefined: For these parts, it is not specified in detail what has to be done at which step, what kinds of information sources have to be considered, and what kinds of data have to be produced – we just know that the task requires and produces certain input and output artifacts, and may have a coarse agenda as an informal guideline for what needs to be done to complete the task. The following scenario illustrates this:

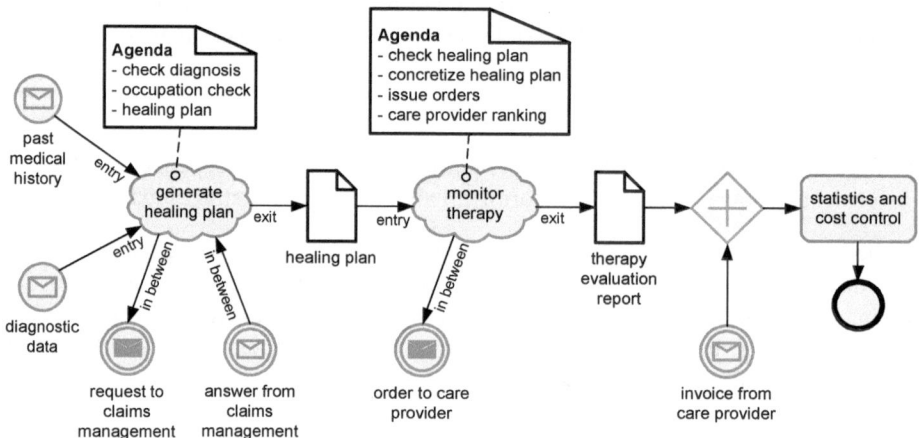

Fig. 1. Structured and unstructured parts in a rehabilitation control process

Case manager Smith, working for the disability insurance company TakeCare, is responsible for the rehabilitation management of the insured person Mr. Miller, who has been injured with a comminuted femoral neck fracture. Supplied with input sources such as Miller's past medical history and diagnostic data, Smith first needs to create a healing plan for this patient. Since there is no pre-defined process for this task, it is represented by the "generate healing plan" element in Fig. 1. As a guideline, Smith only has a coarse agenda at his disposal. Working with this agenda, Smith creates a healing plan, which is the input for the following task "monitor therapy". In the context of both unstructured tasks, Smith has to carry out numerous activities whose execution, repetition or omission is not predefined, but depends on the peculiarities of each case. Smith also has to produce certain deliverables only known within each task, and some deliverables relevant to the surrounding structured process as well (e.g. the therapy evaluation report). To do so, he has to access various external sources, such as rankings of care providers' performance, etc. While working in the unstructured parts, Smith also produces intermediate results that can be processed by activities in the structured process parts, while the unstructured task is still being solved. For example, in the "generate healing plan" task, Smith can issue requests to the claims management department, whose responses are available to Smith as he works on the task. Ultimately, as the therapy is completed, the process continues in a structured way again: Processing the therapy evaluation report and the invoices from care providers are pre-defined standard activities that can be expressed as traditional business processes.

In the following subsections, we will show how case managers can be supporting methodically and technically in working on unstructured, knowledge-intensive tasks like the ones described in the above scenario.

2.1   Agendas

In an unstructured process part, the notion of an *agenda* is of central importance. An agenda (usually derived from a template or best practice for a certain case type) contains links to and names of all entities that a case manager puts on the list of things he plans to consider. If starting from an activity-driven perspective, first-class agenda items are activities that are candidates for execution. If the perspective is more artifact-oriented, then first-class agenda items are any form of artifacts (files, documents, web pages etc.) and other sources of information. If the perspective is more structure-driven, first-class agenda items are key entities of the task's domain. In the given scenario, for example, "healing plan", "care providers" and "therapy" are such key domain entities. This structure-driven perspective allows allocating activities, information sources and documents around key entities of the case domain.

Generally speaking, a case manager does not have a single perspective. Instead, some parts of his work might be activity-driven, others more determined by artifacts, and selected key domain entities might be used for collecting activities and documents under the roof of one keyword. Other ways of structuring items are allowed as well, e.g. along the lines of communication partners, along a timeline, alphabetically or in any other way the case manager considers most useful for organizing his work.

Our notion of an agenda supports these arbitrary mixes of perspectives by just using the general notion of agenda items, and by supporting hierarchies of agenda items in order to support different levels of abstraction. Agenda item hierarchies can be arbitrarily linked with each other. If, for example, an activity such as "generate healing plan" is linked to the domain entity "healing plan", then this may be interpreted to be an activity which works on a healing plan. With this rather broad notion of agendas and agenda items, each case manager can use agendas according to his preferences.

2.2   Artifacts

The solution of an unstructured task rarely can be accomplished by following a purely activity-driven agenda. Rather, one of the most frequently performed activities of a case manager is to search and explore information related to the case. He will access multiple information sources, both internal and external ones, some of which are structured, while others are not. Based on the information gathered, he builds a mental model of the way he wants to solve the case. We collectively call all information that the case manager gathers *artifacts*. He can work with these by combining, evaluating, rating and relating the information they contain. For such situations, the paradigm of working along an agenda is expanded by the paradigm of handling individually assembled universes of information. To support this style of working, we introduce the notion of workspaces.

In the example scenario described before, the healing plan is a key artifact. This artifact may be based on generally accepted therapy plans, but it may also deviate from these mainstream plans due to certain patient-specific risks. In this case, a rationale for the deviation from the plan, or a scientific report describing the alternative therapy, can be researched and attached to the healing plan by the case manager.

The case manager organizes the artifacts for a task in the workspace provided by our toolset, as shown in Sect. 4. He can include information and documents from heterogeneous sources, he can arbitrarily associate artifacts in his workspace with each other, and he can annotate these artifacts and associations with metadata as discussed in the following section. In the course of the task's completion, the case manager will dynamically re-arrange and relate the artifacts in his workspace as he accesses further sources of information and evaluates the collected information.

3   Cloud-Based Artifact Annotations

Actors driving knowledge-intensive processes are known to evaluate gathered artifacts using annotations. They may add metadata such as comments, tags and ratings, and they highlight content. We can find such annotations on two different levels: on the artifact level, where they refer to the complete documents, e-mails, web pages etc. as a whole; and on the content-level, where they concern specific fragments, elements and bits of information that are part of the whole artifact.

In this section, we initially propose requirements and challenges in the context of annotation support. We then propose a system architecture that enables knowledge workers to annotate any kind of document regardless of its format and source.

3.1   Requirements and Challenges

Most commonly available artifact editors and viewers provide basic support for annotating the artifacts they can process. But in some cases, the features differ: Microsoft Word documents, for example, can be annotated on the artifact and content level. Adobe Portable Document Format (PDF) files can also be annotated on both levels, but usually the artifact-level annotation is read-only for PDF viewers. Both file types support similar artifact and content-level annotation features such as tagging, adding comments and highlighting text, although commenting is limited to the content level in PDF files. Limited support for rating artifacts is incorporated in the Microsoft Windows operating system, but only applicable to certain image file types.

In order to make it easier for case managers to apply and work with meta information on the artifacts they deal with, we propose two basic requirements to be satisfied by IT support for case management: Ensuring annotation support for a wide range of heterogeneous artifact types and sources, and providing a homogeneous user experience independently of the artifact type and source. Given these requirements, we can identify several challenges that have to be coped with in order to support annotations of artifacts, which will be discussed in more detail below:

- **Missing write permissions:** Where to store annotation information?
- **Missing annotation support** in artifacts' document/data model: How to store annotation information?
- **Inconsistent user interfaces:** How to provide similar user experience for annotating different types of artifacts with different editors?

Missing write permissions are a problem whenever third-party information shall be annotated or organizational rules prohibit manipulation of data. Sometimes the artifact itself is read-only (e.g. some PDF files). In many cases, the underlying file system may not be accessible, e.g. remote web pages and reference sections in corporate intranets. It is obvious that annotations cannot be stored in such files and locations. Due to the unpredictable and multi-variant nature of data access in knowledge-intensive processes, we assume that issuing general write access – whether pre-arranged or granted on-demand – is not feasible. Therefore, any solution to support annotations of heterogeneous artifacts must involve some external storage for the meta information of sources to which the case manager does not have write access.

While many formats such as Word and PDF documents, as well as JPEG images, provide annotation support, **missing annotation support** in the artifact's file format is a problem for many artifact types. For example, HTML supports only artifact-level annotations (by way of META tags in combination with a profile like the Dublin Core [4]) For images, comments can be stored in the "comment" field as part of text chunks in the PNG specification or with the "COM" marker in the JPEG specification. In addition, there are standards like EXIF and XMP [5] that allow the integration of artifact-level annotation information into media. However, simple artifact types like plain text files do not have such complex data models and hence neither support artifact- nor content-level annotations. It is therefore not yet feasible to store and exchange both artifact- and content-level annotation information within the data models of all kinds of artifacts.

Finally, due to the heterogeneity of file editors and viewers, there are many **inconsistent user interfaces** for annotating artifact – both on the command level (i.e. which menus, buttons or hotkeys are required to annotate an artifact), and on the display level (i.e. how the annotation is represented visually). Consequently, every time the case manager uses a previously unknown viewer or editor for a new artifact type, he has to adapt how he works with annotations, which affects his efficiency – especially since he has to keep switching between many different artifact types in the course of his work.

In the following subsection, we therefore suggest a solution that provides an interface that standardizes the user experience and solves the storage and access problems highlighted above.

3.2   Architecture for Annotation Storage

We believe that the first two challenges mentioned above require a centralized, separate storage space for annotation information and other meta data referring to the artifacts from many decentralized sources.

This storage can be implemented as an annotation service that is part of a private document cloud. The server in the cloud hosts an Annotation Management component and an Annotation Service component. To unify the user experience and smoothly integrate the proposed cloud service into the client, each viewer/editor application must be equipped with an Annotation Add-On that interacts with the server (Fig. 2).

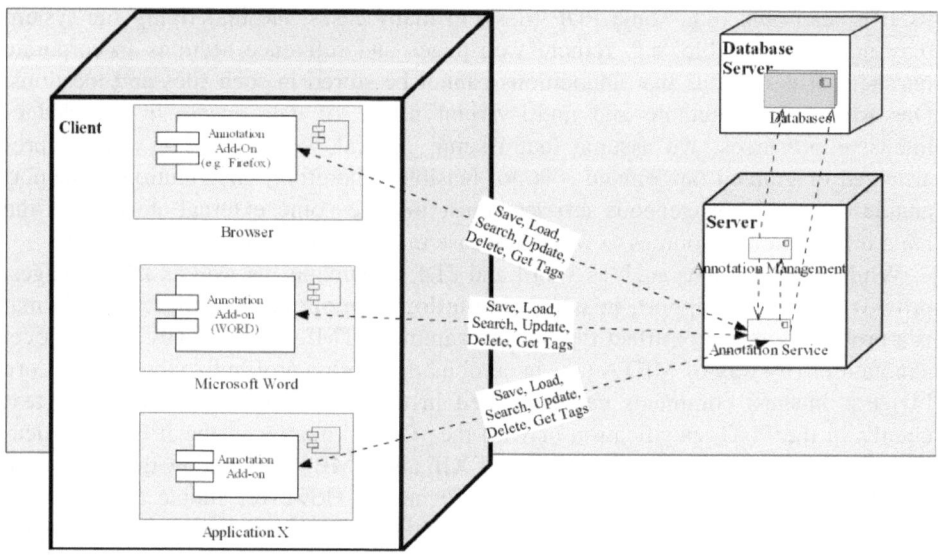

Fig. 2. Coarse architecture of cloud-based annotation service

The Annotation Management component encapsulates the database connection. It also provides annotation management functions to the Annotation Service interface component. The latter component can be a representational state transfer (REST) service providing basic operations to the application add-ons.

For the client side, since different artifacts have to be annotated, an approach using add-ons allows us to extend the basic functionality of different artifact editors and viewers (this approach also solves the problem that some editors or file formats do not have any built-in means for creating annotations). The proposed Annotation Add-On consists of two components, the Annotation Editor and the Annotation Viewer. The Annotation Editor provides a user interface to manipulate annotation information, i.e.

editing properties like comments, ratings, tags etc., while the Annotation Viewer is responsible for showing existing annotations.

In order to integrate the Annotation Add-on into a client application (e.g. a browser or word processor), it binds itself as an event listener in the client application. In order to provide a unified user experience, it is vital that it reacts to the same events in each client application (e.g. a particular keyboard shortcut), in order to open the Annotation Editor from within the web browser or word processor.

To display annotations, each Annotation Add-On uses the capabilities of its surrounding client application to produce visually similar annotation presentations. For this purpose, it may reuse existing annotation capabilities, e.g. the commenting function of PDF and Word documents. If dedicated capabilities are missing, the Annotation Add-On may employ formatting functions to achieve similar representations – for example, highlighting text in a browser could be achieved by formatting the annotated text with a yellow background.

Given such add-ons for the various artifact viewers that case managers work with, they will encounter a unified user interface (and have a central, source-independent storage facility) for all meta data that they associate with the artifacts relevant to their case. This means that they do not need to spend cognitive effort in dealing with the differences of the various viewing tool, or even the impossibility of annotating artifacts from certain sources, and can instead focus on the contents of the artifacts and their meaning and relevance for the case at hand.

4   Case Management Toolset

In this section, we describe a user interface prototype of a case management toolset implementing the concepts introduced in the previous sections. It is currently implemented for the Microsoft Windows platform as a prototype. We will describe it using the example of controlling long-term rehabilitation of patients by insurance case managers.

To help the case managers to keep an overview of their progress on each case, find information relevant to it and work with it, the interface of the toolset provides three window panes, as shown in Fig. 3: The *agenda pane* on the left displays the agenda items that need to be worked on during the traversal of this case. The initial population of the agenda can be managed using templates. So agenda templates are designed to facilitate functions similar to workflow management, but for knowledge-intensive processes.

Clicking on any agenda item will bring up all associated artifacts in the artifacts pane – the main workspace – in the middle. These can be excerpts from or links to web sites or intranet pages, data and documents provided by third-party services, as well as information clipped from spreadsheets, manuals or other documents, e-mails, contacts and appointments, and even notices written by the case manager himself. Artifacts displayed here may have associated annotations that are drawn from the central cloud storage. The source pane on the right provides detailed views of the various artifacts that are relevant to a case, as well as the data sources from which those artifacts are imported – it is essentially a combined web browser and document viewer. To enable case managers to work with a broad range of artifacts, the toolset does not implement editing and viewing logic itself, but instead embeds existing applications such as web browsers or word processors as needed.

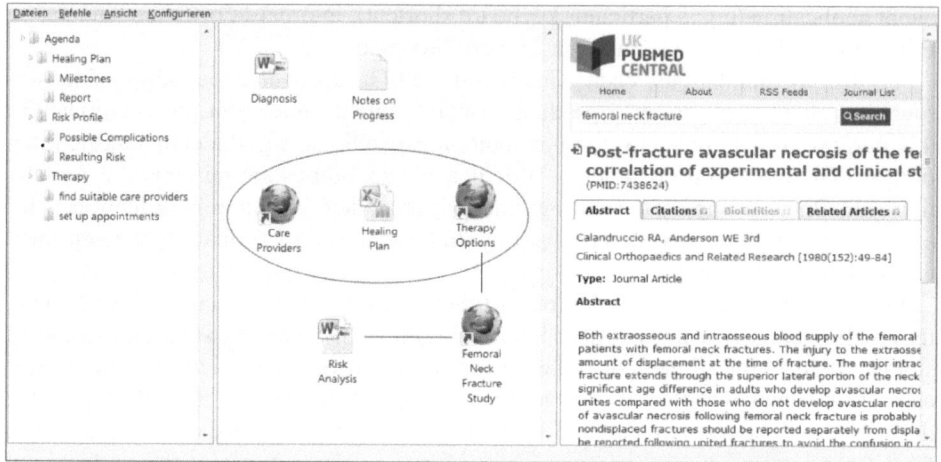

Fig. 3. User interface prototype of toolset with agenda, artifact and source panes

When the case manager picks up a new rehabilitation case, the agenda is filled with the items of the organization's rehabilitation case template, and the artifacts pane is populated with the input artifacts that describe the case, such as the patient's diagnosis. With this, the case manager can begin working on the case: Rather than following a step-by-step process, he may decide upon reviewing the input artifacts that additional agenda items are necessary – for example, obtaining additional information on counter-indications against particular medications, if the patient has other chronic conditions. He can therefore refine the agenda by adding the respective activities to the more coarse key elements found in the template. Since the agenda items aim to guide the task solution, the case manager can annotate them with further attributes he finds helpful to structure his work. For example, the case manager may categorize agenda items by denoting them as "actions", "checkpoints", "reviews" etc., and can attach deadlines, completion percentages or other organizational attributes to them. The tool can use this information to present accordingly filtered views on the artifacts, to remind the case manager of upcoming deadlines, or to give an overview of open issues that will help him to organize and prioritize his work in a particular case.

As the case manager is working on an individual patient's rehabilitation progress, he needs to consider a number of aspects that are prescribed in the agenda, such as the patient's diagnosis and living conditions, recommended therapies, and the location and qualification of suitable care providers. To research information on these and other aspects relevant to the case, the toolset provides a consolidated interface for accessing and excerpting a broad spectrum of data sources: the source pane can embed applications for showing web or intranet pages, spreadsheets, documents, third-party data services, e-mails, etc. A set of bookmarks (defined case-independently in the agenda, but editable and extendable on a case-by-case basis) provides quick access to relevant data sources for the case's domain.

When the case manager finds relevant information while browsing a source document, he can copy & paste excerpts or drag & drop the whole document into the artifact pane. The toolset in that case records the copied information (if feasible for that artifact format), and also a reference of the source where it was found, to enable the case manager to come back to this source for updated information later if necessary. In the source window, the case manager can also use the annotation features provided by the embedded applications' annotation add-ons. Besides ensuring a consistent user experience when working with annotations in the toolset, the cloud-based annotation service also ensures that the annotations will remain associated with the artifacts even when they reside in remote sources that cannot be copied locally.

The extracted artifacts are displayed in the artifacts pane as labeled icons, which can be visually set in relation with each other through free arrangement on the pane. Further visual cues can be added, such as connecting lines for relationships and outlines for groupings. The user interface provides simple drawing tools for this purpose. As shown in the middle of Fig. 3, for example, the case manager might cluster information related to the healing plan, and relate it to specialist information such as medical studies and risk assessments.

To associate the artifacts with agenda items that they are relevant for (e.g. with the agenda item "Therapy"), they can be dragged and dropped onto the respective item. When a user then clicks on an agenda item, only the associated artifacts, their annotations and their mutual relationships will be displayed, in order to introduce some structure into what may grow into a large collection of artifacts.

Ultimately, the case manager will complete the case by creating an "output artifact" that is passed to the following more structured steps of the business process that this unstructured task has been embedded in.

5   Related Work

Our work belongs to the field of approaches that support the execution of those parts of business processes that cannot be reasonably described in a structured manner. The related literature calls these types of processes "semi-structured processes", "knowledge-intensive business processes" [2], "case-oriented business processes" [6], "ad-hoc processes" [7], "weakly-structured business processes" [8], "flexible processes" [9], "knowledge work" [10], "dynamic case management" [11] or "adaptive case management" [12].

A prominent approach to describe processes with a certain variability in following a defined control flow is the "case handling paradigm" as developed by van der Aalst et al. [2]. There, a case consists of activities, data objects, precedence relations between activities, and relations between them, as well as a state space used to describe the current state of activities and data objects. Our approach also focuses on required data objects and defined goals of unstructured activities. However, our approach does not prescribe any precedence relation or control flow definition for their internal steps. Instead, we provide an agenda as a guideline to case managers, but no explicit order of agenda items or control flow elements that would introduce some notion of a structured process.

A further approach called "activity schemes" is presented by Schmidt et al. in [8]. They describe knowledge activities as a graph of knowledge action nodes that includes relations between activities, applications, and resources. In contrast to this knowledge-focused approach, we focus on steps that have to be taken and goals that have to be achieved during a task's completion.

Pinto et al. propose an approach called "goal-oriented and activity-based workflow modeling" [13]. In their domain model for case management, they integrate the data flow with control flow elements and specify the completion of actions through pre- and post-conditions of data objects. The relation of activities is determined by production of data in certain states (the achieved goals). With our use of input and output artifacts for instantiating and completing a task, we apply a similar mechanism. Additionally, our approach strictly separates the structured parts of a business process (ordered activities with control flow and decisions) from the unstructured parts (with defined input and output artifacts, but without any assumptions about the internal control flow), and we do not aim to provide a mixture of both.

To provide support through tools and workspaces, Pinto et al. describe the "work list" concept, which is used to keep goals and activities that have to be considered by the case manager [13]. Holz et al. use a "task list" as a central concept for supporting case management [14]. This task list, particularly, is comparable to our agenda due to the possibility of relating resources to task list items.

Besides the research community, industrial key players also provide solutions to support case management. Forrester [15] identified Pegasystems, IBM, EMC, Appian, Singularity and Global 360 as leading software vendors in this field. However, their tools do not focus extensively on agendas, artifacts and annotations as described here.

6   Conclusion

In this paper, we introduced an approach for handling segments within a structured business process that cannot be executed according to a pre-defined series of steps, but whose completion relies extensively on the expertise and judgment of a case manager who works with a broad spectrum of data sources in order to arrive at results. To support work on such tasks, we introduced the concepts of agendas to guide the case manager, and the artifact workspace in which he can collect, annotate and evaluate relevant data from highly heterogeneous resources. Besides these conceptual foundations, we described mechanisms for storing artifact annotations in a central cloud-based storage, independently of the source artifacts' format and location, and presented a user interface prototype toolset that provides a unified user experience for the case manager's tasks. Since the (possibly quite voluminous) artifact information is stored in the cloud, it enables the case manager to access it from anywhere, even if he is using a portable device in a process requiring mobility (e.g. visiting patients).

By having all relevant information consolidated in one tool, being able to relate them and go back to their sources if necessary, we expect to free case managers from a significant cognitive load that the operational aspects of their research work would otherwise impose on them. A particular feature that supports case managers' natural cognitive processes is that the approach combines the hierarchical thought patterns typically employed in planning activities (i.e. the aspects expressed in the tree of agenda items) with the networked relationships found in knowledge representations such as documents and information sources, which are modeled by the graph structure of the artifact workspace.

In our ongoing work, we are extending our approach with mechanisms for monitoring the steps taken to resolve a case, and deriving agenda template optimizations or data source suggestions from them that can be used in other cases. These steps occur in collaboration with case managers who give feedback on their work patterns and strategies, in order to ensure the effectiveness of the approach in practice. Future research subjects include the explicit support of collaborative work in the completion of unstructured tasks, as well as support for mobile case handlers who directly interact with their environment. For these users, we expect the central cloud storage to also enable much more flexible work since all information and metadata can be retrieved from any place, regardless of its original source location.

Acknowledgments. The authors would like to thank Alexander Kalinowski and Yordan Terziev for their work on the case management toolset prototype.

References

1. Singularity: Case Management – Combining Knowledge With Process. In: BPTrends (July 2009), http://www.bptrends.com
2. van der Aalst, W.M.P., Weske, M., Gruenbauer, D.: Case handling: a new paradigm for business process support. Data & Knowledge Engin. 53(2), 129–162 (2005)
3. Gruhn, V., Köhler, A., Klawes, R.: Modeling and analysis of mobile business processes. Journal of Enterprise Information Management 20(6), 657–676 (2007)
4. Kunze, J., Baker, T.: The Dublin Core Metadata Element Set. RFC 5013 (August 2007), http://www.ietf.org/rfc/rfc5013.txt
5. Adobe, Inc.: Extensible Metadata Platform (XMP), http://www.adobe.com/products/xmp/
6. Reijers, H.A., Limam, S., van der Aalst, W.M.P.: Product-Based Workflow Design. Journal of Management Information Systems 20(1), 229–262 (2003)
7. Vanderfeesten, I., Reijers, H.A., van der Aalst, W.M.P.: Case Handling Systems as Product Based Workflow Design Support. Enterprise Information Systems 12, 187–198 (2008)
8. Schmidt, B., Stoitsev, T., Muehlhaeuser, M.: Activity-centric support for weakly-structured business processes. In: Proceedings of the 2nd ACM SIGCHI Symposium on Engineering Interactive Computing Systems (EICS 2010), pp. 251–260. ACM (2010)
9. Schonenberg, H., Weber, B., van Dongen, B., van der Aalst, W.M.P.: Supporting Flexible Processes through Recommendations Based on History. In: Dumas, M., Reichert, M., Shan, M.-C. (eds.) BPM 2008. LNCS, vol. 5240, pp. 51–66. Springer, Heidelberg (2008)
10. Hädrich, T.: Situation-oriented Provision of Knowledge Services. Dissertation, Martin-Luther-University Halle-Wittenberg (2008)

11. Pegasystems, Inc., `http://www.pega.com/products/case-management`
12. Swenson, K.D.: Mastering the unpredictable: How adaptive case management will revolutionize the way that knowledge workers get things done. Meghan-Kiffer Press (2010)
13. Pinto, B.O., Silva, A.R.: An Architecture for a Blended Workflow Engine. In: Daniel, F., Barkaoui, K., Dustdar, S. (eds.) BPM Workshops 2011, Part II. LNBIP, vol. 100, pp. 382–393. Springer, Heidelberg (2012)
14. Holz, H., Maus, H., Bernardi, A., Rostani, O.: From Lightweight, Proactive Information Delivery to Business Process-Oriented Knowledge Management. Journal of Universal Knowledge Management. Special Issue on Knowledge Infrastructures for the Support of Knowledge Intensive Business Processes 0(2), 101–127 (2005)
15. Miers, D., Le Clair, C.: The Forrester Wave – Dynamic Case Management, Q1 2011 Report, Forrester Research, Inc. (January 2011)

Designing an SLA Protocol with Renegotiation to Maximize Revenues for the CMAC Platform

Adriano Galati[1], Karim Djemame[1], Martyn Fletcher[2,3], Mark Jessop[2,3], Michael Weeks[2,3], Simon Hickinbotham[2,3], and John McAvoy[3]

[1] Collaborative Systems and Performance Group, School of Computing,
University of Leeds, LS2 9JT, UK
[2] Advanced Computer Architecture Group, Department of Computer Science,
University of York, YO10 5DD, UK
[3] Cybula Ltd. R&D Team, Science Park, York, YO10 5DD, UK
{a.galati,k.djemame}@leeds.ac.uk,
{martyn.fletcher,mark.jessop,michael.weeks,simon.hickinbotham}@york.ac.uk,
mcavoy@cybula.com

Abstract. The emerging transformation from a product oriented economy to a service oriented economy based on Cloud environments envisions new scenarios where actual QoS mechanisms need to be redesigned. In such scenarios new models to negotiate and manage Service Level Agreements (SLAs) are necessary. An SLA is a formal contract which defines acceptable service levels to be provided by the Service Provider to its customers in measurable terms. This is meant to guarantee that consumers' service quality expectation can be achieved. In fact, the level of customer satisfaction is crucial in Cloud environments, making SLAs one of the most important and active research topics. The aim of this paper is to explore the possibility of integrating an SLA approach for Cloud services based on the CMAC (Condition Monitoring on A Cloud) platform which offers condition monitoring services in cloud computing environments to detect events on assets as well as data storage services.

Keywords: SLA, WS-Agreement, requirements, renegotiation.

1 Introduction

Cloud computing is emerging as a new computing paradigm and it is gaining increasing popularity throughout the research community. One aspect of the cloud is the provision of software as a service over the internet, i.e. providing applications (services) hosted remotely. Ideally, these services do not require end-user knowledge of the physical compute resource they are accessing, nor particular expertise in the use of the service they are accessing. Whilst the cloud offers opportunities for remote monitoring of assets, there are issues regarding resource allocation and access control that must be addressed to make the approach as efficient as possible to maximize revenues. From the service provider perspective, it is impossible to satisfy all customers' requests and a balance mechanism needs

A. Haller et al. (Eds.): WISE 2011 and 2012 Combined Workshops, LNCS 7652, pp. 105–117, 2013.
© Springer-Verlag Berlin Heidelberg 2013

to be devised through a negotiation process. Eventually, such a process will end up with a commitment between provider and customer. Such a commitment is a commercial contract that guarantees satisfaction of the QoS requirements of customers according to specific service level agreements(SLAs). SLAs define the foundation for the expected level of service agreed between the contracting parties. Therefore, they must cover aspects such as availability, performance, cost, security, legal requirements for data-placements, eco-efficiency, and even penalties in the case of violation of the SLA. The QoS attributes need to be explicitly defined with clear terms and definitions by SLAs which exhibit how service performance is being monitored, and what enforcement mechanisms are in place to ensure SLAs are met [1].

Although the cloud computing research community recognises SLA negotiation as a key aspect of the WS-Agreement specifications, little work has been done to provide insight on how negotiation, and especially automated negotiation, can be realised. In addition, it is difficult to reflect the quality aspects of SLA requirements. The scope of this paper is to present a Cloud environment where an SLA protocol may be designed and developed for the management of the negotiation, the monitoring and the renegotiation phase of the agreed terms. We intend to integrate this tool into the CMAC (Condition Monitoring on A Cloud) platform [7] which offers a range of analytical software tools designed to detect events on assets and complex systems as well as data storage services. For this purpose, we choose the WSAG4J framework [2] which is an implementation of the WS-Agreement standard [6] from the Open Grid forum (OGF). It provides comprehensive support for common SLA management tasks such as SLA template management, SLA negotiation and creation, and SLA monitoring and accounting. In the rest of this paper we present in Section 2 the CMAC platform and in Section 3 general concepts of SLAs. In Section 4 we introduce our SLA protocol and in Section 5 we draw attention to some aspects related to its design and integration in the CMAC platform that we expect to address at this early stage with particular focus on the SLA renegotiation protocol. Finally, Section 6 concludes this paper.

2   CMAC Project

The CMAC project [7] provides a platform that allows people to run compute-intensive research in an ordered manner over the internet. CMAC operates in a manner that combines state of the art pattern search capability with a software-as-a-service platform called YouShare [8]. CMAC brings the process of condition monitoring in cloud computing to deliver an integrated monitoring service that combines monitoring information across geographical locations, thus delivering a powerful solution at low cost. In a typical application, asset data may be distributed across a number of sites, e.g. data from sensors distributed across one or more assets of different sites are uploaded to the cloud computing platform where condition monitoring takes place.

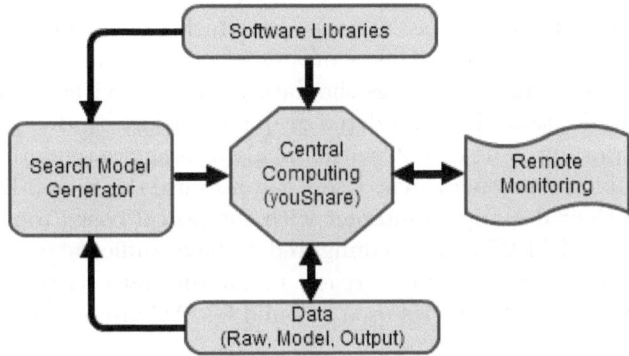

Fig. 1. CMAC top-level overview

Figure 1 presents a top-level overview of the CMAC components. The YouShare platform is the central computing component which heads up the whole system. The *Remote Monitoring* component encapsulates the process of accessing the system to monitor assets from remote locations. The *Software Libraries* provide various services among which monitoring, scheduling, pattern recognition and visualization. The system maintains a scheduler service which recognizes data uploads and deploys pattern search services on the data.

The *Search Model Generator* builds the models that are used to monitor the data. The pattern search services form the core of the condition monitoring process. They are currently developed around the AURA-Alert software [3–5] which is capable of rapidly detecting novel patterns in the data set. Since CMAC is built upon YouShare, any software can be used to process data, as long as the operating system that hosts the software can be run as a virtual machine. The pattern search requires raw data and a search model to generate monitoring data. A model defines the configuration of the pattern search, i.e. search parameters and patterns to be searched for. Models are used to generate monitoring data and visualizations of an asset using the data from the sensors. Then, a specialist engineer can analyse the monitoring data and reconfigure the model. CMAC considers three types of data (not including associated metadata): raw data is the monitoring information and are continuously uploaded to the system from the assets being monitored; model data (search model) is the specification of the pattern search; monitoring (output) data is the result of the monitoring service on raw data by employing a model data. A field engineer uses the monitoring data for managing, maintaining and planning the use of the asset.

The high level picture of the CMAC architecture and its data flows are presented in Figure 2. Currently, the CMAC platform focuses on two main tasks: Pattern Search and Visualization Data Generation. Pattern search is a more intensive service which continuously process data streams uploaded from external assets. In addition, an initial data rendering is carried out with the support of several levels of interactive viewing. The CMAC cloud carries out condition monitoring upon arrival of raw data using the search models. Data processing is carried out on various execution nodes which can be located on different sites. The outputs are archived as a matter of course, but they are also sent to a further service which processes the data for remote visualization on a web browser. A combination of pre-rendered graphical objects and asynchronous requests for raw data provide CMAC with a richly interactive rendering of the data within acceptable processing times. Users can also use the Signal Data Explorer (SDE) [17] on their desktop to interact with the portal based tools for analysis and visualization. CMAC can be configured to have sufficient virtual machines constantly available, with services ready to handle data as soon as it is uploaded. The existing technologies used to build CMAC are presented in the next subsections.

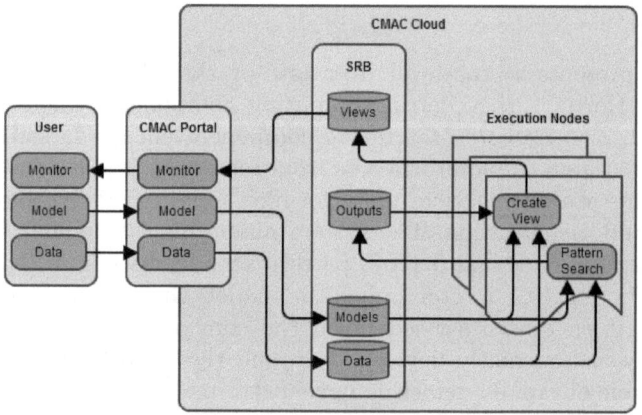

Fig. 2. CMAC architecture and data flows

The project, "Harnessing Large and Diverse Sources of Data Condition Monitoring on a Cloud (CMAC)" Ref: 5175-33386, is funded by the Technology Strategy Board (TSB). The CMAC academic and industrial partners are the University of York, Cybula Ltd. and DTP Group.

2.1   YouShare Platform

The YouShare platform is a development of the CARMEN project (Code Analysis, Repository and Modelling for E-Neuroscience) [9, 14, 15] which provides a unique virtual neuroscience laboratory; an infrastructure for using and sharing data, tools and services that is now in regular use by neuroscientists. The YouShare infrastructure extends the CARMEN platform and allows researchers from any discipline to create collaborative groups through which to work using shared data and software accessed though a web portal.

YouShare allows users to upload raw data sets, to describe them with extensive metadata, to store output data produced by the execution of services, and to apply sharing policies on them. YouShare allows users to create their own services, run these against data held in the system, and to share their services with other YouShare users. The service is then executed and upon completion, the YouShare portal stores the output and displays the result to the user. In YouShare, requests from the portal are handled by servlets, which create YouShare job requests that are passed to the Storage Resource Broker (SRB) and the Service Manager of the execution nodes. YouShare is capable of running any operating system within a virtual machine, and thus any uploaded software application without needing to port it to its appropriate OS, consequently reducing programming overheads. Upon job completions, results are sent back to the servlets where they are packaged and made available through the portal.

The YouShare system operates as Cloud and Software as a Service Model. It is underpinned by compute and data resources at the University of York and forms part of the White Rose Grid computing infrastructure [16].

2.2   Signal Data Explorer

The Signal Data Explorer (SDE) [17] is a Cybula's time-series pattern matching engine and data visualization tool. It is a stand-alone product supported by Windows operating system which provides functionalities for visualization and exploration of time-series data sets, to create and manipulate models as search queries, to search for pattern matches in raw data, to manipulate search models based on the monitoring results. CMAC exploits the components of the SDE and makes its functionalities available for remote monitoring on any HTML5 browser from any operating system.

2.3   AURA-Alert

AURA-Alert [3–5] is a novelty-detection plug-in for SDE, designed to detect anomalies in signal data from complex assets. It is particularly simple to use and set up. It scales very well to large volumes of data and may be used on historical and real-time data feeds. AURA-Alert is also part of CMAC to provide support for remote monitoring services.

3   SLA Concepts

Recently, the WS-Agreement standard has been widely adopted for most of the projects where SLAs are being implemented [2, 10, 11] However, several SLA alternatives have been developed due to WS-Agreement limitations. A comparison of SLA use in six of the european commissions FP6 projects is presented in [18].

In this project our work will be based on the WSAG4J framework which implements an SLA Engine Module. This module provides a generic implementation of a WS-Agreement based SLA engine. WS-Agreement is a Web Service

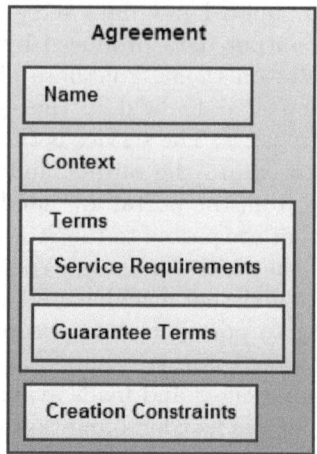

Fig. 3. Structure of a negotiation offer

protocol developed by GRAAP-WG from the OGF[6]. It implements the standard functionality for processing and creating agreement offers, monitoring their states at runtime, and evaluating and accounting agreement guarantees [6]. Besides, agreement acceptance policies and business logic for instantiating and monitoring SLA aware services can easily be plugged in.

Providers produce agreement templates which present the services and describe their properties. The WS-Agreement defines the SLA structure as an agreement template consisting of four blocks (see Figure 3):

- *Name* is optional and contains the name of the agreement. It is not the agreement identifier but just a human identification.
- *Context* contains information on the parties that created the agreement. Thus, the context defines both agreement initiator and agreement responder. Such fields can contain any arbitrary domain specific description of each party in order to resolve them to real world, e.g. an endpoint reference or a distinguished name that uniquely identifies the party. These details are mandatory, but the specification allows the service provider to extend them with more agreement details.
- *Terms* contains all service requirements and guarantee terms related to assurances and commitments between client and service provider (see Section 4 for more details).
- *Creation Constraints* provides some information to control a negotiation process. Negotiation constraints are used by a negotiation participator to define the structure and possible values that can be considered for counter offers. They are a means to express the requirements of a negotiation party. This block only appears on the negotiation template.

SLA negotiation [12, 13], creation and renegotiation of agreements are based on the WS-Agreement specification [6], which follows a Discrete-Offer-Protocol message exchange. The negotiation starts with the client requesting SLA templates from service providers. Once the two parties come to a final agreement it is possible to monitor the quality of the requested services during runtime. By evaluating the individual service term states it can be deduced whether an agreement is fulfilled or risks to be violated if no appropriate countermeasures are taken. Thus, an intelligent resource management mechanism is required for maximizing revenue and minimizing SLA violation penalties. In general, the fulfilment of an agreement is crucial for determining whether a contractual penalty has to be paid from the service provider or the service consumer.

4   SLA Protocol Design

Most research up to now provides little insight on how negotiation, and in particular automated negotiation, can be realised. In addition, it is difficult to define the quality aspects of SLA requirements. Here, we design our SLA protocol which will be integrated in the CMAC platform (see Figure 4). For this purpose, the main challenges that we tackle in order to provide QoS guarantees, are mainly four:

- the identification of the QoS properties, i.e. its requirements and terms, and their publication in the SLA Template;
- the decision process to accept, reject, or renegotiate the counteroffer in the negotiation process;
- SLA violation monitoring based on the agreed terms;
- the contract renegotiation protocol.

With regard to the first issue, it is not currently tackled by the WS-Agreement specification. We have decided to structure the SLA template distinguishing between *requirements* and *terms* (see Figure 3). We assume requirements to describe sufficient conditions required by the service to be executed. More precisely, a service might have need for technical (i.e. a specific operating system, CPU capacity, amount of RAM), syntactical (i.e. defined format of the input data) and ethical requirements (i.e. majority age for the consumer to use the service). Requirements are presented exclusively by the service provider and cannot be negotiable; they are essential for the fulfillment of the service. Our SLA protocol allows negotiation based on the terms presented by the parties. In this context, the agreed terms are necessary conditions, but not sufficient, to reach an agreement; service requirements must be satisfied anyway. It is necessary to determine all the guarantee terms that will be signed by both parties. For the CMAC services we identify some terms which are QoS parameters like the delivery ability of the provider, the performance target of diversity components of user's workloads, the bounds of guaranteed availability and performance, the measurement and reporting mechanisms, the cost of the service, the terms for renegotiation, and the penalty terms for SLA violation. Our SLA protocol defines ad-hoc SLA template structures for each service on the base of its prerequisites.

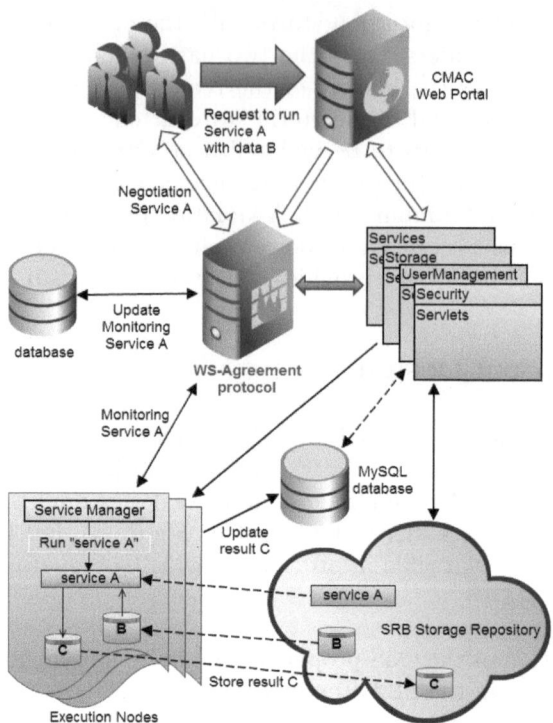

Fig. 4. CMAC architecture with SLA protocol for renegotiation support

Measurement of metrics and definition of all the service requirements and guarantee terms is done by a negotiation process between both parties. Our SLA protocol has been designed to allow renegotiation only for the guarantee terms. In this respect, the WS-Agreement negotiation protocol will decide whether to accept or reject the user's offer, or eventually renegotiate providing a counteroffer, if there are the prerequisites for raising one. The renegotiation is quite flexible; it is based on temporal restrictions, resource constraints, previous offers and a maximum number of renegotiations will be possible. The negotiation constraints in the negotiation template are used to control the negotiation process. Although, virtualization of resources is a prerequisite for building a successful cloud infrastructure, we do not consider it at this stage. In this work we do not address self-renegotiation after system failures, which has currently become an open issue [20]. Once the service requirements and the guarantee terms are met the contract has been stipulated.

At this point the SLA protocol will start monitoring the service and determining whether the service objectives are achieved or violated. Namely, if the provider has delivered the service within the guaranteed terms. The profiling process will record and analyse the service metrics in the local database so as to assess or predict their capabilities. Such a behaviour will assist the job scheduling to be highly optimized in service allocation and resource reservation. The whole system is orchestrated by our WS-Agreement protocol, which acts as broker between the CMAC platform and the users (see Figure 4). At this early stage we assume only the service provider to be in charge of this process, although each party should be in charge of this task and how fairness can be assured between them is an open issue. The monitoring process helps the service provider to prevent violations of the guarantee terms by renegotiating them. In this work, we also intend to design a system able to maximize revenues trying to make services available as soon as possible to clients willing to pay for their immediate availability. This implies postponing queuing requests waiting to use that service and asking their owners whether they accept to renegotiate the agreed terms. Immediate availability is allowed upon their renegotiation. At this stage we assume that renegotiations can be initiated by the service provider upon the customer's consent in the agreement of the initial negotiation. Such acceptance does not imply that the customer will accept the terms of a possible renegotiation.

5   Proposed Approach

Whilst cloud computing technologies offer opportunities for remote monitoring of assets, there are issues regarding resource allocation and access control to optimize their usage. It is not always possible to allocate resources for the deployment of services, thus they have to be concurrently shared between all of the clients' requests. In this context scheduling of incoming data flows might be employed to share resources in the cloud. Time slots might be assigned to data flows uploaded from different assets to be processed. Their frequency and duration is initially agreed during the SLA negotiation phase. Renegotiation requests can be raised for various reasons, e.g. to maximize revenues by reassigning time slots to incoming clients willing to pay for immediate use of the service or to prevent the violation of the agreed terms. At this stage, we assume only the service provider able to initiate a renegotiation. In this work we intend to design and develop an SLA protocol for the management of the negotiation, monitoring and renegotiation phase of the guarantee terms.

5.1   Project Overview

In this section we draw attention to some aspects related to the design and integration of an SLA protocol for the CMAC platform to a point where it can be commercially implemented. Let us consider a service provider using a cluster of several machines and CMAC as his platform to offer a set of services, which can be combined together to form a workflow. In CMAC each service may have multiple instantiations, each with different QoS characteristics.

Multiple users can have access to such an infrastructure, and they can submit workloads consisting of service requests upon the mutual acceptance of an SLA contract initially provided by the service provider. This work focuses mainly on four aspects:

- SLA generation and negotiation control based on service requirements and assets available to satisfy customer requests, as well as admission control to decide which SLAs should be accepted;
- maximising providers' profit implementing an optimal service and resource allocation policy. Profits are recognised when SLA agreements are honored, generally when workload execution completes on time, otherwise penalties are incurred;
- monitoring the agreement compliance during runtime in order to react early to failures and delays;
- renegotiation of the agreed terms to maximize revenues.

To maximise the provider's profit the number of machines allocated to each workload and each service (workflow) might be continuously adapted at runtime. At this early stage we assume that one machine is allocated to each workload whose SLA has been accepted and that the workload arrival rate follows a known distribution. The execution time of each service can be inferred by analysing their performance over time. Probability distributions to estimate service execution times may be derived based on some system characteristics and size of the input data. Such statistical information are useful for a proper server management.

Several metrics employed in QoS control design are being taken into account to monitor a variety of computer systems and software services. These may be classified as *system-level metrics*, such as CPU and memory utilization, cache hit ratio and server queue length, *application-level metrics* such as response time and throughput, or *business-level metrics* such as profits in SLAs. The use of metrics considered as terms on an agreement could be used to take decisions that result in an improvement of QoS in terms of usage (e.g. number of CPUs needed, amount of RAM requested and bandwidth required) or in terms of energy saving among others. Energy saving represents a highly active research topic within GreenIT [19]. In other words, these metrics could be used to achieve a smarter system management along with specific provider objectives.

5.2   Contract Renegotiation Protocol

We aim to integrate the CMAC platform with an SLA protocol which will allow the user to specify and negotiate his requests. During the negotiation phase, the resource orchestrator will be in charge for finding suitable time-slots, namely when all required resources are available at the same time. Once the time-slots are identified and accepted by the client the orchestrator reserves each individual resource. Such a reservation can be expressed as a Quality of Service in an SLA and is a binding agreement for the two parties.

In this project we also intend to design a renegotiation protocol based on the principles of contract law and allows for multi-round renegotiation in CMAC Cloud environments. The protocol allows the initiation of renegotiation through non-binding enquiries to the other party so that an evaluation of the new terms can be conducted before committing to a new contract. Such a new contract is formed once the accepted template is sent by the offeree, i.e. CMAC, and not

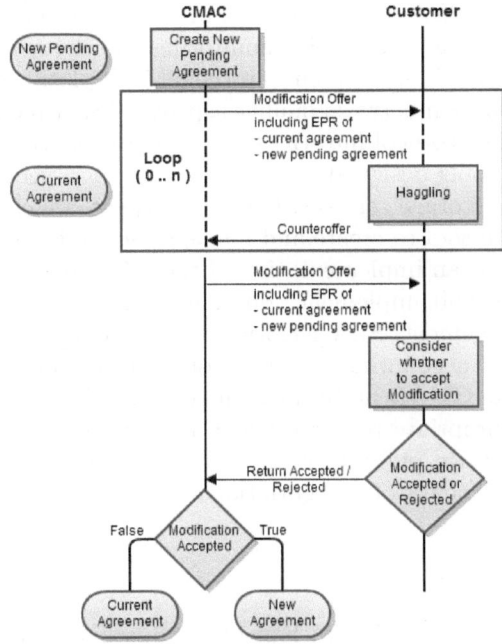

Fig. 5. Process of the renegotiation protocol in the CMAC platform

when the acceptance is received by the offeror, i.e. the customer. This behaviour is meant to prevent the risk of cheating by the customer and minimize disruptions due to possible message lost and delay in the network. In this work we assume the negotiation starting with the client requesting SLA templates from the CMAC service providers. To the contrary, we assume only CMAC able to initiate a renegotiation (asymmetric), upon the customer's consent in the initial agreement. Such a straightforward protocol is described in Figure 5. The renegotiation is initiated by CMAC which sends a modification offer along with the Endpoint Reference (EPR) of the current agreement and of the new pending agreement to the customer. The customer may accept, reject or haggle the modification offer. Offers may be proposed by CMAC based on resource constraints, temporal restrictions, previous offers and a maximum number of sessions, or until a modification offer is accepted and the EPR of the superseding agreement is received by the customer. Once the customer accepts the modification offer CMAC changes the state of the pending agreement to a current agreement. In turn, the customer is referring now to the EPR of the new agreement.

6   Conclusions

Currently, WS-Agreement provides little insight on how negotiation, and in particular automated negotiation, can be realized. In addition, it is difficult to define the quality aspects of SLA requirements. This paper presents an SLA protocol designed to guide the negotiation, the monitoring and the renegotiation phase of the agreed terms to maximize revenues in the CMAC platform. Besides, we also provide a clear distinction between quality aspects of SLA requirements. We aim to integrate our SLA protocol into the CMAC platform which offers a range of analytical software tools designed to detect events on assets and complex systems as well as data storage services. For this purpose, we choose the WSAG4J framework which is a tool to create and manage service level agreements in distributed systems. It is an implementation of the OGF WS-Agreement standard that helps to design and implement common SLA management tasks such as SLA template management, SLA negotiation, creation and renegotiation, and SLA monitoring and accounting. In this work, we intend to develop the first SLA protocol for the CMAC platform with the intention to maximize revenues by designing an appropriate resource allocation process based on time restrictions and related service parameters agreed during the negotiation phase and possibly modified by means of renegotiations.

References

1. Keller, E., Ludwig, H.: The WSLA Framework: Specifying and Monitoring Service Level Agreements for Web Services. Journal of Network and Systems Management 11, 57–81 (2003)
2. WSAG4J - Web Services Agreement for Java, July 3 (2012), https://packcs-e0.scai.fraunhofer.de/wsag4j
3. Austin, J.: Distributed associative memories for high speed symbolic reasoning. International Journal on Fuzzy Sets and Systems 82, 223–233 (1995)
4. Hodge, V.J., Austin, J.: A Binary Neural k-Nearest Neighbour Technique. Knowledge and Information Systems 8(3), 276–292 (2005)
5. Austin, J., Brewer, G., Jackson, T., Hodge, V.J.: AURA-Alert: The use of binary associative memories for condition monitoring applications. In: Proceedings of 7th International Conference on Condition Monitoring and Machinery Failure Prevention Technologies (CM 2010 and MFPT 2010), Stratford-upon-Avon, England, June 2010, vol. 1, pp. 699–711 (2010) ISBN: 978-1-61839-013-4
6. Andrieux, A., Czajkowski, K., Dan, A., Keahey, K., Ludwig, H., Kakata, T., Pruyne, J., Rofrano, J., Tuecke, S., Xu, M.: Web Services Agreement Specification (WS-Agreement), GRAAP-WG, OGF proposed recommendation, October 10 (2011), http://www.ogf.org/documents/GFD.192.pdf
7. Hickinbotham, S., Austin, J., McAvoy, J.: Interactive Graphics on Large Datasets Drives Remote Condition Monitoring on a Cloud. In: The Proceedings of the Open Access Journal of Physics: Conference Series, Part of CMAC Project, for Cybula Ltd. and the University of York, COMADEM 2012, June 18-20 (2012)
8. Austin, J., Fletcher, M., Jackson, T., Jessop, M., Turner, A., Weeks, M.: YouShare, an Online Collaboration Research Environment for Sharing Data and Services, UK e-Science All Hands Meeting (2011)
9. Austin, J., Jackson, T., Fletcher, M., Jessop, M., Liang, B., Weeks, M., Smith, L., Ingram, C., Watson, P.: CARMEN: Code analysis, Repository and Modeling for e-Neuroscience. In: Proceedings of the International Conference on Computetional Science, ICCS 2011, vol. 4, pp. 768–777 (2011)

10. SLA@SOI: Empowering the service industry with SLA-aware infrastructures, FP7-ICT-2007.1.2 Service and Software Architectures, Infrastructures and Engineering, http://sla-at-soi.eu (date of last access: July 4, 2012)
11. Optimis: Optimized Infrastructure Services, FP7-ICT-2007-1 - Objective 1.2, http://www.optimis-project.eu, (date of last access: July 4, 2012)
12. Battré, D., Brazier, F.M.T., Clark, K.P., Oey, M.A., Papaspyrou, A., Wäldrich, O., Wieder, P., Ziegler, W.: A Proposal for WS-Agreement Negotiation. In: Proceedings of the 11th IEEE/ACM International Conference on Grid Computing, pp. 233–241 (October 2010)
13. Waeldrich, O., Battré, D., Brazier, F., Clark, K., Oey, M., Papaspyrou, A., Wieder, P., Ziegler, W.: WS-Agreement Negotiation Version 1.0, GRAAP-WG, OGF proposed recommendation, October 10 (2011), http://www.ogf.org/documents/GFD.193.pdf
14. Fletcher, M., Liang, B., Smith, L., Knowles, A., Jackson, T., Jessop, M., Austin, J.: Neural Network Based Pattern Matching and Spike Detection Tools and Services - in the CARMEN Neuroinformatics Project Special Issue of Neural Networks on Neuroinformatics Journal 9. Elsevier (June 2008)
15. Watson, P., Lord, P., Gibson, F., Periorellis, P., Pitsilis, G.: Cloud Computing for e-Science with CARMEN. In: Proceedings of the 2nd Iberian Grid Infrastructure Conference, pp. 3–14 (2008)
16. Dew, P.M., Schmidt, J.G., Thompson, M.: The white rose grid: practice and experience. In: The Proceedings of the UK eScience-All Hands, Nottingham (2003)
17. Fletcher, M., Jackson, T., Jessop, M., Liang, B., Austin, J.: The Signal Data Explorer: A High Performance Grid based Signal Search Tool for use in Distributed Diagnostic Applications. In: The Proceedings of the Sixth IEEE International Symposium on Cluster Computing and the Grid CCGRID 2006, pp. 217–224 (2006)
18. Parkin, M., Badia, R., Martrat, J.: A comparison of SLA use in six of the european commissions FP6 projects, Institute on Resource Management and Scheduling, CoreGRID-Network of Excellence, Tech. Rep. TR-0129 (April 2008)
19. Laszewski, G., Wang, L.: GreenIT Service Level Agreements. In: The Proceedings of the 10th IEEE International Conference on Grid Computing, Grid 2009 (2009)
20. Brandic, I., Music, D., Dustdar, S.: Service Mediation and Negotiation Bootstrapping as First Achievements Towards Self-adaptable Grid and Cloud Services. In: Wieder, P., Yahyapour, R., Ziegler, W. (eds.) Grids and Service-Oriented Architectures for Service Level Agreements. Springer (2009)

Levi – A Workflow Engine Using BPMN 2.0

Keheliya Gallaba, Umashanthi Pavalanathan, Ishan Jayawardena,
Eranda Sooriyabandara, and Vishaka Nanayakkara

Department of Computer Science and Engineering, University of Moratuwa, Sri Lanka
{keheliya.gallaba,umashanthip,udeshike,
eranda.sooriyabandara}@gmail.com, vishaka@uom.lk

Abstract. Increasing benefits of business process automation and information technology (IT) based governance encourage organizations to model and manage their day to day business activities using business process management systems, in order to achieve increased efficiency and productivity. Many business process languages, such as Business Process Execution Language (BPEL), use a programming oriented view in process modeling as opposed to human oriented view. Recent standardization of Business Process Model and Notation version 2.0 (BPMN 2.0) provides a way to support inter-operation of business processes at user level, rather than at the software engine level. Wide adoption of the BPMN 2.0 standard is limited by the lack of runtimes natively supporting BPMN 2.0. In this paper we discuss about Levi, a cloud-ready BPMN 2.0 execution engine built using the core concurrent runtime of Apache based open source process engine ODE (Orchestration Director Engine), which executes BPMN 2.0 processes natively.

Keywords: Business process, business process management engines, business process modeling, BPMN 2.0, business process execution, workflow engine, business-IT gap.

1   Introduction

Business Process Management (BPM) is a management approach focused on aligning all aspects of an organization with the requirements of its clients. A BPM system can be viewed as a type of Process-Aware Information System (PAIS), which helps an organization to make greater profits by improving the way they do business [1]. The efficiency and productivity enhancement of BPM systems make it suitable for any type of business enterprise [2]. BPM primarily focuses on the comprehensive management and transformation of operations presented in the processes of an organization [3]. A typical organization would have deployed hundreds or thousands of processes most of which controls the main sources of their revenue. Therefore, these processes must be constantly examined and managed on an ongoing basis to assure that they remain as efficient and effective as possible [2]. The performance of these processes must be evaluated to ensure that they meet the organization's business targets, which are based on critical metrics that relate to customer needs and organizational requirements [3].

A. Haller et al. (Eds.): WISE 2011 and 2012 Combined Workshops, LNCS 7652, pp. 118–130, 2013.
© Springer-Verlag Berlin Heidelberg 2013

The concept of BPM has becoming widely adopted since the last two decades. In 2006, Zur Muehlen introduced a Business Process Management life cycle [4] which can be used to improve the way a company conducts its business in the long and short term.

To manage business processes, they have to be modeled and documented. One of the essential parts of business process modeling is choosing the most suitable modeling approach. Among the existing graphical modeling notations prominent modeling approaches are: Petri Nets, UML Activity Diagrams, Role Activity Diagrams (RAD), Data Flow Diagrams (DFDs) and State-Transition Diagrams (STDs) [5], [6], [4]. Notations such as UML and DFDs are focused on the informational perspective of a process (information flow involved in a process) while notations such as RADs and STDs are focused on the behavioral aspect (the behavior of the activities and the actors) of a process. But neither of these is a complete solution, which means that using a model from one perspective will have an opportunity cost of not using the others. In the recent past, there have been efforts in developing web service-based XML execution languages for BPM systems, such as Web Services Business Process Execution Language (WSBPEL/BPEL). But these languages, which were designed for software operations, are not meant for direct humans use. Therefore, only very experienced programmers could work with such languages and business people who do the initial development, management and the monitoring of processes could not take the advantage of these languages. This business-IT gap in the current BPM software does not enable business users to easily model and execute business processes. The reason is the approaches used in building business process engines.

Since business people are more comfortable with visualizing business processes in a flow-chart format there is a human level of "inter-operability" or "portability" that is not addressed by XML execution languages such as WSBPEL. To address this, Business Process Model and Notation (BPMN) was standardized to yield the inter-operation of business processes at the human level, rather than at the software engine level. The first goal of BPMN is to provide a notation that is readily understandable by all business users, from the business analysts who create the initial drafts of the processes to the technical developers responsible for implementing the technology that will perform those processes [7].

BPMN provides a standard visualization mechanism for business processes defined in an execution optimized business process language. Thus, BPMN creates a standardized bridge for the gap between the business process design and process implementation [8], [9]. BPMN enables businesses to model their internal business procedures in a graphical notation and communicate these procedures in a standard manner. It follows the tradition of flowcharting notations for readability and flexibility. The Object Management Group (OMG) is using the experience of the business process notations that have preceded BPMN to create the next generation notation that combines readability, flexibility, and expandability [10], [11].

BPMN 2.0 is a step forward for the whole business process management community because it introduces not only a standard graphical notation, but also concise execution semantics for process execution that can be used to enable the real execution of business processes that are modeled using it [11]. BPMN 2.0 provides a commonly agreed upon formal execution semantics by introducing concise execution semantics, overcoming the major drawback in the earlier versions such as BPMN 1.2 [1]. In addition to that, BPMN 2.0 provides a notation and a model for business processes and an interchange format that can be used to exchange BPMN process definitions between different tools. Diagram interchange format facilitates the exchange of diagrams whereas XML schema interchange allows for easy sharing of model and its attributes. The goal of BPMN 2.0 is to enable portability of process definitions, so that users can take process definitions created in one vendor's environment and use them in another vendor's environment.

By providing a visual modeling language for business processes, BPMN 2.0 enables non-IT experts to communicate and mutually understand their business models. This progress in the area of business process management has resulted in widespread use of BPMN 2.0 as a modeling language [12].

This paper presents Levi, a highly concurrent BPMN 2.0 compatible process engine. Although BPMN 2.0 has achieved reasonable popularity, BPMN 2.0 does not have wider native runtime support yet. We believe that Levi serves as a proof of concept of a native BPMN 2.0 execution engine using existing open source components as the core. Apache ODE, which is among the most influential open source processes engines, provides a BPEL based process execution runtime. Since concurrency and join pattern is one of the key considerations while building a processes execution framework, ODE defines a runtime called JACOB (Java Concurrent Objects), a highly concurrent implementation of join pattern, which support persistent executions. This layer is independent from BPEL, and Levi implements support for BPMN 2.0 on top of JACOB runtime. One of the main challenges of building Levi was mapping BPMN 2.0 constructs to underline JACOB runtime. We will discuss the challenges, design, and solutions we encountered while building Levi, and critically analyze its effectiveness. To make a BPM solution 'Cloud Enabled', it needs to be built focusing on scalability, security and multi tenancy aspects. Levi's runtime architecture is designed considering these features for it to be cloud ready. Since Levi is designed in a way such that its functionality can be exposed as Web services, cloud based business applications can be built on top of Levi with minimum effort, which makes Levi suitable as a business process engine for a cloud enabled environment.

Section 2 of this paper describes the existing approaches of implementing the BPMN 2.0 runtime and the merits and demerits of mapping BPMN 2.0 into different intermediate exchange formats. The reasons for building Levi is explained in section 3. Section 4 discusses the design of Levi and section 5 explains the implementation of BPMN 2.0 runtime in Levi. Section 6 describes the outcome of the work and we explain the future work in section 7.

2   Related Work

The effort of building BPMN 2.0 execution engines has started since the initial release of the BPMN 2.0 beta specification in August 2009. Many vendors considered BPMN as a visual notation to BPEL and started creating BPMN 2.0 execution engines that runs the processes in their existing BPEL engines. Consequently, they tried to map BPMN 2.0 semantics to BPEL semantics which is not straight forward, as we shall discuss later in this section. Some other vendors used other intermediate exchange formats such as jPDL and XPDL to convert the BPMN 2.0 processes and then execute in their engines that does not support BPMN 2.0 process execution natively. At present, with the release of the final version of BPMN 2.0 specification in January 2011, there are several BPMN 2.0 implementers [13] but almost all of them convert BPMN 2.0 processes into some intermediate form even though they claim native execution. Next sections highlight merits and demerits of mapping BPMN 2.0 to various intermediate exchange formats.

2.1   BPMN Runtime through BPEL

Most of the current BPMN 2.0 engine vendors use a BPMN 2.0 to BPEL mapping, which enables a user to first model business processes using BPMN 2.0 constructs. However, at the runtime, those implementations convert the BPMN 2.0 business process into one or more BPEL processes, and execute them using a BPEL engine. The use of such mapping has created many debates among BPM experts. Implementers of the ActiveVOS BPM suite [14] argued that native execution of BPMN 2.0 processes is complex, and that it is simpler to map BPMN processes to BPEL[15].

However, several publications[16], [17], [18], [19] have pointed out that the conceptual mismatch between BPMN 2.0 and BPEL, and discussed the pitfalls of mapping BPMN 2.0 into BPEL. When converting a language to a different language, it is required to measure the feasibility of doing that conversion. Mainly the conversion should minimize if not avoid loss of semantic representation information. That means the transition between languages should establish a high extent of matching of main representation capabilities between the two languages and a matching of control flow support. When converting BPMN 2.0 to BPEL, there exists a significant mismatch of domain representation capability and control flow support as discussed in [18].

Limitations of this mapping have been discussed in academia in a comprehensive manner. Most of the researchers in this field support the argument that BPEL is inherently block oriented, like a computer program, while BPMN is inherently graph oriented, like a flowchart, even though there are minor confusions about the structure of BPEL and BPMN 2.0 [20],[21]. As pointed out by Weidlich et al. [19] this structural incompatibility is the key reason for the pitfalls of the mapping. It further discusses about further reasons for the pitfalls of the mapping and the myth of a straight-forward mapping.

Beside these reasons, the BPMN 2.0 specification itself describes that only a small subset of the BPMN 2.0 constructs are isomorphic with BPEL and can be mapped to BPEL directly. The specification further says that not all BPMN 2.0 processes can be mapped to BPEL in a straightforward manner because BPMN allows the modeler to draw almost arbitrary graphs to model control flow, whereas in BPEL, there are certain restrictions such as control-flow being either block-structured or not containing cycles. The specification [11] essentially says in the "extended mapping" section that engine vendors are on their own, noting "in many cases there is no preferred single mapping of a particular block, but rather, multiple WS-BPEL patterns are possible to map that block to". This contradicts the argument that this mapping is simpler than native BPMN 2.0 execution.

Guo et al. [16] and Indulska et al. [17] argue for the need for bi-directional transformation between BPMN 2.0 and BPEL for a complete such mapping and the limitations of achieving it. [17] uses the Bunge-Wand-Weber representation model to analyze the representational capabilities of BPMN 2.0 and BPEL4WS, and on that basis, argues that the translation between BPMN and BPEL4WS is prone to difficulties due to inconsistent representational capabilities. They also claim that their work serves as a theoretical cornerstone on which the development of better mapping support for BPMN 2.0 and BPEL4WS can be based on.

2.2   BPMN Runtime through jPDL

Similar to the mapping of BPMN 2.0 to BPEL, some argue that BPMN 2.0 to jPDL mapping is suitable for BPMN 2.0 execution engines. jPDL [22] is the jBPM Process Definition Language (JPDL) for jBPM [23], a Business Process Management Suite from the JBoss community. Even though jBPM claims that it support BPMN 2.0 process execution natively, internally it converts the BPMN 2.0 process definition in to jPDL definitions before executing the business process in the existing engine. jBPM implements BPMN 2.0 process execution on top of the jBPM Process Virtual Machine (PVM) which was originally built for executing jPDL processes hence requires a conversion [24]. More over jPDL is not an industry wide standard; it is just the language used only in the jBPM suite and can only be used by it. Hence this conversion is far from being accepted as a standard for executing BPMN 2.0 processes [25].

2.3   BPMN Runtime through XPDL

Some vendors use XPDL (XML Process Definition Language) as the intermediate format to run BPMN 2.0 processes. XPDL [26] is designed to exchange the process definition, both the graphics and the semantics of a workflow business process, among different workflow products [27]. Hence this conversion does not result in native execution of BPMN 2.0 processes.

3   Why Levi?

With the introduction of the operational semantics for BPMN 2.0, it is now possible to build an engine that directly supports BPMN 2.0 – without the intermediate step of generating BPEL. As explained by Leymann [12], no BPEL at all is required to execute process models specified in BPMN 2.0. Levi is designed to be a native BPMN 2.0 execution engine, which can be used to execute business process models that conform to the BPMN 2.0 specification.

Implementing a workflow engine is tantalizing yet a daunting task. There are many non-functional requirements like robustness, efficiency and scalability expected from an enterprise level workflow engine. Open Source WS-BPEL 2.0 implementation like Apache ODE [28] has mechanisms to ensure concurrency, durable continuation, reliability, and recovery. It uses a framework called JACOB, which is a practical combination of ideas from the actor model and process algebra approaches to concurrency and continuation. The implementation of the BPEL constructs is simplified by limiting itself to implementing the BPEL logic and not the infrastructure necessary to support it [29].

Without reinventing the wheel, as ODE's BPEL implementation relies on JACOB framework to implement the BPEL constructs, Levi uses JACOB to implement BPMN 2.0 constructs. Most importantly, it serves as a proof of concept for exploring the possibilities of using Apache ODE and JACOB to execute BPMN 2.0 processes consisting of core BPMN constructs.

4   Architecture of Levi

4.1   Overall Architecture

Major building blocks of a BPMN execution engine are shown in fig. 1. Users first describe their processes using BPMN 2.0 notations and then deploy the processes in Levi. Deployed processes are stored in the process database. Later when users execute the process, the process engine executes the process. Our discussion on Levi will focus on the execution engine that handles the runtime, compared to the build time of a BPMN model. This is due to the fact that for a given BPMN model, the 'build-time' occurs only once, whereas the 'runtime' is expected to be functional each time that model is executed or managed/monitored through the administrator's or any other user's console.

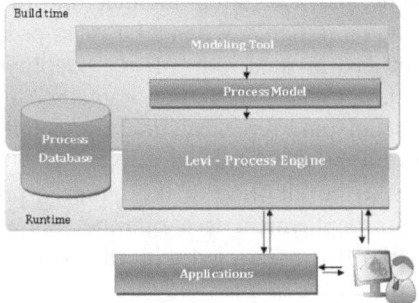

Fig. 1. Deployment and Execution Architecture

Build Time: This is when the user creates a BPMN model for a business scenario to fulfill his requirement. To model a BPMN process, a modeling tool such as BPMN2 Visual Editor for Eclipse [30] can be used. A typical modeling tool supports creating BPMN 2.0 diagrams in a visual editor and generates the corresponding XML representation of that process model. After modeling the basic model in BPMN 2.0, the model must be made into a process archive that can be deployed in Levi. To do that, additional artifacts such as the user input forms, WSDL files, process diagrams etc., must be bundled together with the created BPMN file. Once the process archive is deployed, it is stored in the Process Database of Levi.

BPMN Process Model: A BPMN process model is essentially an XML document that corresponds to the standard BPMN XML Schema document proposed by OMG. The Levi engine expects all the BPMN files to have a '.bpmn' extension and these BPMN files are validated when those are deployed to the system in the form of a business archive.

Format of a Process Archive: The process archive type identified by Levi is called the "Levi Process Archive" type, which is a zip archive renamed to have a '.lar' extension. A valid archive must have a single top most directory in which all the sub directories, BPMN files and other artifacts are included.

Runtime: This refers to two concepts both related to BPMN process execution, depending on the context where those are referred. The first concept is the actual execution time of a deployed business process within the execution engine. The other concept is the subsystem of the execution engine which handles the execution, management, and monitoring of deployed business processes. This is also referred to as the backend of Levi. The frontend of Levi and/or a third party application (web/desktop/mobile) can connect to the backend as shown in the fig. 1 and manage business process via a customized user interface.

For better understanding of the architecture, Levi engine can be partitioned into four functional components: runtime service module, storage service module, user management module and utility module.

4.2   Deployment and Execution Architecture

The high-level deployment and execution architecture of the business process execution engine Levi is shown in fig. 2. When a business process archive is given as the input, a 'deployment' is created out of it. 'Deployment' is the runtime representation of the business process definition contained in the business archive. These process definition details and representation are stored in the Process Engine Database when the business archive is deployed.

There are two concepts to be clarified at this point – process deployment and process instance. A process deployment is a runtime representation of a business process, bundled in a lar file. It is connected with the concept of process definition. Once a lar file is deployed into the engine, only this process deployment is created. When a user wants to execute the business operations in that process, a process instance is created using the object model of the process definition. There can be multiples of process instances created from a single process deployment.

When a user wants to execute a business process, the process definition and object model of that particular process deployment is retrieved from the process engine database and a process instance is created in the runtime, as shown in fig. 2. Properties of the process instance will be persisted in the database. When executing the process, the engine navigates through each BPMN 2.0 element in the process instance, until it reaches the end event. Process instance states are persisted in the database and retrieved when required.

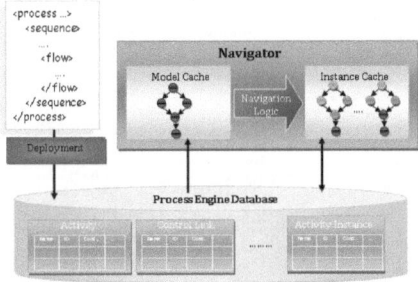

Fig. 2. Deployment and Execution Architecture

4.3   Building Applications Using Levi

The architecture of Levi engine is designed in such a way that real world business applications can be built on top of it with minimum effort. The major building blocks of the engine such as the RuntimeService, StorageService and UserManagmentService are exposed as APIs (Application Program Interfaces), which enables users of the engine to build a customized front end layer according to their business needs. This enables users to build different applications with less effort.

5   Implementation of BPMN2.0 Runtime

The BPMN runtime component handles the execution of BPMN logic within the Levi engine. It acts as the backend for the web user interface where the users interact to deploy, execute, and manage their business processes. The runtime is mainly composed of the runtime abstraction of a BPMN process; the ProcessInstance class, and the data types that represent the set of BPMN 2.0 constructs currently supported by Levi. All these types derive from a single type, called BPMNJacobRunnable and this class, in turn, derives from the JacobRunnable class of Apache ODE. This type hierarchy makes it possible to execute the Levi's representation of BPMN constructs and the process instances on the JacobVPU. Further, XMLBeans was used as the data binding tool to generate Java types from the XML representation of BPMN constructs and these types were used to bring in the definition of elements of the input BPMN documents to the context of the runtime. Each of the BPMN construct types acts as a wrapper for the corresponding XMLBeans generated type. Currently, Levi supports all of the simple BPMN 2.0 constructs as well as UserTask, SendTask and ServiceTask from the descriptive category as shown in fig. 3.

BPMNJacobRunnable defines some common methods related to all construct types and are used by the runtime. JacobRunnable defines an abstract method; run, which must be implemented by all of its derivatives. This method is executed by the JacobVPU and the construct related implementation is written in the run method of each construct type. For example, when implementing the ExclusiveGateway construct, the gateway related logic was written in its run method. Also it has a reference to an object of the type generated by XMLBeans; TExclusiveGateway. This instance brings in all the data present in the original XML element to the scope of the runtime.

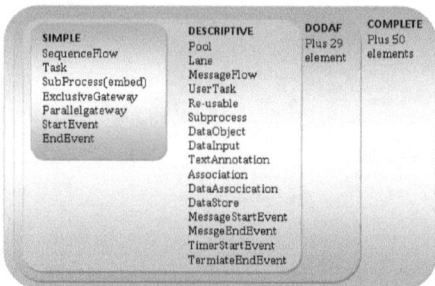

Fig. 3. BPMN 2.0 Constructs

Consider the following XML excerpt from a BPMN document which corresponds to an outgoing sequence flow from an exclusive gateway element. This sequence flow contains the condition upon which is satisfied, the flow takes the path by referring to its target reference. The Levi's implementation of ExclusiveGateway can access the data such as the condition expression "i < 100000" only through the method getConditionExpression of instance of TSequenceFlow.

```
<sequenceFlow id="flow5" sourceRef="exclusiveGw2"
targetRef="exclusiveGw1"> <conditionExpression>
    <![CDATA[i < 100000]]>
</conditionExpression> </sequenceFlow>
```

Similarly, the implementation of ScriptTask accesses the script defined in the BPMN document's ScriptTask element by invoking the getScript method of the instance of TScriptTask. In the run method, it evaluates this script by using the context details of the current process instance, such as the process variables and the script type.

When a process is deployed to the engine in the form of a lar, the runtime constructs the corresponding process definition by parsing and validating the BPMN document together with other dependent entities such as WSDL documents. The process definition (org.levi.engine.impl.bpmn.parser.ProcessDefinition.java) is the static abstraction of a BPMN process inside the Levi engine. It is a data type that aggregates the BPMNJacobRunnable objects which correspond to the BPMN elements of the input BPMN document. When a process instance is created, it is passed with a reference to an instance of the corresponding process definition. The internal design of the process definition has been optimized for efficient navigation of BPMN elements to be used in constructing the process flow during the execution of the process instance.

At the initial stages of the design of the process definition, the iterator pattern was used to navigate the elements of a process instance. This decision was highly influenced by the linear arrangement of BPMN flow elements inside a BPMN document. This lead to incorrect runtime behavior when BPMN documents with elements arranged in a different order other than the order in which the process flow must occur were processed. From this, it was identified that the order of the elements of a BPMN document does not necessarily mimic the actual order of the process flow. BPMN uses an elegant solution to construct the process flow by using sequence flows and setting their source and target reference identifiers. Therefore, after considering all these factors, it was required to come up with a design which had the structure and characteristics similar to those of a graph which enables faster navigation compared to the linear iteration approach. As a solution, the previously described design was proposed in which the sequence flows are grouped into sets of sequence flows based on their target and the source reference IDs separately. These groupings are used as the major data structure in navigating the process elements by the runtime.

Execution of process instances includes starting, pausing, and stopping of process instances of deployed business processes. All these functions are executed when an authorized user gives corresponding command in the frontend. These commands are dispatched to the backend to be executed based on the process parameters. There are two types of executions. First type is executing multiple instances of a same process definition. In this, users can instantiate as many process instances as they wish from a given process definition and the engine is capable of isolating instances from one another and manage the execution. The second type is executing many process instances of different process definitions. The Levi engine supports these two types of process execution. The engine manages multiple instances of the different/same process definitions by resolving the relationship mappings among the users, tasks, process details and other process parameters of the instances accordingly.

6   Results

We have conducted a performance test for Levi with few process scenarios that use the script task construct. We used the following configuration for this purpose: Intel(R) Core(TM)2 Duo CPU T6670 2.20GHz processor with 2.00GB memory, on 32-bit Ubuntu 10.10 operating system.

Fig. 4 shows the test scenario where the top process diagram shows sequential orientation of n ScriptTasks and bottom part shoes the same process oriented in parallel. Fig.5 shows the comparison of running n ScriptTasks oriented sequential and parallel orientation

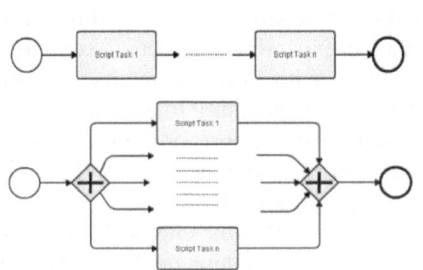

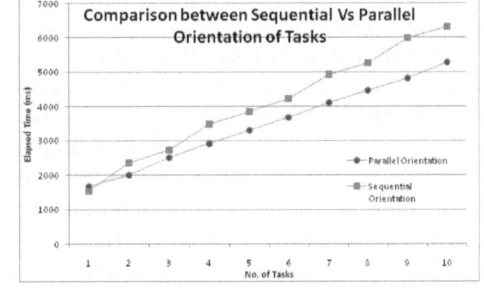

Fig. 4. Sequential and Parallel Orientation of Tasks

Fig. 5. Comparison of Sequential Vs Parallel Orientation of Tasks

As the graph suggests, the sequential approach involves more cost than the parallel approach. The running time of the parallel approach is almost always less than that of the sequential approach. We get this behavior due to the difference of the parallelism involved in each approach. Parallel gateways are used to execute parallel tasks with the help of JacobVPU, in Levi engine. In the meantime for the parallel orientation, the running time is not constant since there is a considerable time delay involved when creating each ScriptTask construct.

Fig. 6 shows the test scenario where a business process of n ScriptTasks is oriented sequentially in the top process diagram and same process is modeled using a loop orientation in the bottom process diagram. Fig. 7 shows the comparison of running time using the sequential and loop orientations as shown in fig. 6.

According to the results, the loop model always takes less time than the sequential model. This is because creating a new object for each ScriptTask in the sequential orientation is an expensive operation to the JVM compared to evaluating conditional expressions in the loop orientation.

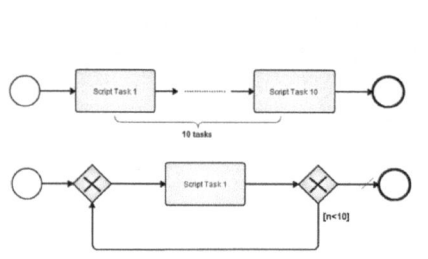

Fig. 6. Sequential and Loop Orientation of constructs

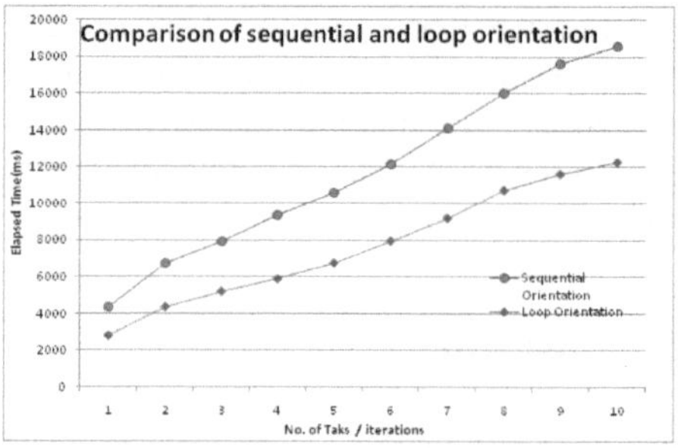

Fig. 7. Comparison of Sequential Vs Loop Orientation of Tasks

7   Future Work

We were successful at implementing the basic BPMN 2.0 constructs in the Levi engine. Since we have designed our engine in such way that addition of new constructs are much simpler, and involves minimum amount of changes, any additional construct can be developed individually as a separate module and integrate to the runtime with less effort. We are working on expanding the set of supported standard BPMN constructs in future together with improvements to the implementations to the existing constructs. Further we are planning to add SOAP web services support for the Service Task and WS-HumanTask support for the User Task.

When we expose Levi engine as a product, performance is one of the most compelling factors to be successful among the competitors in the industry. A proper benchmark does not exist to test the performance of a BPMN engine. Therefore, creating a comprehensive benchmark for Levi is one of our major goals, which will also help improve the industrial value of our project.

A BPMN2 Eclipse plugin [30] is under development to support the full BPMN 2.0 specification. The graphical editor can be integrated with Levi engine to design the processes and can be further improved to support deploying the designed processes through the IDE.

8   Conclusion

We have implemented Levi, a native BPMN 2.0 execution engine by using Apache ODE's JACOB framework. Levi is capable of deploying, persisting, navigating, and executing business processes claiming BPMN 2.0 execution conformance. It serves as a proof of concept for exploring the possibilities of using Apache ODE and JACOB framework to execute processes that consist of core BPMN 2.0 constructs. The implementation of BPMN 2.0 runtime in Levi proves that it is possible to build a BPMN 2.0 execution engine natively without converting into another intermediate representation such as BPEL or jPDL. It also contradicts the debate that converting BPMN 2.0 semantics into BPEL is the simpler way for building a BPMN 2.0 runtime. Further the native BPMN 2.0 runtime feature of Levi enables the rapid support for future expansion of BPMN 2.0 constructs set.

In this paper we have discussed the suitability of BPMN 2.0 to build a business process engine which fulfills both the requirements of business and IT people. The design and implementation of such an engine Levi is described in detail.

References

1. Wohed, P., van der Aalst, W.M.P., Dumas, M., ter Hofstede, A.H.M., Russell, N.: On the Suitability of BPMN for Business Process Modelling. In: Dustdar, S., Fiadeiro, J.L., Sheth, A.P. (eds.) BPM 2006. LNCS, vol. 4102, pp. 161–176. Springer, Heidelberg (2006)
2. Harmon, P.: Business process change: a guide for business managers and BPM and six sigma professionals. Morgan Kaufmann (2007)
3. Michael: Handbook on Business Process Management, vol. 1. Springer (2010)
4. El-Bakry, H.M., Mastorakis, N.: Business process modeling languages for information system development, pp. 249–252 (August 2009)
5. Arkin, A.: Business Process Modeling Language (2003)
6. Cull, R., Eldabi, T.: A hybrid approach to workflow modelling. Journal of Enterprise Information Management 23(3), 268–281 (2010)
7. Owen, B.M., Raj, J.: BPMN and Business Process Management - Introduction to the New Business Process Modeling Standard (2003)
8. Badica, C., Badica, A.: Businss process modelling using role activity diagrams
9. Recker, J., et al.: Do Process Modelling Techniques Get Better? A Comparative Ontological Analysis of BPMN. In: 16th Australasian Conference on Information Systems (2005)
10. Allweyer, T.: BPMN 2.0 Introduction to the Standard for Business Process Modeling (2010)
11. Object Management Group, Business Process Model and Notation specification(BPMN 2.0) (2011)
12. Leymann, F.: BPEL vs. BPMN 2.0: Should you care? In: Mendling, J., Weidlich, M., Weske, M. (eds.) BPMN 2010. LNBIP, vol. 67, pp. 8–13. Springer, Heidelberg (2010)
13. BPMN Supporters - Current Implementations Of BPMN, http://www.omg.org/bpmn/BPMN_Supporters.htm (accessed: October 01, 2011)
14. BPM, Business Process Management Software, Business Process Management Suite | ActiveVOS BPMS from Active Endpoints, http://www.activevos.com/ (accessed: October 01, 2011)

15. BPMN or BPEL: which is simpler to understand? | VOSibilities, http://www.activevos.com/blog/bpel/bpmn-or-bpel-which-is-simpler/2009/11/19/ (accessed: October 10, 2011)
16. Gao, Y.: BPMN - BPEL Transformation and Round Trip Engineering (2008)
17. Indulska, M., et al.: Are we there yet? Seamless Mapping of BPMN to BPEL4WS. In: 13th Americas Conference on Information Systems, pp. 1–11 (2007)
18. Recker, J., Mendling, J.: On the Translation between BPMN and BPEL: Conceptual Mismatch between Process Modeling Languages. In: The 18th International Conference on Advanced Information Systems Engineering. Proceedings of Workshops and Doctoral Consortium, pp. 521–532 (2006)
19. Weidlich, M., Decker, G., Großkopf, A., Weske, M.: BPEL to BPMN: The myth of a straight-forward mapping. In: Meersman, R., Tari, Z. (eds.) OTM 2008, Part I. LNCS, vol. 5331, pp. 265–282. Springer, Heidelberg (2008)
20. BPMN vs BPEL: Are We Still Debating This? BPMS Watch, http://www.brsilver.com/2009/11/19/bpmn-vs-bpel-are-we-still-debating-this/ (accessed: October 01, 2011)
21. Dumas, M., García-Bañuelos, L., Polyvyanyy, A.: Unraveling unstructured process models. In: Mendling, J., Weidlich, M., Weske, M. (eds.) BPMN 2010. LNBIP, vol. 67, pp. 1–7. Springer, Heidelberg (2010)
22. Chapter 18.jBPM Process Definition Language (JPDL), http://docs.jboss.org/jbpm/v3/userguide/jpdl.html (accessed: October 01, 2011)
23. jBPM - JBoss Community, http://www.jboss.org/jbpm (accessed: October 01, 2011)
24. jBPM BPMN | JBoss Community, http://community.jboss.org/wiki/JBPMBPMN (accessed: October 01, 2011)
25. What's in the Architecture?: XPDL,BEPL,JPDL,BPMNS,BPDM et al. Standards and More Standards, http://rabisblog.blogspot.com/2007/04/xpdlbepljpdlbpmnsbpdm-et-al-standards.html (accessed: October 01, 2011)
26. XPDL, http://www.xpdl.org/ (accessed: October 01, 2011)
27. White, S.A.: XPDL and BPMN. Management, 221–238
28. Apache ODE, http://ode.apache.org/ (accessed: October 01, 2011)
29. InfoQ: An Introduction to Apache ODE, http://www.infoq.com/articles/paul-brown-ode (accessed: October 01, 2011)
30. Eclipse Modeling - BPMN2 Eclipse Plugin, http://www.eclipse.org/modeling/mdt/?project=bpmn2 (accessed: October 01, 2011)

Proceedings of the 1ˢᵗ International Workshop on Engineering Semantic Enterprise (ESE) 2012

Maciej Dabrowski[1], John Breslin[1], Alexandre Passant[2],
and Eric Gordon Prud'hommeaux[3]

[1] DERI, National University of Ireland Galway, Ireland
`firstname.lastname@deri.org`
[2] Seevl.net, Ireland
`alex@seevl.net`
[3] W3C, USA
`eric@w3.org`

Preface

Two main trends emerged in the enterprise in the past years. On one hand, Web 2.0 tools such as blogs, microblogs and wikis for enterprise-scale collaboration and information management became widely used for information management, leading to a move to "Enterprise 2.0". At the same time, Semantic Web technologies have emerged allowing enterprise users to transparently provide structured and meaningful data in the enterprise thanks to vocabularies representing social data. These technologies enhance information management and sharing with an enterprise, leading to the "Social Semantic Enterprise", where both Semantic Web technologies and Social Web principles converge. Although the adoption of the Semantic Web technologies in enterprises is growing very fast, some major challenges related to their exploitation in enterprises remain, including personalization, near real-time integration or scalability. For this workshop, we have selected four papers that addressed the above topics.

First paper by Del Nostro et. al. entitled "A Semantic-Based Architecture for Collaborative Enterprise Management: The ARISTOTELE Platform" presents a platform that supports creativity knowledge-intensive organizations using semantic technologies enabling self-organizing acquisition, processing and sharing of new information.

Baclawski et. al. in their paper "ICOM: A Framework for Integrated Collaborative Work Environments" address the important issue of collaborative work environments. They discuss ICOM, a proposed OASIS, which enables integration of enterprise collaboration platforms and standalone services across enterprise boundaries.

The third paper "Knowledge-Based Semantification of Business Communications in ERP Environments" by Meimaris and Vafopoulos discuss the importance of information about communications and propose a set of semantic components extending ERP systems to provide a basis for intelligent knowledge management.

The fourth paper "Ubiquitous Service Capability Modeling and Similarity Based Searching" by Gao and Derguech proposes a capability model for ubiquitous devices to describe the dynamics and relevance of services and a generic similarity metric for service capabilities that increases quality of information search.

We hope that this workshop will stimulate further research in this area and the value of the semantic technologies will be further exploited by knowledge-intensive organizations – a topic that contributes to system engineering research area of WISE.

Maciej,
John,
Alex,
Eric

A Semantic-Based Architecture for Managing Knowledge-Intensive Organizations: The ARISTOTELE Platform

Pierluigi Del Nostro, Francesco Orciuoli, Stefano Paolozzi, Pierluigi Ritrovato, and Daniele Toti

CRMPA, MOMA S.p.A., Italy
{orciuoli,ritrovato}@momanet.it, {delnostro,paolozzi,toti}@crmpa.it
http://www.aristotele-ip.eu

Abstract. We present the semantic-based architecture of the ARISTO-TELE platform, which is founded on the definition and development of models, methodologies, technologies and tools to support the emergence of competences and creativity within workers by self-organizing acquisition, processing and sharing of new information inside knowledge-intensive organizations. ARISTOTELE's architecture relies on semantic data by means of a number of conceptual models, which define the context of interest for an enterprise via a set of concepts and relationships among them. Instances of these models are used to annotate content data, thus creating a semantic network of information that actualizes the Linked Data paradigm within the information space of an organization. In this paper we describe the building elements of the ARISTOTELE platform, the conceptual models which lie behind them and the core Linked Data Layer component responsible of managing information for the whole system.

Keywords: Semantic Web, Semantic Enterprise architecture, Knowledge Modeling, Knowledge Management, Human Resource Management, Collaboration, Innovation, Learning, Training.

1 Introduction

Semantic Web technologies have simplified knowledge-intensive applications, by enabling interoperable and machine-readable data in software systems. However, a wide scale adoption of these systems, especially in the enterprise world, is still far from being achieved. In this paper we present the semantic-based architecture of the ARISTOTELE platform [5], designed as an integrated set of online services that might provide workers, managers, trainers and other stakeholders involved in an organization with tools and resources to support and enhance the emergence of their competences and creativity. Its functionalities are based upon a wide range of methodologies and strategies, along with a set of models for representing the adopting organization's domain.

A. Haller et al. (Eds.): WISE 2011 and 2012 Combined Workshops, LNCS 7652, pp. 133–146, 2013.

The architecture of the platform allows in fact for the semantic representation of the information within the adopting organization, enabling open innovation through the Linked Data paradigm, thus allowing the organization to exploit resources that are outside its boundaries and contributing to the expansion of the knowledge available over the Web. Tools and services cooperate with one another, taking advantage of the defined models to store, extract and handle the organizational knowledge in order to fulfill such tasks.

Here, we proceed to describe the building elements of the ARISTOTELE platform, the conceptual models which lie behind them and the core component responsible of managing information and knowledge for the entire system.

The paper is structured as follows. In Section 2, related work is discussed. In Section 3, we present an overview of ARISTOTELE's architecture, listing its services and tools and describing the structure of its data layer. In Section 4, we introduce the conceptual models actually exploited by the platform for representing semantic-based information, with special emphasis on the Knowledge Model. Section 5 is then devoted to provide a deeper insight into the Linked Data Layer component, which is responsible of handling and encapsulating information for the whole system. And finally, in Section 6 we draw our conclusions.

2   Related Work

The ARISTOTELE platform, thanks to its innovative nature of a comprehensive solution for knowledge management within an enterprise, feature few competitors to be actually used for a sensible comparison. On the other hand, a certain number of research and industrial proposals possess a partial subset of ARISTOTELE's capabilities, since each of those proposals focuses on specific aspects of enterprise-related knowledge.

For instance, PROLIX [20] features a process-oriented learning approach built with a flexible architecture based on the SOA (Service-Oriented Architecture) paradigm. Its core objectives revolve around filling competence gaps for workers, tailoring learning material to their needs (by using a proper methodology for matching required competences with workers' profiles) and providing ad-hoc fruition of such a material. As it stands, PROLIX focuses on coupling business processes with learning processes in corporate environments and on providing competence-based decision support for those processes, whereas it does not specifically take into account social relationships among workers within an enterprise.

A similar consideration applies to APOSDLE [16], since the latter focuses on self-directed learning integrated within the working environment, as well as competence modeling and development, but does not consider aspects related to social relationships among workers.

MATURE [19], instead, is made up of a Conceptual model of enterprise knowledge, an employee-level collaborative platform to be integrated into the working environment of learners, and an organization-level platform to encourage innovation and collaborative activities.

As far as specific tools are concerned, Cogito Answers [3] is a semantic search platform with a Natural Language interface (NLI) providing users with access to unstructured information via natural language queries. In this regard, it serves a similar purpose to some of the ARISTOTELE's services (like the Knowledge Building Shared Services and the Recommender System, which will be explained in Section 3).

3   Overview of ARISTOTELE's Architecture

The design methodology for the ARISTOTELE platform and its underlying architectural choices are based upon a well-known framework for the definition of Enterprise architectures for business organizations: MIKE2.0 [11]. This framework has a specific focus on information management and provides solutions for Semantic Enterprise, Enterprise 2.0 and Knowledge Management, which are key aspects for the ARISTOTELE platform.

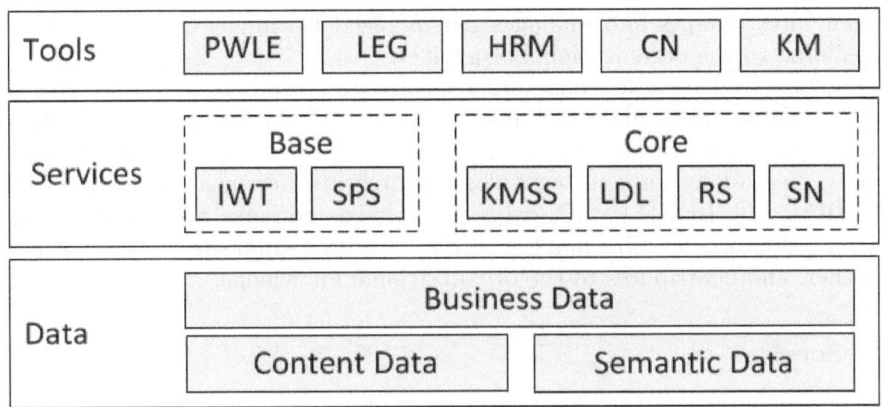

Fig. 1. ARISTOTELE's architecture

To provide a first big picture of ARISTOTELE's architecture it is possible to identify three logical layers, as depicted in Fig. 1: (i) tools; (ii) services; (iii) data. These layers are presented below.

3.1   Tools

Tools build up the presentation layer: a tool is a software entity conceived and designed to provide a set of conceptually-related functionalities in order to solve some kind of real-world problem for a user. These tools draw upon functionalities offered by services and provide the final users with a web-based interface to access them.

Specifically, the ARISTOTELE platform provides the following tools, each related to one or more of the key areas of the project:

- Knowledge Management (KM): this is a set of tools exploiting methodologies and techniques which support organizational knowledge building and maintenance (merging, matching, versioning) in a semi-automatic way, via the definition of knowledge extraction techniques as well;
- Human Resource Management (HRM) Tools: they take advantage of methodologies and techniques supporting competence gap analysis, team and group formation, competence development for internal resources, and recruitment;
- Social Collaboration and Networking (CN) Tool: it exploits methodologies and techniques to combine adaptive learning strategies with "non-adaptive" emergent competence change, based on serendipitous exploitation of other people's knowledge made available in social networks;
- Learning Experience Generation (LEG) Tool: it monitors a worker's behavior, provides workers with suitable contents and didactic approaches, and generates, adapts and manages personalized learning experiences custom-tailored to the organizational objectives.
- Personal and reliable Work and Learning Environment (PWLE): this tool is available online and accessible by the organizational user base, with the purpose of assisting a worker in his/her daily working activities, by acting as a methodological and technological integrator for most of the other tools (HRM, CN, LEG). PWLE helps "knowledge workers" achieve their objectives, supports learning and knowledge activities, connects workers with each other, and contributes to the organizational knowledge.

3.2 Services

Services implement the business logic and data access. In ARISTOTELE they can be distinguished into Base Services and Core Services.

Base Services encompass all the basic services from the reference frameworks and systems adopted by the platform that are relevant to it as a whole (e.g. User Authentication, User Profile Services, Business Connectivity Services etc.). Specifically, we refer to the following systems:

- Intelligent Web Teacher (IWT)
- SharePoint Server 2010

IWT is an innovative, extensible and open platform for Learning and Knowledge Management conceived for delivering user-tailored and scalable e-Learning solutions [2].

SharePoint Server 2010 is Microsoft's business collaboration platform for the enterprise and the Web. It is an extensible solution creation platform, particularly suitable for the realization of enterprise collaboration applications.

Core Services, instead, are domain-specific and implement the actual business functions of the platform, and are instantiated to serve the upper-level Tools. ARISTOTELE's Core Services are the following:

- Knowledge Building Shared Services (KMSS)
- Linked Data Layer (LDL)
- Recommender System (RS)
- Social Networking (SN)

The purpose of the KMSS services is to perform all the knowledge building capabilities of the platform, like text analysis, conceptualization, ontology building and ontology merging. Besides, KMSS implement the services related to specific business functionalities for the tools (i.e. expert finding, tag suggesting, competence gap computation), which leverage on the knowledge building capabilities themselves.

The LDL service enables data access for tools or other services, by allowing the aggregation of information distributed over heterogeneous sources and offering a set of storage-independent classes (see the following definition of Business Entities). By means of a Web Service the LDL exposes a set of methods for managing and querying data that can reside on different repositories.

The RS provides information filtering in order to recommend information items that are likely to be of interest to ta worker.

SN has the goal to promote a collaborative approach within the organization, following the principles of Enterprise 2.0.

3.3   Data

The data layer of ARISTOTELE is made up of data coming from disparate sources, including relational databases, SharePoint objects, RDF/OWL graphs and additional external or legacy data. Information stored in those repositories is interlinked by exploiting the ARISTOTELE models and instances coming from the Semantic Repository of the platform, where an RDF/OWL-based representation brings the Linked Data paradigm to fruition within the information space held by the organization.

More specifically, we can identify three "macro"-types of data managed by the ARISTOTELE platform: (i) content data, (ii) semantic data and (iii) business data.

Content data are documents or any kind of digital information, accessible by the members of the organization; they can be structured or un-structured, can reside in heterogeneous data-sources and can be managed by different applications. In the context of the ARISTOTELE platform, content data is mainly stored in SharePoint 2010 and IWT databases.

Semantic data are used to represent ARISTOTELE models and instances. ARISTOTELE models define the context of interest for the project by means of a set of concepts and relationships between them, specified by means of ontologies; more details regarding these models are described in Section 4. Instances of semantic data are then used to annotate content data, thus creating a semantic network of information that brings the Linked Data paradigm to fruition within the information space of the organization. Semantic data are stored in an open-source Semantic Repository, Sesame [1], widespread in the Semantic Web community for its performance and scalability capabilities. Standard formalisms (RDF, RDFs, OWL) have been adopted for the implementation of semantic data.

Data managed by the tools are built by merging information from content and semantic data: we call these sets of classes Business Entities, and they basically build up the business data of the platform. A Business Entity has the role of decoupling the communication and the management of data — at a business level — from the nature of the actual data sources. For this purpose, the Linked Data Layer component of the ARISTOTELE platform comes into play when other components within ARISTOTELE need to access and handle content or semantic data, by providing them with proper means for interacting with the needed underlying data repositories and storage. In Section 5, we will see how data is handled by means of Business Entities by such a core component.

4   Models for the Representation of Knowledge-Intensive Enterprise

After providing an overview of the whole architecture, in this section we briefly present the conceptual models used to semantically represent the enterprise knowledge, as we earlier said in Section 3.

The purpose of the ARISTOTELE models is actually two-fold. First of all, they need to represent the enterprise assets in an effective way. Secondly, they need to support the day-by-day activities of the so-called *knowledge workers*, i.e. workers whose main capital with respect to the organization they work for is knowledge [10], inside knowledge-intensive organizations. Four models have thus been defined:

- Knowledge Model (KM) - This model focuses on three aspects of knowledge:
  - knowledge concerning enterprise assets, e.g. strategies, processes, activities, documentation, and, in general, all the output generated by workers in their activities, including social interactions with their peers;
  - knowledge concerning organization-specific information, in the shape of lightweight ontologies meant to provide a shared classification of the organization's resources and entities;
  - knowledge concerning training, related to the educational domain that may be exploited by a company.
- Competence Model (CM) - This model provides a representation of competences and their relationships with other concepts such as context, evidences and objectives. With this model it is possible to represent:
  - competences of a worker and the evidence of their acquisition;
  - competences needed to fulfill a particular role/job inside the organization;
  - competences needed to achieve a particular goal or objective that is important for the organization [6].

- Worker Model (WM) - This model provides a representation of the knowledge workers inside their organization, including their personal information, social activities and relationships, learning preferences and needs, working activities and involvements.
- Learning Experience Model (LEM) - This model provides a representation of the elements required to support the generation of personalized learning experiences in order to satisfy specific learning needs of the workers, let them achieve a new competence or fill a competence gap.

These four models are interrelated and can cooperate with each other in order to cause a breakthrough in current practices related to competence development, human resources management, knowledge management and learning. The joint adoption of all of these models has therefore been designed to improve the current organizational practices and support innovation in organizations.

All of these models are formalized using RDF/XML syntax. This way, it is possible to support a concrete enforcement of such models on semantic web stacks, their mututal interoperability, as well as their easy linkage with other models emerging within the semantic web community. These models are already interconnected and their entities are interlinked with each other, as shown in Figure 2. Here, for the scope of this paper we focus on one of these models, the Knowledge Model, for which we provide a more detailed description below.

4.1 Knowledge Model

The Knowledge Model (shown in Figure 3) represents the explicit knowledge produced within organization. It was designed to support the following tasks:

- assigning an activity, when a new activity is created and it is assigned to a working group with the goal of creating a new product or service;
- describing a job, when a new technical figure is introduced in the organization, requiring the addition of new competences to the company's profile. In most cases this is equivalent to the assignment of a role in the organizational chart;
- addressing a new challenge, when the organization means to identify the emergence of new challenges or needs from the market itself, like a new product released by a competitor, or a new technology that can drive potential innovations within the business area of the organization. Linking enterprise goals to competences is required to address new challenges, and the Knowledge Model was designed to support the representation of these links.

The starting point for the definition of this model consists of a set of specifications analyzed throughout the inception of the platform, such as Dublin Core [17], SKOS [15], SIOC [14], MOAT [12] and FOAF [4], and a set of enterprise ontologies and models that can be mutually interlinked, such as the REA ontology [8], the oXPDL ontology [9] and the I* model [13,18].

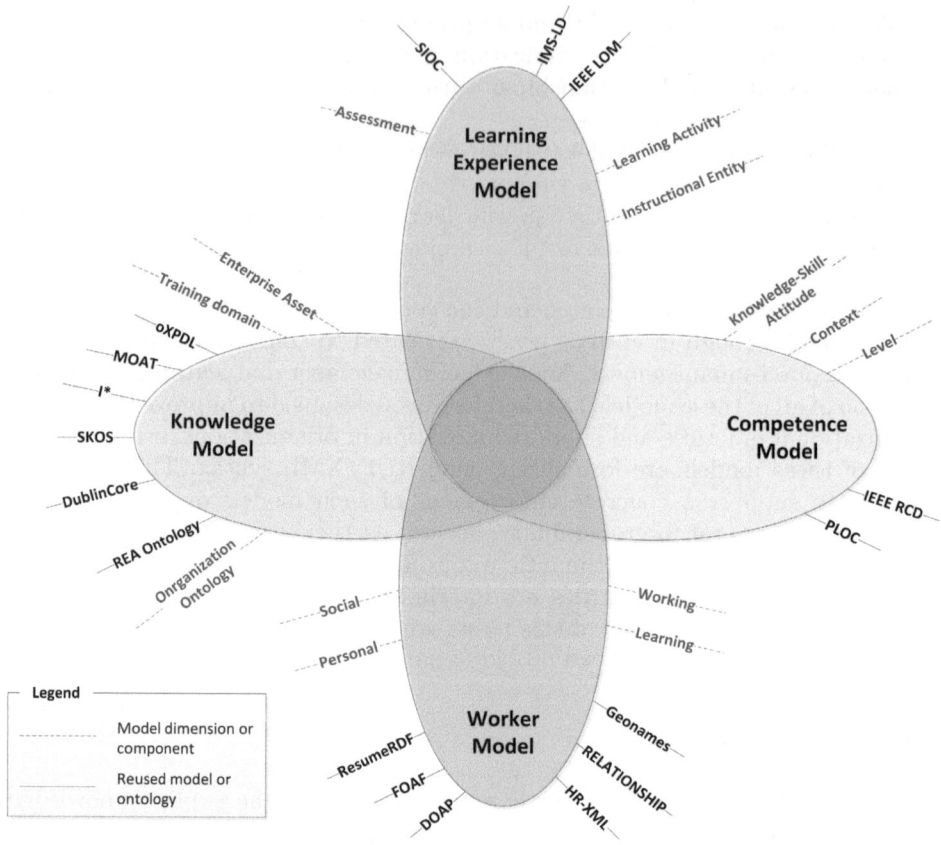

Fig. 2. Relationships among the ARISTOTELE models

Knowledge Concerning Enterprise Assets. This portion of knowledge needs to represent typical enterprise-based assets such as processes, activities, goal, objectives etc. This information is intended to be general and exploitable by any enteprise. In this part of the model we include some platform-specific concepts as well as several concepts coming from the other well-known standards we earlier mentioned (such as FOAF, SIOC, I*, REA etc.), in order to ensure high-level interoperability with existing processes, projects and practices inside the adopting enterprise.

The main concepts of this part of the Knowledge Model are the following:

- Activity, which is a central notion of the Knowledge Model, and is used to represent both the everyday activities and the ARISTOTELE-enabled activities. It is also important for the Goal management processes;
- Agent, used to represent a person, group, software or physical artifact. Typical agents are workers; other agents include organizations and groups;
- Asset, used to represent one of the assets available in the enterprise (e.g. structured and unstructured documents, tools and workers as well);

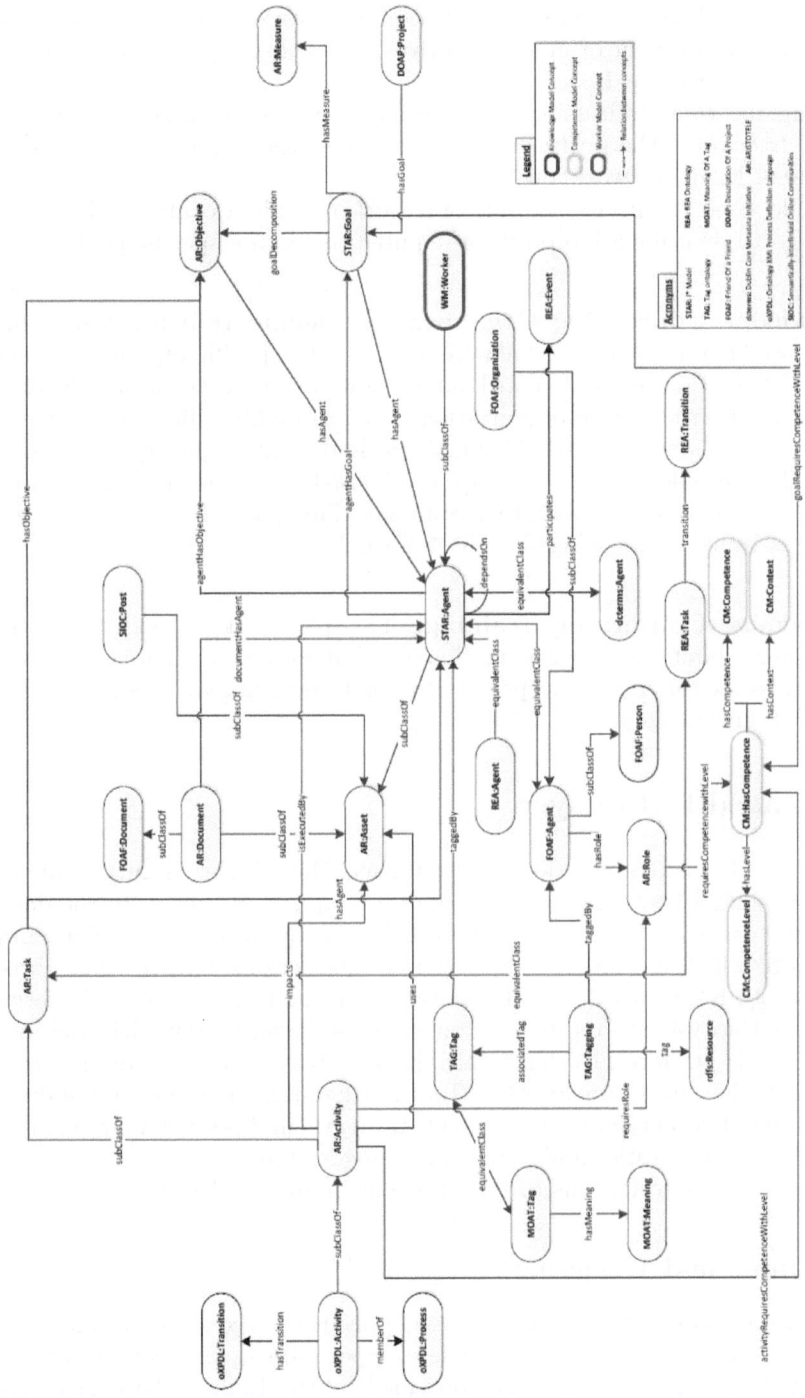

Fig. 3. The Knowledge Model

- Goal, used to map both organizational or worker-level goals. It can be decomposed into one or more Objectives, which also can be further decomposed into one or more Tasks;
- Organization, addressing the information required for scheduling activities or to express the position of an activity in the organization, respecting to the order in the workflow;
- Role, used to specify the role of a worker inside a company. It can also be used to describe a job position required to reach a specific goal.

Knowledge Concerning Organization-Specific Information. This part of knowledge represents the knowledge about the specific organization in which ARISTOTELE is adopted. This kind of knowledge reflects the specific classification of information of an organization that can highly differ among companies. To address this aspect ARISTOTELE exploits lightweight organization ontologies. These ontologies conceptualize the know-how of the enterprise in terms of areas of interest for the adopting company. This part of the Knowledge Model is representing by adopting the SKOS formalism.

Knowledge Concerning Training. This part of knowledge represents the knowledge about the educational domain that may be exploited by a company to address the different concepts to be taught and learned. This is done by using the IWT knowledge model [2,7].

5   Linked Data Layer

As we briefly said in Section 3, the Linked Data Layer (LDL) component of ARISTOTELE has the responsibility of decoupling the access to the data layer from the other components of ARISTOTELE, by declaring a unified interface thanks to which it allows for entity-based interactions with the data. Basically, its purpose is to provide a data source-independent abstraction, in terms of a Business Entity-based representation, over the underlying models and instances of the whole platform. In this regard, it mediates between the data source layer and the higher-level Core Services/Tools, by handling querying and update operations over different data sources, aggregating them if needed, mapping retrieved data to domain entities and persisting changes, if any.

Let us now describe this core component in greater detail.

5.1   Input and Triggering

Every time a component (tool or service) needs to access the data layer for any CRUD (create, read, update and delete), operation against content or semantic data, it relies upon the interface provided by the Linked Data Layer in order to query the data layer itself. Therefore, the LDL is triggered whenever such a CRUD operation is to be performed.

Upon invocation of one of its interface methods by a component of the platform, the LDL takes care of forwarding the call received to the specific module related to a determined business entity to be manipulated. For instance, if a component requests the list of Tasks associated with an Objective O, such a request is forwarded to the internal LDL component which is responsible of handling Tasks.

In this context, each business entity can be populated by integrating data from different sources; as in the example above, the information related to a Task might be found both in the RDF/OWL Triplestore and within the SharePoint content database. Nevertheless, data access requests are formulated in a data source-independent way, by invoking the LDL methods and passing parameters (if necessary), so that those data sources needed to retrieve all the requested information are completely transparent with respect to the invoking component. The LDL is thus responsible of translating such requests by selecting the corresponding language to be used to interact with a specific data source.

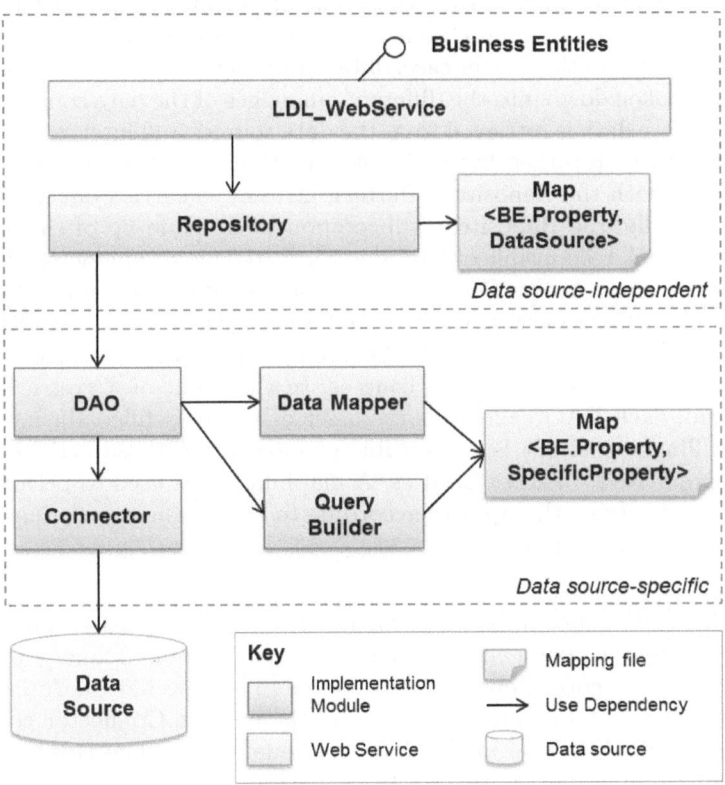

Fig. 4. Architecture of the Linked Data Layer component

5.2 Architecture of the Linked Data Layer

The architecture of the module is depicted in Figure 4; the components building it up are described in greater detail below.

LDL_WebService. The functionalities of LDL are made accessible by means of a standard, Web accessible interface via a Web Service. This is the only way to access the LDL module from the other components of the platform: the methods receive and transmit data in the form of Business Entities and/or additional standard-typed parameters and values (integer, strings) accordingly. As it stands, this is the outermost layer of the module; any call to its methods is handed over to the Repository subcomponent immediately below in the module's hierarchy.

Repository. This subcomponent receives calls from the LDL_WebService, and is assigned the duty of handling them, by storing/modifying data within the underlying data sources involved (in the event of "create/update/delete" (CUD) operations) or integrating data from those data sources (in the event of a "retrieve" operation). In the former case, data is passed as Business Entities, and appropriately broken down into the different languages of the data sources, whereas in the latter case data is retrieved from the data sources and integrated into Business Entities to be returned by the invoked method(s). This process is designed in accordance with the Repository Pattern [21] and is carried out in the following way. Actually, the Repository subcomponent is made up of single, specific Repositories, each responsible of handling the CRUD logic of a determined Business Entity. A Data Access Object (DAO) for each data source is called by the repositories in order to access the required data sources. As far as CUD operations are concerned, the required DAOs are called in order to correctly modify the data within the required data sources. In the event of a "retrieve" operation, instead, each DAO called returns a set of partially-filled Business Entity, and the calling repository is responsible of integrating those entities returned into fully-populated Business Entities. A mapping file is used to perform such a merging, by specifying the priority according to which using the values returned from one data source or the other.

DAO. The DAO subcomponents, when called by a repository, interacts with the QueryBuilder component in order to get the required query to be performed against its corresponding data source. The selected query is then executed against the data source by invoking the specific Connector component, and its results are provided in the data source language. The DAO component subsequently relinquishes control to the corresponding DataMapper component, which takes the result of the query as input and gives Business Entities in return. As stated earlier, the outcome of this mechanism is a set of partially-filled Business Entities, to be subsequently handled and integrated by the upper-layer repository.

Connector. This component provides the lower-level communication channel with a specific data source.

Map <BE.Property, DataSource>. This is a mapping file, in XML format, where mapping information between properties of a Business Entity and their corresponding data source to be used to retrieve such a property is stored, for each Business Entity handled by the platform.

Map <BE.Property, SpecificProperty>. This is another mapping file, in XML format, where mapping information between properties of a Business Entity and their corresponding representation in the data source specified by the previous mapping file is stored.

QueryBuilder. This subcomponent is accessed by the DAO subcomponents, and is responsible of translating of high-level requests coming from the DAOs into data source-specific queries according to the language of its related data source (i.e. SQL, SPARQL, CAML). Basic CRUD operations (simple retrievals, creations, updates and deletions) are automatically generated by resorting to the mapping file Map<BE.Property, SpecificProperty>, whereas custom queries are manually composed in order to provide advanced functionalities where needed.

DataMapper. This subcomponent is responsible of translating the data source-specific result of a query executed by a DAO (e.g. a SPARQL ResultSet, a SQL ResultSet, a SharePoint list etc.) into a set of Business Entities.

6 Conclusion

In this paper we have presented the architecture of the ARISTOTELE platform, a semantic-based collaborative system for enterprise management. We have listed the tools and services it includes, as well as its underlying data layer made up by lower-level content and semantic data and higher-level business data. We have then presented ARISTOTELE's conceptual models used for representing semantic-based information coming from the adopting organization's domain of interest, with special attention upon the Knowledge Model of the platform. And finally, we have detailed one of its core component, the Linked Data Layer, which is used to manage the knowledge thus represented and make it accessible by both the internal components and external tools.

Acknowledgments. The research reported in this paper is partially supported by the European Committee under the Collaborative Project ARISTOTELE "Personalized Learning & Collaborative Working Environments Fostering Social Creativity and Innovations Inside the Organization" under Grant Agreement n. 257886.

References

1. Broekstra, J., Kampman, A., van Harmelen, F.: Sesame: A Generic Architecture for Storing and Querying RDF and RDF Schema. In: Horrocks, I., Hendler, J. (eds.) ISWC 2002. LNCS, vol. 2342, pp. 54–68. Springer, Heidelberg (2002)
2. Capuano, N., Miranda, S., Orciuoli, F.: IWT: A Semantic Web-based Educational System. In: IV Workshop of the AI*IA Working Group on Artificial Intelligence & e-Learning held in conjunction with the XI Int. Conf. of the Italian Association for Artificial Intelligence (AI*IA 2009), pp. 11–16. Reggio Emilia, Italy (2009)
3. Expert System Introduces Cogito Answers to Take Enterprise Search and Self-Help Solutions to the Next Level, `http://www.expertsystem.net/news/press-releases/expert-system-introduces-cogito-answers-to-take-enterprise-search-and-self-help-solutions-to-the-next-level`
4. Friend of a Friend (FOAF), `http://www.foaf-project.org/`
5. Gaeta, A., Gaeta, G., Orciuoli, F., Ritrovato, P.: Managing Semantic Models for Representing Intangible Enterprise Assets: The ARISTOTELE Project Software Architecture. In: Sixth International Conference on Complex, Intelligent and Software Intensive Systems (CISIS), pp. 1024–1029 (2012)
6. Gaeta, M., Orciouli, F., Fenza, G., Mangione, G.R., Ritrovato, P.: A Semantic Approach for Improving Competence Assessment in Organizations. In: IEEE 12th International Conference on Advanced Learning Technologies (ICALT), pp. 85–87 (2012)
7. Gaeta, M., Orciuoli, F., Paolozzi, S., Salerno, S.: Ontology extraction for knowledge reuse: The e-learning perspective. IEEE Transactions on Systems, Man, and Cybernetics Part A:Systems and Humans 41, 5765718, 798–809 (2011)
8. Geerts, G., McCarthy, W.: An Ontological Analysis of the Economic Primitives of the Extended - REA Enterprise Information Architecture. International Journal of Accounting Information Systems 3, 1–16 (2002)
9. Haller, A., Marmolowski, M., Oren, E., Gaaloul, W.: oXPDL: a Process Model Exchange Ontology. Technical report, DERI – Digital Enterprise Research Institute (2007)
10. Jemielniak, D.: The New Knowledge Workers. Edward Elgar Publishing, Cheltenham (2012)
11. MIKE2.0 Governance Association, `http://mike2.openmethodology.org/wiki/MIKE2.0_Governance_Association`
12. Meaning Of A Tag (MOAT), `http://moat-project.org/index`
13. Mylopoulos, J., Chung, L., Yu, E.: From Object-Oriented to Goal-Oriented Requirements Analysis. Communications of the ACM 42(1), 31–37 (1999)
14. Semantically-Interlinked Online Communities (SIOC), `http://sioc-project.org/`
15. Simple Knowledge Organization System (SKOS), `http://www.w3.org/2004/02/skos/`
16. The APOSDLE Project, `http://www.aposdle.tugraz.at/home`
17. The Dublin Core Metadata Initiative, `http://dublincore.org/documents/dcmi-terms/`
18. The I* Wiki page, `http://istar.rwth-aachen.de/tiki-index.php`
19. The MATURE Project, `http://mature-ip.eu/`
20. The PROLIX Project, `http://www.prolixproject.org/`
21. The Repository Pattern, `http://msdn.microsoft.com/en-us/library/ff649690.aspx`

ICOM: A Framework for Integrated Collaborative Work Environments

Kenneth Baclawski[1], Eric Chan[2] Laura Drăgan[3], Patrick Durusau[4],
Deirdre Lee[3], Peter Yim[5], and Yuwang Yin[1]

[1] Northeastern University, Boston, MA 02115, USA
kenb@ccs.neu.edu, ywang.yin@gmail.com
http://www.ccs.neu.edu/home/kenb
[2] Oracle Corporation, Redwood Shores, CA 94065, USA
eric.s.chan@oracle.com
[3] Digital Enterprise Research Institute, Galway, Ireland
firstname.lastname@deri.org
[4] Consultant
patrick@durusau.net
http://tm.durusau.net
[5] CIM Engineering, Inc. ("CIM3"), San Mateo, CA 94402, USA
peter.yim@cim3.com
http://ontolog.cim3.net/cgi-bin/wiki.pl?PeterYim

Abstract. Collaboration is an important activity that is increasingly
using technology to improve the productivity of the participants. The
Integrated Collaboration Object Model (ICOM) is a proposed OASIS
standard for interoperation among collaboration services. ICOM is intended to be a framework for integrating a broad range of domain models
for collaboration environments. The intention is to encourage independent software vendors and open source communities to create common
collaboration clients that interoperate with integrated collaboration platforms and standalone collaboration services across enterprise boundaries.
This paper provides an overview of ICOM that covers the high-level concepts, directory, space, access control, metadata, content management,
and unified message models. ICOM is represented in several formats: the
Java Persistence API, XML Schema, RDF and OWL. We also describe
an example application based on ICOM.

Keywords: JPA, collaborative work environments, Semantic Web.

1 Introduction

The Integrated Collaboration Object Model (ICOM) for Interoperable Collaboration Services defines a framework for integrating a broad range of domain
models for collaboration activities in an integrated and interoperable collaborative work environment. The framework is not intended to prescribe how applications or services conforming to its model implement, store, or transport the

A. Haller et al. (Eds.): WISE 2011 and 2012 Combined Workshops, LNCS 7652, pp. 147–158, 2013.
© Springer-Verlag Berlin Heidelberg 2013

data for objects. It is intended as a basis for integrating a broad range of collaboration objects to enable seamless transitions across collaboration activities. This enables applications to maintain a complete thread of conversations across multiple collaboration activities. The model integrates a broad range of collaboration activities, by encompassing and improving on a range of models which are part of existing standards and technologies. The model is modular to allow extensibility. The fundamental concepts, metadata concepts, and their relations are included in the core part of the model, while the specific concepts and relations for each area of collaboration activities are defined in separate extension modules.

In the next section, we define the ICOM Core model. It defines the classes that bring together the model of directory management, identity management, and content management in a framework with a common access control model and metadata model. We continue with the description of the extension modules, which extend the core to define the specialized models for different collaboration activities. The range of collaboration models include content sharing and collaboration, asynchronous communication, instant communication, presence awareness, moderated group discussion, time management, coordination, etc. The core defines three distinctive branches of concepts, which complement each other: Artifact, Subject, and Scope.

The Subject and Artifact branches support separation of concerns for user administration and content management. The Subject branch includes the model of actors, groups, and roles. These concepts typically appear as the subject in the (subject, privilege, object) triples of an access control model. The Artifact branch includes the model of content and metadata produced by actors.

The Scope branch defines the model of communities and spaces that contain subjects and artifacts. Communities and spaces join the subjects and artifacts in a role-based access control model where a role is assigned to an actor in a specific scope. Thus Scope, Subject, and Artifact form a framework for applications to integrate and interoperate with directory, identity management, content management, and collaboration services.

The model specified in ICOM is based on existing standards and technologies. ICOM core model encompasses LDAP Directory Information Models [1]. The extension modules integrate models from Content Management Interoperability Services [2], Java Content Repository API [3], Web Distributed Authoring and Versioning (WebDAV) [4], Internet Message Access Protocol (IMAP) [5], Simple Mail Transfer Protocol (SMTP) [6], Extensible Messaging and Presence Protocol (XMPP) [7], XMPP Instant Messaging and Presence [8], vCard MIME Directory Profile [9], Internet Calendaring and Scheduling Core Object Specification (iCalendar) [10], and Calendaring Extensions to WebDAV (CalDAV) [11]. ICOM is open for extensions with additional domain models to enable seamless integration with business processes and social networks: for example in process integration domain which includes Business Process Model and Notation [12], Web Services Business Process Execution Language [13], WS-BPEL Extension for People [14], and Web Services for Human Task [15]; in social networking domain, which includes Friend of a Friend (FOAF) [16], Semantically-Interlinked Online Communities (SIOC) [17], Open Social[1], and Facebook Platform Open Graph[2].

ICOM uses the Content Management Interoperability Services (CMIS) [2] grammar to define classes and properties.

The OASIS ICOM TC Wiki[3] provides supplemental information, including overview, primer, extensions, use cases, and mappings to various standards and proprietary data models. The integrated model can be the foundation for defining the application programming interfaces (API) for integrated collaboration applications to interoperate with collaboration services. A service provider interface (SPI) can be specified to support interchangeable and interoperable services that conform to the ICOM application framework. ICOM does not prescribe how applications or services conforming to its model implement, store, or transport the data for objects.

ICOM is represented in several formats: the Java Persistence API, XML Schema, RDF and OWL. The UML and XML Schema are derived from the Java classes, while the other specifications are directly derived from the authoritative specification.

2   ICOM Modular and Extensible Framework

ICOM specifies a set of concepts in a collaboration environment, in terms of class and property definitions. An ICOM object may be composed of information from multiple repositories or collaboration services. All objects in the ICOM framework must be instances of at least one class. The class and property definitions correspond to the UML meta-model, which is an OMG Meta Object Facility (MOF) M2-model. The UML diagrams for ICOM were generated from Java classes, which were directly translated from the authoritative ICOM specification. Some of these diagrams are shown below. The full set of UML class diagrams is in [18].

2.1   ICOM Core

The ICOM Core model has three branches in its class hierarchy: Scope, Subject, and Artifact. The Scope branch includes the model of communities and spaces which are containers of subjects and artifacts. This branch is concerned with directory management, providing hierarchical classified listings of Role, Group, and Actor for administration, search and indexing, and uniform reference. Community and space list the resources as directory entries [1]. Although spaces contain both subjects and artifacts, the membership of subjects in a space is administered separately from management of artifacts in the space. Discretionary

---

[1] http://opensocial.org
[2] https://developers.facebook.com/docs/opengraph
[3] https://wiki.oasis-open.org/icom

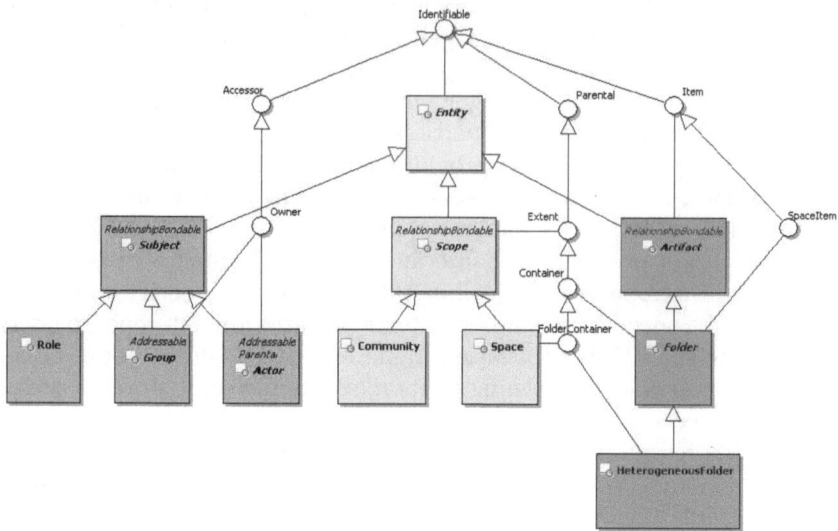

Fig. 1. The Main Branches of the ICOM Core Model

access control (DAC) and role-based access control (RBAC) policies are defined in terms of the subjects from the directory.

The Subject branch includes the model of actors, groups, and roles. This branch is concerned with the identity of actors in collaboration. The Artifact branch includes the model of folder, heterogeneous folder, and content produced by actors.

The Core model also defines the Metadata and Access Control models in separate namespaces. The Metadata is concerned with annotations. The Access Control defines the discretionary access control (DAC) model for entity-level granularity and role-based access control (RBAC) model for scope-level granularity.

The Subject and Artifact branches support the separation of concerns of user administration and content management. Typically subjects and artifacts are joined in the (subject, privilege, artifact) triples of the access control model. Some of the triples are derived from the scopes of the role assignments and the artifacts contained by the scopes.

Figure 1 depicts the top-level abstract classes forming the main branch and the Scope, Subject, and Artifact classes that represent the roots of the three major sub-branches of the ICOM class hierarchy. To deal with the fact that some programming languages, such as Java, do not support multiple inheritance, the model defines two types of classes: ordinary and mixin. The mixin classes are represented in Java as interfaces rather than classes. The mixin classes are shown as circles rather than rectangles in the class diagram. It also shows the core classes in the Scope, Subject, and Artifact branches. It only shows the

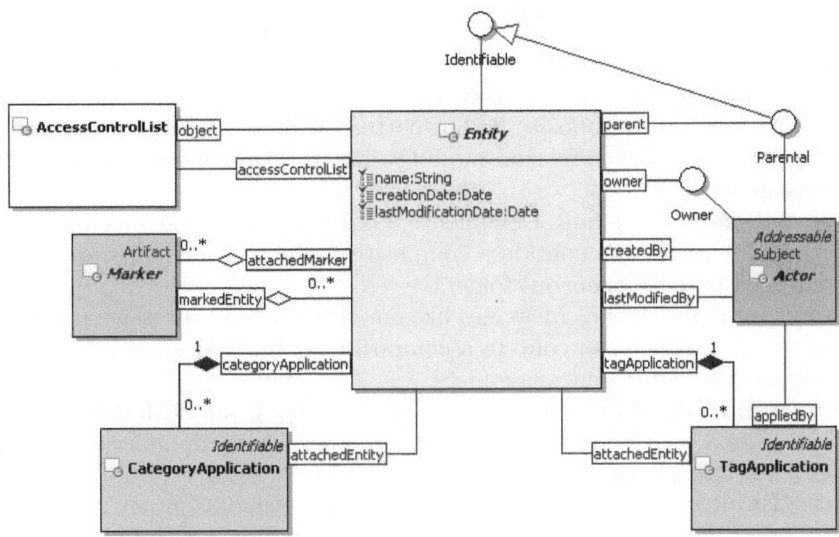

Fig. 2. ICOM Entity Classes

subclass relationships, not the attributes or the associations. Figure 2 depicts the Entity class in more detail, showing both its attributes and associations with other classes.

2.2   ICOM Extensions

Each ICOM extension module defines a model of a collaboration activity. The specification includes models for content creation, communication, coordination, discussion forum, and conference. Most of the extension modules in this section introduce specialized subclasses of Artifact and Folder of Artifact Branch. In addition to the extension modules described here, the ICOM framework allows additional extension modules. For example, applications can adopt a model for the CMIS Policy base type as a new extension module, which can be used to integrate with BPMN or BPEL processes outside the ICOM domain. An ICOM space can provide a durable context for continuity of conversations and activities related to a business process type or process instance. Some new extension modules may import the models from related standards.

ICOM defines containers that provide contexts and structures for specific areas of collaborative activities. For example, a Space is a container that contains HeterogeneousFolder, AddressBook, Calendar, TaskList, Forum, and Conference. These subcontainers are briefly described below in the corresponding extension modules. HeterogeneousFolder (defined in Core) is a general purpose container that can list any type of artifacts, and therefore, can serve as a library of documents and wiki pages to support content sharing and co-creation, an inbox or outbox for communication, or a trash folder to archive all types of artifacts deleted from a space.

The following modules are specified as extension modules of ICOM:

Content module defines classes for Content, MultiContent, and SimpleContent. Content represents a piece of data in a document or message. The module uses the Composite design pattern to form objects. The module is referenced by other extension modules, like the Document module and the Message module.

Document module defines Document, WikiPage, and a model for version control. A document can contain a composite content. Documents are typically contained by heterogeneous folders.

Message module defines Messages like emails or instant message, and related classes. A message can contain a composite content. Messages are typically contained by heterogeneous folders.

Presence module defines Presence, Activity, and ContactMethod. Presence represents a watchable state of an actor. Presence state is derived using an actor's subscriptions.

AddressBook module defines AddressBook and PersonContact. AddressBook is a specialized container to manage contact or personal information, such as addresses, phone numbers, birthdays, anniversaries, and other entries. A person contact can reference a person in an ICOM community as well as information about a person who may not be in any ICOM community.

Calendar module defines Calendar, Occurrence, and OccurrenceSeries. Calendar is a specialized container to support time management. Occurrence artifacts are used to resolve the free-busy times of participants for scheduling of meetings and booking of resources.

FreeBusy module defines the FreeBusy class. FreeBusy is a view derived from occurrences in a calendar or a set of calendars using an actor's privileges.

TaskList module defines TaskList and Task. TaskList is a specialized container to support task coordination. Tasks are used to coordinate the assignment and progress of work.

Forum module defines Forum, Topic, Announcement, and DiscussionMessage. Topics, announcements, and discussions are used for threaded conversations. Moderators of a forum can prune, merge, or fork the discussion threads. Forum is a specialized container to support Topic sub-containers and Announcement sub-containers for time-sensitive communication.

Conference module defines Conference and related classes. Conference is a specialized container that provides a durable context for real-time interactions. A conference can contain visual, audio, and chat transcripts of the conference sessions. It also contains the current status, conference settings, past sessions, active session, and activity logs.

3   Features of the Model

3.1   Persistence

Maintaining data persistently is a necessity for nearly all software applications. Since the predominant storage technology is the relational model but the predominant software languages are object-oriented, it is necessary to have an object-relational mapping. An increasing number of applications are using the Java Persistence API (JPA) for this task. Accordingly, one of the first mappings for ICOM was a mapping to JPA. This mapping consists of Java annotations added to the POJO classes. The java.net OpenICOM project [19] is incubating a JPA framework which manages ICOM POJO objects. OpenICOM framework emulates the JPA programming model but is not limited to object-relational mapping. It supports pluggable data access connectors that can be implemented using proprietary APIs for collaboration platforms, standard protocols such as LDAP, WebDAV, IMAP, SMTP, XMPP, CalDAV, etc., or NoSQL databases.

The ICOM specification defines a class called Entity which is the superclass of any class that supports a persistent identifier, a change token for optimistic locking, and an access control list. The object identifier and change token are annotated, respectively, by javax.persistence.Id and javax.persistence.Version, matching the ICOM concept of Entity with the JPA concept of Entity. ICOM Entity has another fundamental dimension for access control list, which together with JPA Id and Version, defines a unit of persistent information for concurrency and access control. The generation of object identifiers is implementation dependent; however, ICOM recommends that the object identifiers should be globally unique to support permanent references to the entities that may migrate amongst interoperable ICOM repositories. An object identifier is read-only (immutable) once it is assigned and should never be duplicated or re-used for more than one object. The UML diagram in Figure 2 depicts the Entity class, properties, and cardinality of the properties. Entity's properties include name, created by, creation date, last modified by, last modification date, owner, parent, attached markers, category applications, tag applications, and access control list.

Figure 3 shows some of JPA annotations for the Entity class. For the sake of brevity, the import statements were not shown.

3.2   Interoperability

One of the goals of ICOM is seamless interoperability between different collaborative work environments. This requires an interchange mechanism between such environments. Since these environments may use different programming languages, the interchange format must be language-independent. The most commonly-used interchange format today is XML, and ICOM has been expressed in terms of XML Schema. This allows one to exchange data via SOAP/REST web services. The XML Schema specification was derived from the POJO classes using the JAXB schemagen processor[4]. Figure 4 shows the XML Schema representation of the Entity class.

---

[4] http://docs.oracle.com/javase/7/docs/technotes/
   tools/share/schemagen.html

```
package icom;

@Entity
@XmlType(name="Entity", namespace="http://docs.oasis-open.org/ns/icom/core/201008")
@XmlRootElement(namespace="http://docs.oasis-open.org/ns/icom/core/201008")
@XmlAccessorType(FIELD)
public abstract class Entity implements Identifiable {
 @EmbeddedId
 protected Id objectId;
 @Version
 protected ChangeToken changeToken;
 @ManyToOne
 Actor createdBy;
 @ManyToMany(targetEntity=Marker.class)
 @XmlElement(name="attachedMarker",
  namespace="http://docs.oasis-open.org/ns/icom/metadata/201008")
 Set<Marker> attachedMarkers;
 ...
}
```

Fig. 3. Some of the fields and their JPA annotations for the Entity class

```
<xs:complexType name="Entity" abstract="true">
 <xs:sequence>
  <xs:element name="objectId" type="icom_core:Id" minOccurs="0" />
  <xs:element name="changeToken" type="icom_core:ChangeToken" minOccurs="0" />
  <xs:element name="name" type="xs:string" minOccurs="0" />
  <xs:element name="creationDate" type="xs:dateTime" minOccurs="0" />
  <xs:element name="createdBy" type="icom_core:Actor" minOccurs="0" />
  <xs:element name="lastModificationDate" type="xs:dateTime" minOccurs="0" />
  <xs:element name="lastModifiedBy" type="icom_core:Actor" minOccurs="0" />
  <xs:element name="parent" type="xs:anyType" minOccurs="0" />
  <xs:element ref="icom_ac:owner" minOccurs="0" />
  <xs:element ref="icom_ac:accessControlList" minOccurs="0" />
  <xs:element ref="icom_meta:attachedMarker" minOccurs="0" maxOccurs="unbounded" />
  <xs:element ref="icom_meta:categoryApplication" minOccurs="0" maxOccurs="unbounded" />
  <xs:element ref="icom_meta:tagApplication" minOccurs="0" maxOccurs="unbounded" />
 </xs:sequence>
</xs:complexType>
```

Fig. 4. XML Schema for the Entity class

3.3   Semantic Representation

ICOM also has representations in RDF and OWL. The semantic representation is modular, reflecting the UML model, consisting of a core ontology, and extensions. The ICOM ontologies were defined through a direct translation from the authoritative specification.

Providing a semantic representation further accelerates the communication and interconnection of data between different collaboration tools. Other benefits include: access to the growing amount of Linked Data that is available, and inference. With inference, we can enrich the data, but also check for consistency when using ICOM with other ontologies. ICOM data with a seamless programming model like JPA and a concomitant RDF representation will lower the barrier for applying inference engines. Figuratively speaking, a rich vocabulary of "nouns" in ICOM makes up for the strong "verbs" in service interfaces. A well-defined set of classes of ICOM makes the API amenable for rule-based applications and declarative inference. ICOM containers are active or reactive entities, for example conference and chat rooms are highly active while outbox, calendar, and task list are reactive. Their behavior can be augmented by applications.

ICOM Calendar

August 2012		◀ Today ▶		August ▾ 2012 — Go to month		
Sunday	**Monday**	**Tuesday**	**Wednesday**	**Thursday**	**Friday**	**Saturday**
29	30	31	1	2	3	4
5	6	7	8	9	10	11
12	13	14 PSMW Meeting Aug 15 2012	15	16	17	18
19	20 PSMW Meeting Aug 20 2012 — PSMW Meeting Aug 20 2012 Afternoon	21	22	23	24	25
26	27 PSMW Meeting Aug 28 2012	28	29	30	31	1

Fig. 5. Example of a calendar module that integrates SMW and ICOM

4   A Use Case — Integration with Semantic Media Wiki

A *wiki* is a website which allows its users to collaborate to produce and edit content via a web browser. The resulting content is easily understood by users, but computers cannot understand or evaluate it. A *semantic wiki* allows a user to add semantic annotations that allow the wiki to function as a knowledge base which supports inference and semantic query capabilities. The most popular wiki software is MediaWiki[5], the underlying software for Wikipedia. The most popular semantic wiki software is Semantic MediaWiki (SMW)[6].

Unfortunately, just having semantic annotations does not mean that different individuals or communities will specify annotations that are compatible with other individuals or communities. Ontologies can provide standard terminologies for annotations that allow for semantic interoperability within an SMW site as well as between sites. The ICOM ontology (available in both RDF and OWL) is especially appropriate for SMW because both ICOM and SMW are

---

[5] http://www.mediawiki.org/

[6] http://semantic-mediawiki.org/

Page   Discussion   Read   Edit with form   Edit ˅ ⌐

PSMW Meeting Aug 15 2012

location	Boston, MA
organizer	YuwangYin
participant	YuwangYin, Admin, Kenb
priority	Low
editMode	AttendeeCopy
startDate	Aug 15 2012 20:00:00
endDate	Aug 15 2012 21:00:00
fromRecurringOccurrenceSeries	True
exceptionToOccurrenceSeries	True
occurrenceStatus	Tentative
occurrenceType	Meeting
attendee	
attendeeParticipantStatus	

Category: Event

Fig. 6. An example of a customized Form used for a meeting

concerned with supporting collaboration. We have developed a proof of concept for the use of ICOM in a semantic wiki by developing SMW modules for calendars, which make use of the powerful semantic search feature of SMW. Figure 5 illustrates one view of the calendar module. This module is publicly avalable at http://psmw-test.cim3.net/w/index.php/ICOM_Calendar.

Other capabilities enabled by the integration of ICOM with SMW include the ability to export data to different formats and to mash-up calendar data from multiple sources. Our SMW modules illustrate how the ICOM RDF and OWL ontologies can enhance interoperability of collaboration tools.

Our Calendar module is a customizable Template that is written in the calendar format of the Semantic Result Formats[20]. This Template uses ICOM properties. Instantiating the Template produces a reusable Form. If one clicks on the meeting on August 15 in Figure 5, one can see an example of such a Form, which is shown in Figure 6. This form shows the ICOM properties of the meeting along with their values. The values can be modified by clicking on the

"Edit with form" tab. The meetings to be included in a particular Form are determined by a SPARQL[21] query. A user can either customize the Template or reuse an existing Form to create their own Calendar.

5   Conclusion and Future Work

We have described the ICOM framework for Interoperable Collaboration Services. Like most standards, the ICOM is specified in an authoritative document using a specification language (CMIS) that is designed to be easily readable by humans. However, ICOM is intended to be mainly used as an interoperability framework within a collaborative work environment for such tasks as storing data, sending data to other collaboration services as well as interacting with human collaborators. As a result, the ICOM framework has been translated to a number of other languages to facilitate these many purposes. Inasmuch as many other kinds of applications have these same requirements, it would be useful for their standards to be translated to other languages as well. Accordingly, we plan to use our translation software with other standards.

Acknowledgements. The authors wish to extend their thanks to colleagues who have helped with different domain expertise. Especially, we thank Rafiul Ahad and Stefan Decker for supporting this work.

References

1. Zeilenga, K.D.: Lightweight Directory Access Protocol (LDAP): Directory Information Models. IETF RFC 4512 (Proposed Standard) (June 2006)
2. Brown, A., Gur-Esh, E., McVeigh, R., Müller, F.: Content Management Interoperability Services (CMIS), Version 1.0. OASIS Standard (May 2010)
3. Nuescheler, D.: Content Repository for Java^{TM} Technology API Version 2.0. (August 2009), http://jcp.org/en/jsr/detail?id=283
4. Dusseault, L.: HTTP Extensions for Web Distributed Authoring and Versioning (WebDAV). IETF RFC 4918 (Proposed Standard) (June 2007)
5. Crispin, M.: Internet Message Access Protocol Version 4rev1. IETF RFC 2060 (Proposed Standard) (December 1996)
6. Klensin, J.: Simple mail transfer protocol. IETF RFC 5321 (Draft Standard) (October 2008)
7. Saint-Andre, P.: Extensible Messaging and Presence Protocol (XMPP): Core. IETF RFC 3920 (Proposed Standard) (October 2004); Obsoleted by RFC 6120, updated by RFC 6122
8. Saint-Andre, P.: Extensible Messaging and Presence Protocol (XMPP): Instant Messaging and Presence. IETF RFC 3921 (Proposed Standard) (October 2004) Obsoleted by RFC 6121
9. Dawson, F., Howes, T.: vCard MIME Directory Profile. IETF RFC 2426 (Proposed Standard) (September 1998)
10. Desruisseaux, B.: Internet Calendaring and Scheduling Core Object Specification (iCalendar). IETF RFC 5545 (Proposed Standard) (September 2009)

11. Daboo, C., Desruisseaux, B., Dusseault, L.: Calendaring Extensions to WebDAV (CalDAV). IETF RFC 4791 (Proposed Standard) (March 2007)
12. Business Process Model and Notation (BPMN) Version 2.0. (January 2011), http://www.omg.org/spec/BPMN/2.0
13. Jordan, D., Evdemon, J.: Web Services Business Process Execution Language (WSBPEL), Version 2.0. OASIS Standard (April 2007)
14. Ings, D.: WS-BPEL Extension for People (BPEL4People) Specification, Version 1.1. OASIS Committee Specification (August 2010)
15. Ings, D.: Web Services — Human Task (WS-HumanTask) Specification, Version 1.1. OASIS Committee Specification (August 2010)
16. Brickley, D., Miller, L.: FOAF Vocabulary Specification 0.98 (August 2010), http://xmlns.com/foaf/spec/
17. Berrueta, D., Brickley, D., Decker, S., Fernndez, S., Grn, C., Harth, A., Heath, T., Idehen, K., Kjernsmo, K., Miles, A., Passant, A., Polleres, A., Polo, L., Sintek, M.: SIOC Core Ontology Specification. W3C Member Submission, W3C (June 2007)
18. Chan, E.S., Durusau, P.: Integrated Collaboration Object Model (ICOM) for Interoperable Collaboration Services, Version 1.0. OASIS Committee Specification (August 2012)
19. Chan, E.S.: OpenICOM: A JPA Framework for Integrated Collaboration Environments, Part 1 (March 2011), http://today.java.net/article/2011/03/21/openicom-jpa-framework-integrated-collaboration-environments-part-1
20. Hong Kong, J., Koren, Y., De Dauw, J.: Semantic Result Formats (2012), http://semantic-mediawiki.org/wiki/Semantic_Result_Formats
21. Prud'hommeaux, E., Seaborne, A.: SPARQL Query Language for RDF. W3C Recommentation, W3C (January 2008)

Knowledge-Based Semantification of Business Communications in ERP Environments

Marios Meimaris and Michalis Vafopoulos

School of Electrical & Computer Engineering, National Technical University of Athens
{m.meimaris,vaf}@medialab.ntua.gr

Abstract. The complexity of information in business environments is increasing in rates that are difficult for companies to handle using traditional solutions. The diversity of information and communications that stem from remote and heterogeneous sources creates the need for information management in multidimensional contexts and platforms, such as web sources, mobile devices and databases. As web technologies evolve, the volume, temporality and heterogeneity of available data exhibits a degree of dynamicity that is beyond manual control. Without these limitations, the aggregation of this information can provide new value-driving layers in business environments. Moreover, when this data is semantically structured, knowledge sharing and rule-based management allow for new ways of controlling the flow of information, both internally and externally. Focusing on the extraction of *information about communications* and its exploitation in platform-independent manners, we propose, design and implement a set of semantic components extending ERP systems in order to assist semantic web interoperability, to provide a basis for intelligent knowledge management and to unify communication level platforms under shared sets of principles.

Keywords: Semantic Web, Linked Open Data, Business Intelligence, Knowledge Management, Ontologies.

1 Introduction

Modern technologies and design approaches make the web an exceptionally dynamic environment, where data and information is created, consumed, edited and deleted within very small periods of time. This information can be in the form of communication between people and organizations, as in the example of social networking, or in the form of electronic publishing of facts and otherwise static data with varied time scopes. The dynamicity of these types of information makes it necessary for people and systems to be capable of being constantly up to date and ready to respond to challenges, problems and difficulties that result from change.

In the web business world, when it comes to responsiveness such aspects are crucial. Within business contexts, these types of information form a level that will be called the *Business Communications Level*. They can stem from both internal (e.g. databases, employee mobile phones etc.) and external (e.g. social media, web

A. Haller et al. (Eds.): WISE 2011 and 2012 Combined Workshops, LNCS 7652, pp. 159–172, 2013.

repositories etc.) sources. In the traditional sense, such data is usually neither reusable nor easily interchangeable between heterogeneous and remote sources. It is even more difficult to take advantage of such data sources in cases where response times have to be minimal, as is the case in the business world.

Aggregating these types of data and information and combining them through a unified schematic structure will become essential for business entities in order to survive and adapt to the required responsiveness. Moreover, the use of data that stems from mobile devices can provide and quantify contextual and operational information that was not taken into consideration before. For this reason, it is within our belief that providing platforms with the ability to support interoperability through interchanging diverse communicational data is a step toward an evolved business environment. We have designed and implemented a set of components that follow knowledge-based, semantic web and open data directives, as well as facilitate the extraction and reuse of mobile phone data under a shared ontological substrate, as an extension to traditional Enterprise Resource Planning (ERP) systems, in order to propose an ontology for communications and demonstrate how semantic extensions thrive in environments that are rich in diverse information.

This paper is organized as follows: section (2) will provide the background, section (3) will describe the conceptual modelling, section (4) will describe the system implementation and sections (5) and (6) will provide an evaluation as well as present the results and a brief discussion of this work's contribution.

2   Background

2.1   Knowledge Management and Business Intelligence

Generally, Knowledge Management (KM) can be considered as a loosely defined set of methodologies and techniques that aim at externalizing what is considered as internal, or tacit knowledge. A particular subset of Knowledge Management, as is suggested in [1] is *Business Intelligence* (BI), which is considered to be a knowledge level that provides analytical tools to support decision-making. With the incorporation of new technologies, classical BI functionality is extended to include context identification and socially-aware components, this way taking advantage of the richness of information that stems from social web as well as mobile environments.

Within corporate contexts, information can be found both internally and externally. Internal sources include databases, company knowledge, mobile communications and any other types of information that are directly controlled by the company. External sources include web 2.0 (social media, blogosphere etc.), semantic web services, open data repositories, the linked open data (LOD) cloud and so on.

2.2   Internal Sources

Databases and ERP Systems
It is usual for companies to hold their data in local databases. More often than not, these are traditional relational databases that use schemata designed and created in

company-specific manners. In many cases, they are incorporated in Enterprise Re-source Planning (ERP) or similar systems. ERP systems provide a framework for managing business-related resources and assets (both human and non-human) with respect to projects, processes and specific tasks that need to be accomplished. This way, inter-departmental cooperation and coordination can be achieved. Successful deployment of ERP systems reduces overall operating costs and improves enterprise data management and interdepartmental information exchange [2].

Smaizys & Vasilecas [3] argue that the absence of widely accepted standards and formalisms on business rules is what leads ERP system designers to time-consuming solutions which can often be both misleading and discouraging. It is therefore necessary for this data to be structured in ways that allow for sharing and combining with external data under shared and loosely coupled principles. Furthermore, deploying semantic technologies on ERP systems is essential for the development of web service oriented architectures in the semantic web [4].

Mobile Devices

Diverse smartphone functionalities have led companies to incorporate them into their operation management models. Graf and Tellian [5] talk about the benefits of using smartphones as trackers in Supply Chain Management, through the use of data connections and GPS. Furthermore, the app market has led companies to develop applications for managing human resources. Modern smartphones permit context-aware application development. For instance, they include metadata such as geospatial and chronological usage data. Furthermore, in the case of open operating systems, data extraction is feasible, making usage contexts and communication patterns identifiable and tractable.

Little work has been done to assess the use of mobile communication data, and it is within our hypothesis that its aggregation leads to a level of information, called the *Communication Level*, which makes possible the derivation of more complex information regarding the user and the developing communicational contexts they participate in. We build *communication profiles* of users and assess the derived networks.

2.3   External Sources

External data can generally be considered as dynamic or static. Dynamic data originates from social media and otherwise dynamic sources, such as Google+, Twitter and Facebook, web 2.0 platforms in general and so on, whereas static data includes unchanging (or slowly changing) information, such as business registries and other repositories. Both are powerful sources of information, but must be handled differently.

Social Media

We consider social data to be of high business value, because, if interpreted correctly, they hold key information and knowledge about public sentiment. Social media provide added value in many ways. Two main value drivers associated with them are *social media marketing* and *sentiment mining*. Sentiment analysis can provide valuable insight on the public opinion, as well as provide useful results in e-commerce [6].

(Linked) Open Data

Open Data refers to the notion of openness (i.e. free public access) in certain kinds of data which can range from crime reports to government decisions and from bus schedules to registry information about corporate entities. Open datasets are often designed to be machine-understandable and for this reason it is common for them to be shared through open document formats such as JSON, RDF and CSV. A particular case of data openness is that of *Linked Open Data*, which refers to a set of methodologies and guidelines for interlinking data on the web. The advantages of using open datasets such as OpenCorporates.com are that there is no need to store data locally, or tend to the needed updates, since such larger data catalogues publish and curate updated data.

The shift to semantic web technologies and (linked) open data paradigms in the business world is guided by the provision of unique descriptions of corporate entities as reference, publishing of information relevant to company-related affairs, such as accounting facts and figures, operating decisions and board changes and so on (e.g. *OpenCorporates.com*), as well as publishing offers for business transactions that include products and services (e.g. GoodRelations). For a discussion of the implications of an integration of linked open data in business environments, see [7].

3   Conceptual Design

3.1   Test Scenario

The domain of sales was considered as a test domain for the system. The graph-like nature of sales is based on people (actors in general) and their communications, thus making it a suitable test bed for our implementation. It is expandable, ranging from the presales to the after-sales service levels and many different tasks can be identified within it. The advantage of using semantic technologies in this approach is that a common substrate of meaning is employed in a context and platform independent way, for both incoming and outgoing data.

Communication instances stem from diverse sources, such as mobile phones, computers, landlines etc. Especially in large enterprise networks, this results in a highly interconnected graph of dynamic nature. Furthermore, the created network is not isolated from the external world and sometimes the boundaries are not distinct. The scenario is defined as follows:

> *The implemented system is used by a sales company. The combination of external and internal information triggers sales threats regarding prices and offers of particular products/service. The system notifies with alerts when threats are triggered. These might be concerning the same goods offered by competitors, or similar goods with matching functionality (e.g. a competitor company issues a better offer about a particular product, thus threatening our company's existing clientele). The implemented system handles the alerts in two ways:*

a. Automatic suggestions for handling a threat. These include identification of a subset of employees that is suitable to handle the threat, assignment of particular clients/client sets to these employees, suggestion of suitable communication strategies and so on.

b. Provision of related information and statistics for the particular threat for educated manual handling.

Past information drawn from the system is shared among salespeople within the company. The system automatically uses a set of classifications on the market and clients in order to outline the conditions, the results of which are made available to the salesperson. Finally, the salesperson responds to the threat by adapting her operational plan according to these. The results of her attempts are quantified and measured before they are made part of the shared knowledge, thus rebooting the cycle.

The purpose of the system is to demonstrate how corporate decision making can be assisted. The system's input is made up of the sources that were described in section II. The company's salespeople handle both new and existing clients. The system builds *communication profiles* for both clients and employees, based on past data. Therefore, when an existing client needs to be approached, the salesperson can consult the client's profile in order to select a strategy of approach. In the case of new clients, the market profile shows what communication strategy has been successful product-wise in the past.

3.2   Ontologies and Vocabularies

The *local* or *communications* ontology was created mainly to represent the *Communications* domain, which forms a large part of our approach. This has been created from scratch, always having in mind the principle of *minimal ontological commitment* [8], which simply states that in ontology design, the representational claims should be exactly as many as needed for the description of the domain of interest, in order to avoid knowledge redundancy. Classes can be seen in figure 1, with simplified names to assist readability. However, reusing existing ontologies is generally preferred over creating new ones. In our approach, parts of existing ontologies have been employed in order to describe high-level concepts and relationships such as people, organizations and companies, products and services and so on. The reuse of existing ontologies is a default recommendation, as it is quicker, less expensive and drives information sharing between diverse sources.

In our implementation, several external ontologies have been reused, such as the *GoodRelations Ontology* [9] and the *Friend-of-a-Friend* (FOAF) vocabulary. *GoodRelations* provides a conceptual framework for representing e-commerce offers of goods in a way that can be processed by web services. It is adopted by large corporations such as Google, Yahoo! and BestBuy. *FOAF* [10] is an ontology for the description of individuals and organizations, and the relationships between them. It is widely accepted as a reference model for the description of people and groups. Class equivalence between ontologies ensures the interlinking of these concepts in the structural level (e.g. lo cal:Agent is equivalent to foaf:Agent).

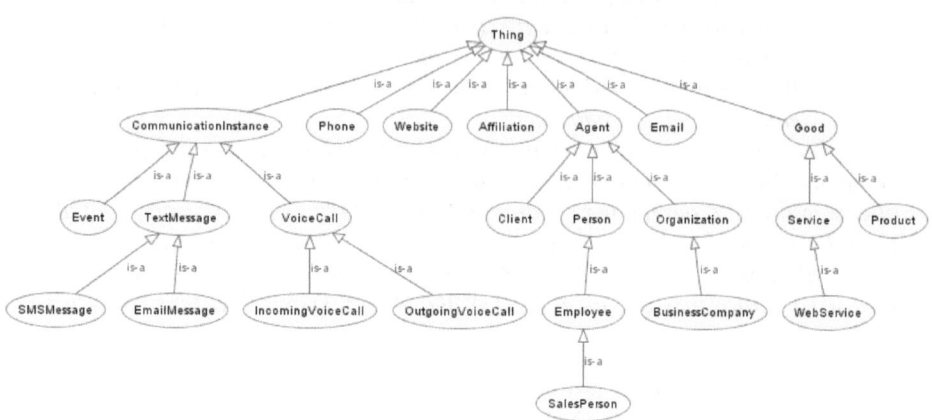

Fig. 1. Local and Communications Ontology

4   System Implementation

4.1   Technological Choices

The implemented system and all associated methods and classes are written in Java. Semantic web and ontology integration are done with the use of the Jena Semantic Web Framework [11]. Demonstration is implemented in the form of JSP (JavaServer Pages) on top of simple HTML, resulting in a web environment capable of providing basic demonstrative functionality.

We have chosen to expand on OpenERP [12], which has rounded capabilities and small size. Thomas Herzog [13] shows that OpenERP (cited as TinyERP at the time of writing) covers all necessary functionalities with a smaller set of database tables than the competing systems. For the mobile communications part, we use Android as the scenario's mobile environment. Android's programming framework is built on top of Java classes, making the incorporation of external Java libraries natural.

Data is converted and stored in the RDF format. Storing and linking data with semantic (ontological) structure is particularly easy with the use of RDF, a w3c recommendation. We have selected *OpenLink Virtuoso* to store RDF triples. The choice was based on the fact that it has rounded capabilities and good querying performance, as suggested by standardized benchmarking tests [14]. Furthermore, Virtuoso contains java libraries for connecting with the Jena Semantic Web Framework, making it less costly for the implemented system to perform RDF data transactions between the server and the various external platforms.

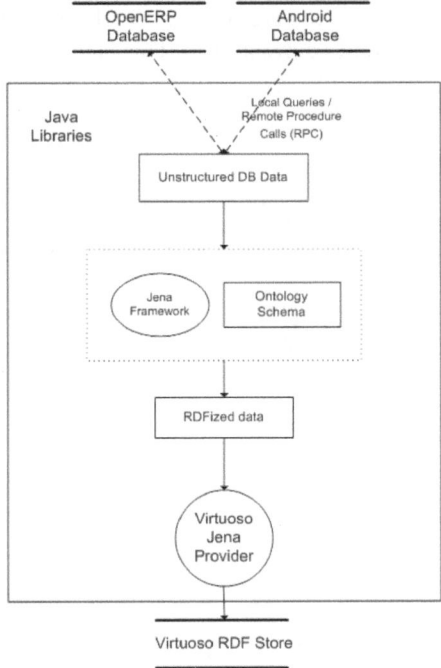

Fig. 2. Data extraction, RDF conversion and storing

4.2   Information Retrieval and Semantification

Various data sources are used for information retrieval in this scenario. Each of them is queried differently, according to the technologies it is subject to. They are converted to RDF using the Jena framework, the system's ontologies and a set of conversion rules (figure 2). After the conversion, they are stored in the quad store. RDF data is then offered depending on the context, and the use of negotiation rules ensures that the availability of the data is limited and controlled accordingly. In practice, this would mean that different subsets of the data will be made available to different parties at different contexts.

4.3   Information Aggregation and Utilization

Content Aggregation
As has already been established, ERP systems form the centre of the scenario, in the sense that the ERP is the main context driver. For this reason our implementation stands as an extension to its functionality, providing several layers, both structural and functional, around the domain of interest, mainly focusing on the benefits of the communicational approach. The structural layer is a consequence of the provision of

semantics, while the functional layers are the sum of all extra functionalities that are implemented on top of the system's current ones. This can be seen in figure 3.

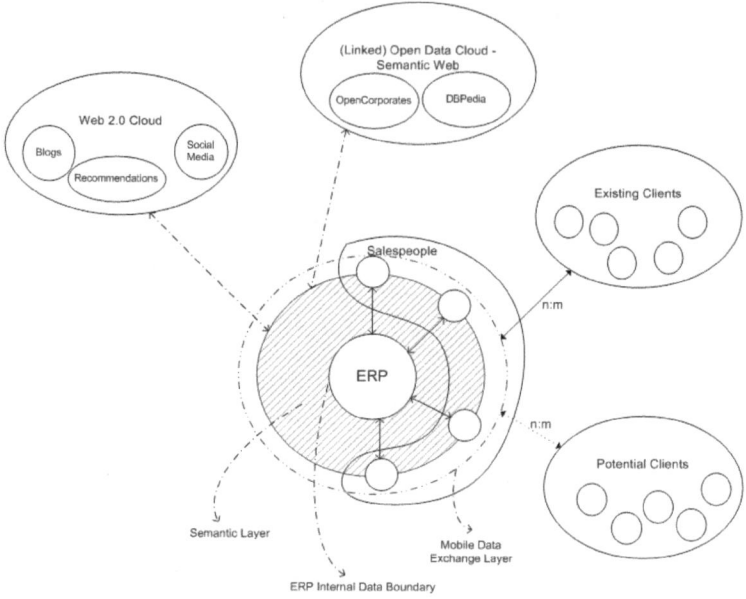

Fig. 3. Abstraction of the information layers

Information is shared within the dashed layer throughout the extended system. Practically, the data that exists within the information layer is in RDF form, stored in the dedicated quad store. The mobile data is exchanged at the boundary of the semantic layer. Finally, external information that stems from web sources such as social media and (linked) open data is exchanged through the boundaries of the semantic layer as well. Within these, data is stored in aggregation, under common semantic structure.

Web Interface

A web interface has been created for demonstration. This is a web environment where employees with managerial status are notified when price alerts are triggered. The implemented version includes a browsing environment of employees, clients and products/services, with respect to various sets of restrictions. Furthermore, the user can read descriptions and metadata of incoming alerts, get automatically-generated handling strategies, assign employees to clients and products, see communication profiles of employees and clients and read several data-centric analyses of cross-patterns between the variables.

a) Alert Notifications

Upon request of this webpage, the server is queried in order to return a list of the un-handled alerts, along with a subset of their non-functional properties. It is assumed that the alerts are created from external sources, mainly social media. However, the creation of such alerts lies beyond the scope of this paper.

By selecting an incoming alert notification, the web environment shows the user the alert's profile. As each price alert is created in reference to a particular product with a particular business function[1], the alert profile page will go on to provide an initial analysis of the alert based on these constants. Based on this design, an alert profile page offers department, employee and client rankings with respect to the alert.

b) Employee, Client and Product/Service profiles

Employee and Client views provide insights to the associated agent, including their static as well as dynamic information profiles. Employee and Client profiles are simi-lar for the most part, with the differences being in the types of associations that are produced. Their conjunction is made up of the following information:

- Static information that defines the employee/client as an ERP entity, including contact details drawn from the Contacts Level. In the case of employees, their de-partment associations are further provided. In the case of clients, a link to data drawn from OpenCorporates is also present.
- Top products/services associated with the particular agent, along with processed information. This includes transaction statistics, rankings by business function, prominent communication strategies and communication type statistics, paired em-ployee-client rankings and so on.
- Rankings of communication types associated with the agent, including communi-cation statistics by type or supertype, (e.g. email vs. text messaging in general), communication patterns that arise, preferred methods of communication by product or by client and so on.
- Rankings of business functions independently of the product/service they refer to, including total number of transactions and total income. This highlights the top-selling business functions for each employee/client/good.

c) Threat handling

Incoming threat alerts are ranked and classified based on their description. The system offers its users the possibility to assess the threat automatically, by suggesting the best means to handle it, based on predefined rule sets. Given that incoming threats are always in reference to a particular product/service, it is easy to identify the threatened clients and outline the context. This way we can evaluate historical data and suggest communication strategies based on the identified context. The overall alert resolution process can be seen in figure 4.

Evaluation of incoming alerts is followed by resolution. The alert metadata helps classify and categorize it, and assign it to appropriate handlers. Technically this can

---

[1] *Business functions* have to do with the different ways a product/service can be commercial-ized, e.g. Sell, Repair, Maintain etc. They are formally represented with GoodRelations.

be achieved as follows: the web environment is user-dependent (controlled with dynamically triggered ontology rules based on the user and their profile/history) and identification of the user restricts the query results accordingly.

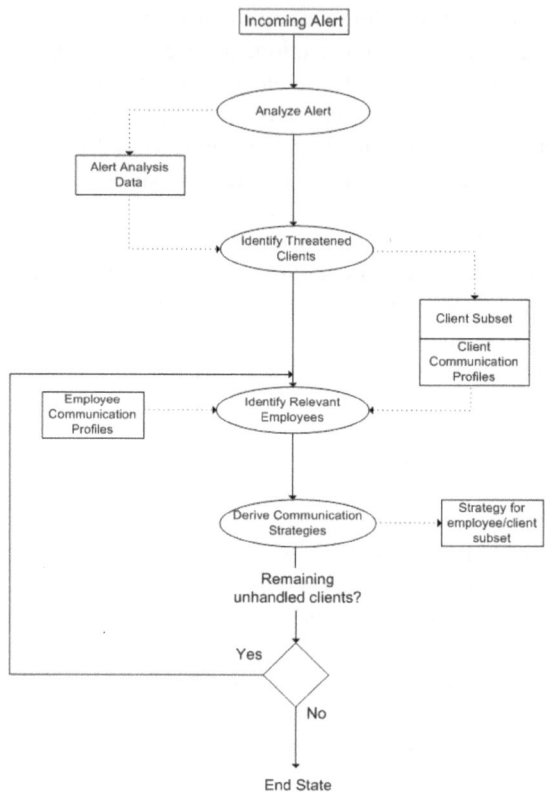

Fig. 4. Alert handling process

5   Evaluation

The main purpose of this work is to highlight the richness of information that business communications have to offer, and this is shown through a corporate use-case scenario. Furthermore, it is demonstrated how various components can be deployed both server-side (ERP extensions) and client-side (Android apps) in order to convert and publish data as RDF, with the use of ontologies. In this section we will focus on qualitative evaluation of the conceptual approach, as well as discuss implementation implications.

Fig. 5. Screenshots from demonstration interface

5.1   Communications Ontology

Our ontology captures the knowledge that is related to communications. These stem from mobile phones, landlines, the world wide web, and so on, but they all belong to the same communications taxonomy. As smartphones and modern computing solutions are able to capture metadata of specific communication instances, we show how this can be performed on the fly and in an ontology-based manner.

Heterogeneous sources are aggregated under a common denomination, creating value as super-layers of information are brought forth as a result of this mash-up. For instance, combining the knowledge of how a particular employee usually handles (communication-wise) clients of a particular nature, along with the knowledge of how successful in terms of sales this employee has been, we outline market conditions and client behaviors on the fly. This is particularly useful in environments where employees need to handle threats with minimal response times, as in the scenario. Combining this information helps companies in more than one ways. To name a few:

- Improved HR management: the communication profiles of employees create an in-house network of nodes and edges that show how employees communicate with each other. Sinks and sources can be identified, case-specific or overall.
- Educated threat-handling: when clients are threatened, we can identify the best strategy to approach the case at hand, depending on their communication profiles.
- Unknown market analysis: when entering new markets, communication patterns of goods in known markets are consulted and approach strategies are adapted accordingly.

5.2   Semantic Business Information Management

The added functionality from the superimposing of diverse data lies within the derivation of new information. The new information, because of the use of knowledge-based technologies, creates dynamic and on-demand rule-driven derivations that help classify, categorize and annotate entities on custom classifications. Knowledge about communications becomes a manageable and tractable asset and ontology revisions can be performed without any downtime.

Sharing and exchanging chunks of data within the company or with the outside world is subject to rules that trigger dynamically. For instance, depending on the status and history of an employee, she can have access to past historical data concerning the threat at hand. The temporality of the access is dependent on the context under which she consumes the requested information. SPIN [15] is an example of how ontology rules can be managed and used programmatically.

With the use of semantic web standards, the system is suited for integration with external semantic web resources. In the business environment, web services are emerging as ways of incorporating knowledge exchange in the markets, making the sharing of information, both inbound and outbound, a task that can be handled by automated software agents.

5.3   Implementation

Even though the implemented system serves to demonstrate some of the functionalities of the conceptual approach, several of the limitations that were imposed during the development reflect issues that should arise in commercial design and development. We will focus on the extraction, conversion and storing of internal data.

Quad Store

It was necessary to deploy a dedicated quad store that would act as the centralized store for all platforms. Being able to deploy such a store is a strong assumption and not a viable commercial solution. The need for two separate stores is a weakness, but necessary for platform-independent interoperation. Quad stores also offer ready-to-deploy data endpoint functionality (e.g. SPARQL) that are customizable (security-wise) and web-accessible. However, ERP systems that are designed to function on top of RDF databases will exhibit the benefits of this approach, and have built-in RDF creation without having to depend on conversion extensions.

Mobile Data

With this approach, mobile data extraction and conversion is inexpensive and decentralized, as each corporate phone is responsible for converting and uploading its own data. Because android phones cannot yet act as quad stores with SPARQL endpoints, RDF are stored in files in the phone and deleted after uploading. However, a more proper mechanism for storing should be deployed, without having to rely on the phone's file system.

ERP Data

The component for converting ERP data to RDF is deployed server-side, running in parallel to the ERP services and acting as a bridge between the ERP internal store and the central quad store. The conversion process is not costly and is triggered by SQL transactions with the use of event handlers at run-time.

6 Conclusions

This work provides a conceptual framework and a demonstrative implementation[2] for taking advantage of the business communication layer of information, which has grown to be a diverse environment of heterogeneous sources and causality relationships. Communications cover a large range of information flows from and to companies, both internal and external.

The proposed extensions form a unification layer around the company's data core that can exchange data on demand and under rule-based control. The implications as far as business management improvements are concerned, have to do with the following points.

1. Minimization of response time
2. Enrichment of internal context to that of an *open business environment*
3. Shared meaning and data, creation of *super-layers* of information that create value bigger than the summation of their constituents

---

[2] The code was part of the author's MBA thesis and can be found in
https://github.com/mmeimaris/android-rdfizer.
It is our intention to keep it updated.

4. Exploitation of mobile information, as mobile devices provide an abundance of usage data and metadata, thus defining contexts (and sub-contexts) on their own.

References

1. Herschel, R., Yermish, I.: Knowledge Management in Business Intelligence. In: King, W.R. (ed.) Knowledge Management and Organizational Learning, Annals of Information Systems, vol. 4, pp. 131–143. Springer US (2009)
2. Choi, H.R., Kim, H.S., Park, B.J., Park, N.-K., Lee, S.W.: An ERP approach for container terminal operating systems. Maritime Policy & Management 30(3), 197–210 (2003)
3. Smaizys, A., Vasilecas, O.: Business Rules Based Agile ERP Systems Development. Informatica 2009 20(3), 439–460 (2009)
4. Tjoa, A.M., Shayeganfar, F., Anjomshoaa, A., Karim, S.: Exploitation of Semantic Web Technology in ERP Systems. In: Tjoa, A.M., Xu, L., Chaudhry, S. (eds.) Research and Practical Issues of Enterprise Information Systems. IFIP, vol. 205, pp. 417–427. Springer, Boston (2010)
5. Graf, H., Tellian, N.: Smartphones as enabler of Supply Chain Event Management. In: Grzybowska, K. (ed.) Management of Global and Regional Supply Chain – Research and Concepts, pp. 133–143. Publishing House of Poznan University of Technology, Upper Austria University of Applied Sciences, Poznan (2011)
6. Davies, A., Ghahramani, Z.: Language-independent Bayesian sentiment mining of Twitter. In: Workshop on Social Network Mining and Analysis (2011)
7. Frischmuth, P., Klímek, J., Auer, S., Tramp, S., Unbehauen, J., Holzweißig, K., Marquardt, C.M.: Linked Data in Enterprise Information Integration (2012)
8. Gruber, T.R.: Toward Principles for the Design of Ontologies Used for Knowledge Sharing. International Journal of Human Computer Studies 43.5, 907–928 (1995)
9. Hepp, M.: An Ontology for Describing Products and Services Offers on the Web. In: Gangemi, A., Euzenat, J. (eds.) EKAW 2008. LNCS (LNAI), vol. 5268, pp. 329–346. Springer, Heidelberg (2008)
10. Brickley, D., Miller, L.: FOAF vocabulary specification 0.98. In: Namespace Document 9 (2010), http://xmlns.com/foaf/spec/
11. Carroll, J.J., Dickinson, I., Dollin, C., Reynolds, D., Seaborne, A., Wilkinson, K.: Jena: implementing the semantic web recommendations. In: Proceedings of the 13th International World Wide Web Conference on Alternate Track Papers & Posters, pp. 74–83 (2004)
12. Pinckaers, F., Gardiner, G.: A Modern Approach to Integrated Business Management (2009)
13. Herzog, T.: A Comparison of Open Source ERP Systems (2006)
14. Mironov, V., Seethappan, N., Blonde, W.: Benchmarking triple stores with biological data. In: Semantic Web Applications and Tools for Life Sciences (SWAT4LS), Berlin, Germany (2010)
15. Fürber, C., Hepp, M.: Using SPARQL and SPIN for Data Quality Management on the Semantic Web. In: Abramowicz, W., Tolksdorf, R. (eds.) BIS 2010. LNBIP, vol. 47, pp. 35–46. Springer, Heidelberg (2010)

Ubiquitous Service Capability Modeling and Similarity Based Searching

Feng Gao and Wassim Derguech

Digital Enterprise Research Institute,
National University of Ireland, Galway, Ireland
`firstname.lastname@deri.org`
`http://www.deri.ie`

Abstract. The proliferation of the Internet-of-Things, which is a part
of the envisioned future internet, is about to significantly enrich the interactions between people and physical environments, making computing
pervasive in our daily lives. To deliver such ubiquitous data as services
over the internet, it is essential to describe their capabilities and enable
the discovery of the capabilities. However, current service descriptions
lack the ability to describe dynamic information, which is crucial for
ubiquitous device capabilities. Moreover, they do not provide explicit information on how services are related to each other. In this paper, we
propose a capability model for ubiquitous devices (sensors) based on the
capability meta-model to describe the dynamics and relevance of sensor
services. We also define a generic similarity metric for service capabilities
and develop a heuristic searching algorithm based on this metric to find
similar sensor services. .

Keywords: ubiquitous service, capability model, service dynamics, service relevance, heuristic search.

1   Introduction

Today computing does not just happen in labs, data centers or desktop computers. The proliferation of sensor networks and mobile devices makes computing pervasive in our daily lives. These techniques are key enablements of the
Internet-of-Things, which is a part of the envisioned future Internet that will significantly enrich the interactions between internet users and the physical world.
To deliver such pervasive data over the Internet as services, it is required to uniformly describe their service capability, so that the device and data heterogeneity can be resolved. However, from our observations, current service description
mechanisms cannot address the following requirements for describing ubiquitous
services.

- **Service Dynamics.** Ubiquitous services are inherently dynamic. From the
  functional perspective, the data observed by a sensing device can vary over
  time. A sensor service providing a concrete observation value at real-time

A. Haller et al. (Eds.): WISE 2011 and 2012 Combined Workshops, LNCS 7652, pp. 173–184, 2013.

can be seen as an *Service Offer* (same as the concept introduced in [8]), and
users may have interests in such service offers while building applications on
top of them. On the other hand, unlike services such as online payment procedures, taking a sensor measurement does not imply a state change of the
real world. Therefore, it is safe to fetch sensor service offers for the users before they actually invoke the service and make use of the sensor data in their
applications [8]. From the non-functional perspective, it is also important
to model the service dynamics. Non-functional features of a sensing device
may vary depending on real-time conditions of physical environments. For
example, the accuracy of a thermal sensor could be lower under extreme
temperatures. Therefore, it is useful to describe such non-functional service
dynamics so that users find information on these non-functional properties
at run-time.
- **Service Relevance.** Current service description methods provide little information on how services with similar capabilities are related. Most approaches merely assign a category for a certain service. We argue that hierarchical service categories based on "sub-class-of" relations are not sufficient to describe service relevance. Providing details of service relevance is
particularly important for sensor applications. The number of sensors and
ubiquitous devices used in *Situation Awareness* applications such as cata-
clysm detections and homeland security programs can scale up to hundreds
of thousands. The users of such systems are interested in both reusing existing sensor capability models to create many functionally similar ones, as well
as finding a similar replacement sensor service in an affordable time period
among a huge amount of service descriptions, when, e.g., one of the sensor
services is mal-functional or offline.

In this paper, we present our solutions to address the above requirements. We
leverage the *Capability Meta-Model* described in [1] to define an upper level
sensor capability ontology. We then propose a generic metric to evaluate the
similarity of service capabilities and use this metric to develop an efficient capability discovery algorithm using heuristic search.

The remainder of the paper is organized as follows. Section 2 describes the capability meta-model briefly as a preliminary work to this paper, then it presents
our upper level ontology for sensor capability. Section 3 elaborates how capability
similarity is defined and how we use this similarity to find relevant capabilities.
Section 4 discuss the related work before we conclude in Section 5.

2   Sensor Service Capability Description

In this section we first introduce the service capability meta-model in [1] and
show how it captures service dynamics and relevance, and then we demonstrate
how we use this meta-model to describe sensor service capabilities.

2.1   Service Capability Meta-model

A service capability is about what a service does. The concept of service capability plays a central role in service description. We use the capability meta-model described in [1] to create a capability model for sensor services using semantic data. The meta-model is modeled with Resource Description Frameworks (RDF). A capability is defined by an *action verb* a set of *attribute* and *attribute-value* pairs. An *action verb* can be defined in a capability domain ontology, where the required attributes and possible values are specified for this action verb, and the fine grained semantics in terms of preconditions and effects are defined [2]. However in this paper, we do not further discuss the use of action verb, since we consider sensor services have only one "sensing" action and there's no preconditions or effects for this action. Rather, we focus on modeling the service dynamics and relevance within and between sensor service capabilities. We use the human-readable Turtle[1] format for the semantic representation throughout the paper, Listing 1.1 represents the basic triples of the capability meta-model, in which the terms *Capability, Attribute, AttributeValue* and their relations are defined.

Listing 1.1. Snippet of Capability Meta-Model

```
@prefix  rdf:  <http://www.w3.org/1999/02/22−rdf−syntax−ns#>
@prefix  rdfs:  <http://www.w3.org/2000/01/rdf−schema#>
@prefix  owl:  <http://www.w3.org/2002/07/owl#>
@prefix  :  <http://.../capability#>

:Capability  a  rdfs:Class ,  owl:Class .
:Attribute  rdfs:subClassOf  rdf:Property ;
        rdfs:domain  [  owl:unionOf  :AttributeValue ,  :Capability ];
        rdfs:range  :AttributeValue .
:AttributeValue  owl:equal  owl:Thing .
```

Service Dynamics. As we discussed earlier, there are various reasons to have dynamic attribute values in service descriptions, e.g.: attribute dependency, inherent dynamicity, etc. Attribute values in the capability meta-model are categorized into *EnumerationValue, RangedValue, ConstrainedValue, ConditionalValue* and *DynamicValue*. Among which, we use *ConditionalValue* and *DynamicValue* to model the values of dynamic attributes of sensor services. Basically, what we can specify statically for service dynamics are: (i) how to compute/assign concrete values based on real-time inputs/conditions using rules, or, (ii) if there are no static rules or they cannot be specified, then where to retrieve concrete values in real-time. The rules are defined with the *Expression* class, and they are in the form of SPARQL segments. Input attribute values for the

---

[1] Turtle format. http://www.w3.org/TeamSubmission/turtle/

rules are specified with *hasRequired* property. Locations to retrieve concrete values are defined using *ServiceEndpoints*, along with a message mapping schema to transform XML messages into RDF triples, if necessary. Exactly how these rules expressions are reorganized into SPARQL queries and evaluated, or how data-fetching is performed using service endpoints are determined by a discovery engine that makes use of such static information.

Service Relevance. The capability meta-model is able to capture parental relationships between capabilities, so that they can be modeled on different abstraction levels and abstract capabilities can be reused to create concrete ones. In particular, three kinds of relationships are defined, including *is-variant-of*, *specify* and *extend*. For 2 attribute values v_1 and v_2, v_1 *isVariantOf* v_2 iff v_1 is more concrete (a sub-class or an instance of) than v_2. For 2 capabilities C_1 and C_2, C_1 *specify* C_2 iff C_1 and C_2 has the same set of attributes but C_1 has more concrete attribute values; C_1 *extend* C_2 iff C_1 has more attributes than C_1 and their shared attributes have the same values.

More formally, given two Capabilities C_1 and C_2, we denote A_1 and A_2 as the sets of attributes of C_1 and C_2 (including attributes for attribute values), respectively. For any attribute a in Capability C, we denote $C(a)$ as the attribue-value of a in C. We provide the following definitions using these notions.

$$C_1 \; extend \; C_2 \iff A_1 \supset A_2 \land \forall a \in A_2[C_1(a) = C_2(a)] \tag{1}$$

$$C_1 \; specify \; C_2 \iff \forall a_1 \in A_1, a_2 \in A_2$$
$$[a_1 \in A_2 \land a_2 \in A_1 \land \exists a^{'} \in A_1[C_1(a^{'}) \; isVariantOf \; C_2(a^{'})]] \tag{2}$$

$$C_1(a) \; isVariantOf \; C_2(a) \iff$$
$$C_1(a) \; rdfs{:}subClassOf \; C_2(a) \lor C_1(a) \; rdf{:}type \; C_2(a) \tag{3}$$

Using the above mentioned relations we can construct a hierarchy of capabilities. We denote C_1 is a parent of C_2 (and C_2 is a child of C_1) iff C_2 *specify* $C_1 \lor C_2$ *extend* C_1. However, this is not sufficient to directly compare two arbitrary capabilities unless one of them is a direct parent of another. We define more complicated relations based to compare arbitrary capabilities. This will be discussed in details later in this paper.

2.2   Sensor Service Capability Ontology

Using the capability meta-model described above, we are able to create an ontology for sensor capabilities. Our ontology reuses some terms and relations from the Semantic Sensor Network (SSN) ontology[2]. Classes and properties for the

---

[2] SSN ontology. http://www.w3.org/2005/Incubator/
ssn/wiki/Semantic_Sensor_Net_Ontology

sensor capability ontology is shown in Figure 1. In the sensor capability ontology the namespace *"cap:"* refers to the capability meta-model ontology, *"ssn:"* refers to the SSN ontology, *"sc:"* is the target namespaces, which is the sensor capability ontology.

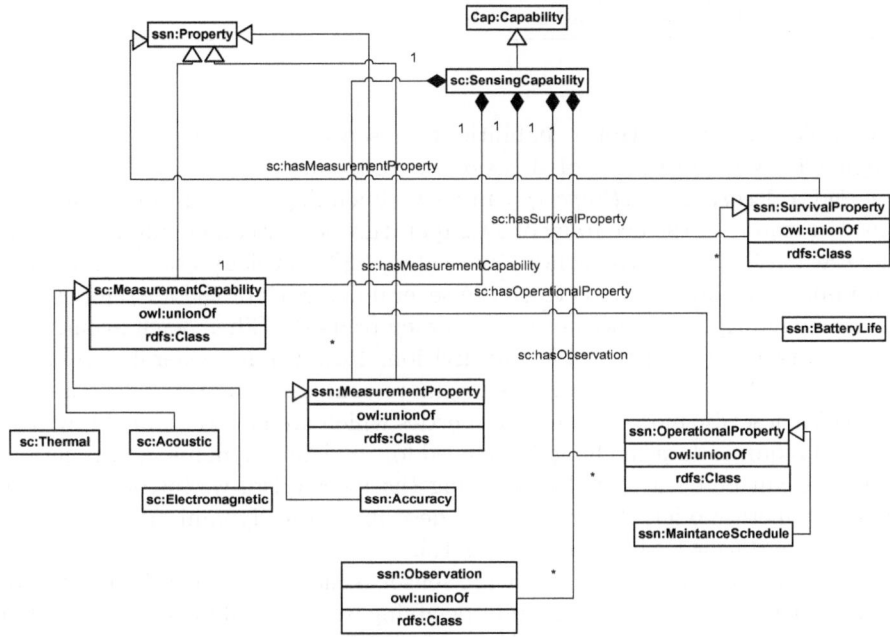

Fig. 1. Sensor Capability Ontology

As shown in the UML diagram, the *sc:SensingCapability* is a sub-class of capability. Five attributes (sub-properties of *cap:Attribute*) are used to describe a sensor capability: *hasMeasurementCapability, hasMeasurementProperty, hasOperationalProperty, hasSurvivalProperty* and *hasObservation*. Listing 1.2 shows a snippet of the upper ontology for sensor capability. Note that the domains and ranges of these attributes are unioned with *rdfs:Class* to allow both instances and sub-classes as domains and ranges so that we can identify the refining relations we previously defined. Attribute values of these attributes are sub-classes of *ssn:Property*. Measurement, operational, survival properties and observations are reused from SSN ontology and represent the same aspects of sensors as they do in the original SSN ontology, and these 4 attributes are "many-to-one" relations.

sc:MeasurementCapability is similar to the original *ssn:MeasurementCapability* which describes the functional properties of a sensor, i.e., what does a sensor observe. However the *hasMeasurementCapability* attribute is a bijective relation between a sensor and its measurement capability in our ontology, as opposed to the "many-one" relation defined in the SSN ontology. The reason that

Listing 1.2. Snippet of Upper Sensor Capability

```
sc : TopSensorCapability  a  sc : SensingCapability ;
        sc : hasMeasurementCapability  sc : MeasurementCapability ;
        sc : hasMeasurementProperty  ssn : MeasurementProperty ;
        sc : hasOperationalProperty  ssn : OperationalProperty ;
        sc : hasSurvivalProperty  ssn : SurvivalProperty ;
        sc : hasObservation  ssn : Observation .
```

SSN ontology needs multiple capabilities for a sensor is that there is a one-to-one mapping between instances of *MeasurementCapability* and *MeasurementProperty*, and a *MeasurementProperty* can vary depending on environmental conditions. For example, the accuracy of a temperature sensor can be influenced by air temperature. To model the conditional values SSN ontology uses ranged values with upper and lower bounds to create several divisions of the air temperature, and create multiple instances of the accuracy property. Then, each accuracy instance is assigned to a temperature division. However no general approach is described in SSN ontology for modeling more complex conditions. It is also impossible to describe more dynamic properties which are not constrained by static rules and requires data-fetching. Such need for modeling dynamic non-functional properties can be easily fulfilled by using the mechanism we introduced in the capability meta-model. An example of describing the dynamic accuracy for a temperature sensor is shown in Listing 1.3.

In the above example, a temperature sensor is modeled as a variant of the top sensor capability by extending and specifying some attributes. The attribute value of *sc:hasObservation* is, and always will be, a dynamic value. A service endpoint is specified for performing the datafetching task and a lifting schema is used for data transformation. The accuracy of the measurement is defined with a conditional value. The accuracy has a required parameter, which is the air temperature measured by the sensor. Two conditions are specified for the accuracy value, if it is between 0 and 10 degrees the accuracy will be 80%, if it is above 10 the accuracy will be 90%. Note that multiple conditions can be used to assign different values to a condition attribute value, but these conditions must be disjoint. When the discovery engine encounters a query on the accuracy, it will recognize the accuracy as a conditional value and checks if there are any required attributes for evaluating the condition. If there is a dynamic required attribute the engine will perform the datafetching task to retrieve the concrete value for the dynamic attribute as well as create a query for the conditional value. The *cap:asVariable* property will specify a variable for the data fetched from the dynamic value, the *cap:hasValue* in the condition indicates the value to be bound for the accuracy, the *cap:hasExpression* specifies the filter for the query and the triple pattern for the *WHERE* clause is derived from the lifting schema. Multiple queries will be created and evaluated for multiple conditions. There will be at most one positive result since all conditions are disjoint. A sample query for the real-time sensor accuracy is shown in Listing 1.4.

Listing 1.3. Snippet of A Specific Sensor Capability

```
sc : Sensor1  a  ssn : SensingDevice , sc : SensingCapability ;
    sc : hasMeasurementCapability  sc : Thermal ;
    sc : hasMeasurementProperty  sc : Mp_1 ;
    sc : hasOperatingProperty  sc : Op_1 ;
    sc : hasSurvivalProperty  sc : Sp_1 ;
    sc : hasObservation
        [ rdfs : subClassOf  ssn : Observation , cap : DynamicValue ;
        sc : hasEndpoint  "www. deri . org / sensor1 "^^ xsd : URI ;
        sc : hasLifting  "www. deri . org / lifting . xsd "^^ xsd : URI ] .

sc : Mp_1  a  sc : Accuracy ,  cap : ConditionalValue ;
    cap : hasRequired  [ cap : hasAttribute  sc : hasObservation ;
        cap : asVariable  "?temp" ] .
    cap : hasCondition  [  cap : hasValue  "0.8"^^ xsd : decimal ;
        sc : hasExpression  " Filter (?temp<10&&?temp>0)"^^ cap : SPARQL ];
        [ cap : hasValue  "0.9"^^ xsd : decimal ;
        sc : hasExpression  " Filter (?temp>10)"^^ cap : SPARQL ] .

sc : Sp_1  a  ssn : BatteryLife ;
    cap : hasValue  "1"^^ xsd : decimal ;
    cap : hasUnit  dbpedia : Month .
```

Listing 1.4. Sample Query for Accuracy

```
SELECT ?acc
WHERE { _x  a  ssn : Observation ;
            cap : hasValue  ?temp
BIND ( 0.8 ,? acc )
FILTER (?temp<10&&?temp>0)}
```

3   Capability Similarity and Heuristic Search

In this section we first introduce the relations we define for service capabilities and how similarity is calculated based on these relations, then, we present our heuristic search algorithm using the similarity metric.

3.1   Enriched Capability Relations

The *specify* and *extend* relation discussed earlier in the capability meta-model were binary relations (*specify*, *extend* $\subset (C \times C)$ where C is the set of capabilities) defined on the capability level to capture parenting relations between capabilities. To discover more detailed relations between capabilities, we need to define relations on the attribute level. To fully compare two capabilities, we first need to compare their sets of attributes and distinguish between shared attributes and different attributes. Then we compare the attribute values of the shared attributes, the attribute values for a shared attribute can be same, different or one of them is a variant of the other. Thus, we define the following 3-ary relations: *ShareSame, ShareDiffer, Specify, Generalize, DifferMore* and *Differ-Less* $\subset (C \times C \times A)$ where C is the set of all capabilities and A is the set of all attributes.

Given two Capabilities C_1 and C_2, we denote A_1 and A_2 as their set of attributes, respectively; and for any attribute a in Capability C, we denote $C(a)$ as the attribue-value of a in C. A relation from C_1 to C_2 is denoted $R(C_1, C_2)$, which is composed of several *atomic relations* $R_a(C_1, C_2)$. A *complete relation* $R_c(C_1, C_2)$ contains all atomic relations bettwen C_1 and C_2. We define six types of *atomic* relations, i.e., *Specify, Generalize, ShareSame, ShareDiffer, Differ-More, DifferLess* based on the following definitions.

- **ShareSame:** we say C_1 *ShareSame* C_2 on a *iff*:
  1. the attribute a is an attribute of both C_2 and C_1,
  2. the attribute-value of a in C_1 is equal to that in C_2,
  Or more formally:

$$C_1 \; ShareSame \; C_2 \; on \; a \iff a \in A_1 \wedge a \in A_2 \wedge (C_1(a) = C_2(a)). \quad (4)$$

- **ShareDiffer:** we say C_1 *ShareDiffer* C_2 on a *iff*:
  1. the attribute a is an attribute of both C_1 and C_2,
  2. the attribute-value of a in C_1 is different to that in C_2,
  Or more formally:

$$
\begin{aligned}
&C_1 \; ShareDiffer \; C_2 \; on \; a \iff \\
&(a \in A_1 \wedge a \in A_2) \wedge (C_1(a) \neq C_2(a) \wedge \\
&\neg(C_1(a) \; isVariantOf \; C_2(a)) \wedge \neg(C_2(a) \; isVariantOf \; C_1(a)).
\end{aligned}
\quad (5)
$$

- **Specify:** we say C_1 *Specify* C_2 on a *iff*:
  1. the attribute a is an attribute of both C_1 and C_2,
  2. the attribute-value of a in C_1 is *variantOf* the one in C_2
  Or more formally:

$$
\begin{aligned}
&S_1 \; Specify \; S_2 \; on \; a \iff \\
&(a \in A_1 \wedge a \in A_2) \wedge (C_1(a) \; isVariantOf \; C_2(a))
\end{aligned}
\quad (6)
$$

- **Generalize:** we say C_1 *Generalizes* C_2 on a *iff*: C_2 *specifies* C_1 on a. Generalize is a reverse relation of specify.
- **DifferMore:** we say C_1 *DifferMore* C_2 on a *iff*: the attribute a is an attribute of C_1 but not of C_2. Or more formally:

$$C_1 \; DifferMore \; C_2 \; on \; a \iff a \in A_1 \wedge a \notin A_2 \quad (7)$$

- **DifferLess:** we say C_1 *DifferLess* C_2 on a *iff*: C_2 *differMore* C_1 on a. *DifferLess* is a reverse relation of *DifferMore*.

When capability engineers reuse existing capabilities to create new ones, the above relations can be derived between the newly created capabilities and existing ones. It is safe to assume that the new capabilities are functionally *simliar* to the old ones used as templates, i.e., they will have many (or at least a portion of) shared attributes and values. This assumption is important for the efficiency of our searching algorithm, which we will discuss in Section 3.3.

3.2 Capability Distance

To give an heuristic to the searching algorithm, we borrow the concept of *distance* from the prototype theory [5] to denote the difference/similarity between two capabilities. However, it is not trivial to accurately define the quality dimensions for the attribute-value featured concepts [4]. Therefore, it is hard to build the conceptual space and define convex regions for categorizing natural concepts. In such conceptual space (e.g., RGB color space), distance between capabilities can represent their semantic similarity. In our approach, categorizing concepts based on semantic distance is not our main purpose. Instead, we define "approximate" quality dimensions with the number of shared/different attributes and values to compute their distance and guide the search algorithm.

Given a *complete relation* between 2 capabilities, we denote *DL, DM, SS, SD, SP* and *GE* as the number of attributes in *DifferLess, DifferMore, ShareSame, ShareDifferent, SPecify* and *GEneralize* relations, respectively. A distance in attribute d_A is the quotient of the number of the different attributes divided by the number of the same attributes, formally:

$$d_A = \frac{(DL + DM)}{(SS + SD + SP + GE)} \tag{8}$$

Similarly, A distance in value d_V is defined as:[3]

$$d_V = \frac{(SD + 0.5 * SP + 0.5 * GE)}{SS} \tag{9}$$

Both d_A and d_V are symmetric distances and have a range of $[0, \infty)$. Apparently 2 capabilities can be considered as identical when both distances are 0. Furthermore we define the distance of two capabilities d as: $d^2 = d_V^2 + d_A^2$.

3.3 Heuristic Searching on the Capability Cloud

Our heuristic searching algorithm is a variation of the *Simulated Annealing* (SA) [6]. Simulated annealing is a generic probabilistic meta-heuristic for the global optimization problem of applied mathematics, namely locating a good approximation to the global optimum of a given function in a large search space. Simulated annealing may be more effective than exhaustive enumeration provided

---

[3] Relations of SP and GE are given a weight of 0.5 just to indicate that they are less "far" from SS than those of SD, but not necessarily to be exactly half the distance.

that the goal is merely to find an acceptably good solution in a fixed amount of time, rather than all the best possible solution.

The heuristic searching algorithm takes as inputs (i) a capability cloud with descriptions of sensor capabilities and the relations between them, (ii) an arbitrary node in the graph that indicates the current capability that the user is looking at (but not satisfied or no longer available) as the starting point, and (iii) a description for the target capability that the user is looking for. In our approach, we use the distance d as the time-varying temperature and the change of distance Δd as the Δt in the original SA. The distance between two capabilities can be derived by either examing all attributes and values directly or computing a relation path that exists between them in the capability cloud. We describe the steps in our algorithm as follows:

1. Evaluate the initial node. set the initial "Temperature" d as the distance from the target to the initial state. Add the initial state to the current node set (CNS for short).
2. For each node in CNS, check if it is the target node (with a distance of 0). If yes, add it to the CNS and result set.
3. Get all the unvisited neighbors of the current state, compute their distance to the target, derive $\Delta d = (distance-of-neighbor)-(distance-of-current)$.
4. If the Δd is negative, accept; otherwise accept the neighbor with a probability of $EXP(-\Delta d/d)$.
5. Replace the current node with the set of accepted neighbors in the CNS. Label the current node and all the evaluated neighbors as visited.
6. Repeat 3-5 for all the nodes in the CNS for a predefined L iterations.
7. Update "temperature" d, according to the annealing schedule.
8. Repeat 3-7 until temperature is 0.

The idea of the SA algorithm come from the analogy of the manner in which liquids freeze or metals recrystallize in the process of annealing. Comparing with a *greedy search*, the crucial improvement is the possibility p of accepting a "worse" state than the current state: when it is "hot", i.e., the temperature is high at the starting phase, p is relatively large, and it is likely that the algorithm will wander randomly and avoid being stuck. When the temperature falls down, p will get smaller, in particular, when it hits the freezing point of 0, the algorithm becomes a greedy search. Notice that even the original algorithm is an analogy to the physical system, the p given by the *EXP* function is unrelated to the specific problem in annealing domain and is often hard-coded in many SA implementations.

The cooling procedure is controlled by a user defined function that will update the temperature according to the time, namely *Annealing Schedule*, normally the *Annealing Schedule* will be a decreasing mathematical function that reaches 0 when the maximum allowed time has expired. However in our case, as the start node is arbitrary and we cannot ensure a high initial temperature for the sake of result completeness, we propose to use a decreasing function that the absolute value of its first derivative is increasing, e.g.: $d = log(-x + 1 + t)$, in which d

is the time-varying temperature, t is the total time allowed. Therefore, we can have more time to have the high acceptance possibility.

To improve the efficiency of the algorithm, it is also required that neighbors are quite similar. So that Δd tends to be quite small in each iteration. The rationale behind this is that it will avoid exploring "very good" states as well as "very bad" states, however, there are much more bad states than the good ones, thus the algorithm will achieve better efficiency in general. This is related to the assumption that we made earlier, which indicates neighboring nodes have a relatively small distance.

4   Related Work

In this section, we discuss some related work in semantic service description and similarity-based search.

Service description is essential to enable service discovery, invocation and composition. Existing semantic service descriptions, e.g., OWL-S[4], SA-WSDL[5], WSMO[6] describe service functionalities as a set of inputs, outputs, preconditions and effects (IOPE) but not as attribute-featured capabilities. Using such IOPE based service description, it is possible to automatically compose web services and achieve a goal when there's no direct matching service. However, these approaches cannot express the dynamic aspects of services nor do they provide explicit information on how similar web services are related. In our approach, we can capture such service dynamics using the capability models, we are also able to facilitate the creation of new service capabilities by specifying the relations explicitly between the new capabilities and the old ones used as templates.

Similarity-based service discovery has been studied extensively for various kinds of services. In [3], multiple evidences are considered to evaluate similarity for service operations and inputs/outputs based on *clustered concepts* extracted from WSDL documents. In [7], a hybrid service matchmaking is proposed for OWL-S services. Among the hybrid filters presented in [7], the *subsumed-by* and *nearest-neighbor* filter leverage different Information Retrieval (IR) similarity metrics. In [9], a *replacement degree* is computed for two *service protocols* based on how their sub-protocols can replace each other in the context of mediated service interactions. Comparing to the above approaches, (i) we are more focused on providing an easy way for the users to specify their requirements to find a replacement for an unsatisfiable service capability using explicit description on their relevance and (ii) we do not use the similarity metric to "recommend" semantically similar substitutes but to guide our heuristic searching algorithm to find exact (and potentially plug-in).

---

[4] OWL-S. http://www.w3.org/Submission/OWL-S/

[5] SA-WSDL. http://www.w3.org/2002/ws/sawsdl/

[6] WSMO. http://www.wsmo.org/

5   Conclusions and Future Work

In this paper, we elaborate the need of capturing the dynamics and relevance in ubiquitous service capabilities. We discuss how capability meta-model can be used to fulfill such need. We adopt the capability meta-model and create an upper level ontology for sensor capabilities. The upper sensor capability is aligned with Semantic Sensor Network (SSN) ontology created by W3C SSN incubator group. Then, we derive a concrete sensor service capability from the upper level sensor capability, which is able to express the service dynamics that cannot be captured using the original SSN ontology. We also enriched the relations defined for capabilities to give details on how two capabilities are related to each other regarding to their attributes and values. A capability cloud can be constructed using these relations and we propose an efficient searching algorithm to find functionally similar replacement for a certain node in the cloud. The algorithm is based on *Simulated Annealing* and it uses the *distance* we define between capabilities as a heuristic.

As an on-going research, there are still a lot of work to do such as evaluating the searching algorithm. Also, our current searching mechanism requires a full description on the target capability. We expect to handle with partially described targets as well to enable a *plug-in* match in the future.

Acknowledgments. This work is supported by the Science Foundation Ireland under Grant No. SFI/08/CE/I1380 (Lion-2).

References

[1] Bhiri, S., Derguech, W., Zaremba, M.: Web service capability meta model. In: Krempels, K.H., Cordeiro, J. (eds.) WEBIST, pp. 47–57. SciTePress (2012)

[2] Derguech, W., Bhiri, S.: Capability modelling - case of logistics capabilities. In: PALS (2012)

[3] Dong, X., Halevy, A., Madhavan, J., Nemes, E., Zhang, J.: Similarity search for web services. In: Proceedings of the Thirtieth International Conference on Very Large Data Bases, VLDB 2004. VLDB Endowment, vol. 30, pp. 372–383 (2004)

[4] Gardenfors, P.: Conceptual spaces as a framework for knowledge representation. Mind and Matter 2, 9–27 (2004)

[5] Johnson, R.: Prototype Theory, Cognitive Linguistics and Pedagogical Grammar. In: Working Papers in Linguistics and Language Training (8), pp. 12–24 (1982)

[6] Kirkpatrick, S., Gelatt, C.D., Vecchi, M.P.: Optimization by simulated annealing. Science 220, 671–680 (1983)

[7] Klusch, M., Fries, B., Sycara, K.: Automated semantic web service discovery with owls-mx. In: Proceedings of the Fifth International Joint Conference on Autonomous Agents and Multiagent Systems, AAMAS 2006, pp. 915–922. ACM (2006)

[8] Zaremba, M., Vitvar, T., Moran, M.: Towards optimized data fetching for service discovery. In: Fifth European Conference on Web Services, ECOWS 2007, pp. 191–200 (2007)

[9] Zhou, Z., Gao, F., Shu, L.: Service protocol replaceability assessment in mediated service interactions. In: 2011 IEEE International Conference on Communications (ICC), pp. 1–5 (June 2011)

Introduction to the Proceedings
of the Workshop on Social Web Analytics
for Trend Detection (SoWeTrend) 2012

Athena Vakali[1] and Hakim Hacid[2]

[1] Department of Informatics, Aristotle University, Greece
avakali@csd.auth.gr
[2] Bell Labs France
hakim.hacid@alcatel-lucent.com

Introduction

Social networks on the Web are growing rapidly and offer a fertile ground for data mining and analysis. So many social networks flourish in an unforeseen manner and social data are evolving from various social streams. Such data embed human interactions and interests and their analysis is valuable for detecting trends, events and phenomena on today's Web reality. Methodologies and techniques of information retrieval, databases, preference modeling, graph theory, etc. have been leveraged and adapted into the social web dynamic environment but still open questions remain in terms of introducing new methodologies which will facilitate trend detection. SoWeTrend workshop has addressed such issues by providing a forum for researchers and practitioners to discuss the social data analytics relevant topics with emphasis on the critical actors in emerging online social networks.

The workshop attracted interested work in the areas of social networks mining, analysis, topic and trend detection. The accepted papers placed emphasis in the areas of predictive social networks analysis and clustering with respect to social networks recommending along with work focusing on the data itself from the perspective of querying but also from the view of crowdsourcing. An invited presentation on "Aiding the Detection of Fake Accounts in Large Scale Social Online Services" enhanced Workshop's focus with emphasis on the issues of privacy and security. SoWeTrend has promoted a fruitful discussion among participants in order to capture major topics raised and to trigger conclusions and future work roadmaps.

Invited talk given by prof. Michael Sirivianos (from the Cyprus University of Technology) was on the fake accounts detection in large social online services and the presenter has attracted a live discussion with the audience. The topic has major impact in trends formation since false opinions and malicious activities can be generated from such fake accounts and this is a major issue for both online social networks users and online social services operators. The invited speaker indicated a rather important methodology and implemented framework to tackle with such fake accounts detection and future topics of research were identified and summarized.

The paper entitled "Associating items with scenes identified in social Q&A data" has focused on content and items associations in order to identify scenes via utilizing characteristic items on social query and answer systems. The proposed scene mining and document clustering has been shown to be quite effective for such content associations. Such an approach is important for social networks trends detection.

The paper entitled "An Epistemic Equivalence for Predictive Social Networks Analysis" presented a semantic and dynamic model for predictive recommendations and trend predictions. This work utilizes a social graph along with knowledge engineering practices and its results have been highlighted in accordance to a particular ongoing research project.

The paper entitled "Dynamic clustering process to calculate affinity degree of users as basis of a social network recommender" has presented a recommender which utilizes a graph model to extract knowledge on users interactions and information trends in a social network setup. Performance issues were also raised and future work ideas were summarized and highlighted.

The paper entitled "Data Leak Aware Crowdsourcing in Social Network" proposed a particular clustering based approach for discovering collaborative and competitive team and trends formation in social networks. Specific implicit social network interactions were identified via a probabilistic approach and its future work ideas are important for qualitative trends detection.

In summary SoWeTrend has promoted discussion on recent and significant developments in the area of social networks trend detection via cross-fertilization of techniques in the areas of mining, graph and probabilistic modeling as well as on crowdsourcing and content associations. Methodologies and techniques identified enabled participant researchers to understand how trends in social networks and social media, can be detected and predicted. Such an understanding of trends capturing and mining in social networks is rather important for several areas such as marketing, security, and Web search.

We sincerely appreciate and thank the SoWeTrend 2012Program Committee Members for their valuable contribution in paper reviewing, selection and evaluation and we also thank WISE 2012 Conference chairs and organizers for their continuous support to our SoWeTrend 2012 workshop.

Athena Vakali,
Hakim Hacid

SoWeTrend 2012 Workshop Chairs

Associating Items with Scenes Identified in Social Q&A Data

Shin-ya Sato, Masami Takahashi, and Masato Matsuo

NTT Network Innovation Laboratories
3-9-11, Midori-cho, Musashinno-shi, Tokyo, Japan

Abstract. We discuss the problem of discovering associations between typical situations (scenes) in our daily lives and their characteristic items, which refer to anything from real objects to imaginary beings or abstract concepts. Once scenes are associated with items, the scenes can be further computationally analyzed (e.g., compared, tracked) on the basis of their associated items. In our approach for mining such associations, a list L of items and a set D of Web documents, in which scenes are identified, are first prepared. Next, D is divided using latent Dirichlet allocation (LDA) into clusters, each of which can be regarded as corresponding to a distinct characteristic scene. Then, the relevance between the scenes and items in L is estimated on the basis of the statistical significance of occurrence of items in the clusters. We developed two simple techniques for improving the quality (consistency) of the clustering result obtained using LDA with the expectation that the improved clustering result yields better performance in revealing item-scene associations. The most effective of the two techniques, PACA, purifies original clusters (i.e., eliminates unwanted elements in each cluster) created using a clustering algorithm by using the outcome from another clustering algorithm. Through an experiment using pages in a social Q&A site, we verified the effectiveness of the cluster purification techniques and the total effectiveness of our approach of associating items with scenes.

Keywords: social Q&A, scene mining, document clustering, agglomerative clustering, K-means, community detection, LDA.

1 Introduction

The Web has long been considered an important source of information. With the growth of social media, which do not only bring people together but also data representing the ways people behave and interact with each other, the value of the Web as an information/knowledge source has become much more significant.

We propose an approach for analyzing textual data generated by social media for determining typical context in the data, which is more appropriately called "scenes". The goal of our analysis is to find a set of items that characterize each scene, where an item refers to anything from a real object to an imaginary being or abstract concept. Taking the scene of Halloween as an example, characteristic items include pumpkin and witch. Once scenes are associated with

A. Haller et al. (Eds.): WISE 2011 and 2012 Combined Workshops, LNCS 7652, pp. 187–200, 2013.

items, they can be compared on the basis of their associated items. For example, the more items two scenes share, the more likely they are similar to each other. Furthermore, the differences between two scenes can be explained by the differences in their associated items. The proposed approach takes a set of documents and a list of items as inputs. It first identifies scenes in the document set. Next, the document set is divided into clusters so that each cluster can be regarded as corresponding to a distinct characteristic scene. Then, the relevance between the scenes and items in L is estimated on the basis of the statistical significance of occurrence of items in the clusters. We developed two simple techniques for improving the quality (consistency) of the clustering result obtained using LDA with the expectation that the improved clustering result yields better performance in revealing item-scene associations. The most effective of the two techniques, PACA (purification by auxiliary clustering algorithms), purifies original clusters (i.e., eliminates unwanted elements in each cluster) created using a clustering algorithm by using the outcome from another clustering algorithm.

We conducted an experiment to confirm if the proposed approach works using data (pages) of social Q&A, where each page contains answers to a problem that a questioner is facing in her/his life. Thus, in this case, scenes to be identified in the data are situations where people are facing typical problems. For items, we prepared a list of places, which includes names for locations, buildings, institutions, shops, and different types rooms. It can be said that associating these items with scenes is equivalent to finding places related to each problem (e.g., places where the problem will likely occur).

This paper is organized as follows. First, we review related work in Section 2. Then, we describe our approach for solving the problem of discovering scene-item association in Section 3. We devote Section 4 to explaining our techniques of cluster purification. We present the results of an experiment using data of a Q&A site for verifying the effectiveness of our approach.

2   Related work

2.1   Topic/Trend Detection

One of the fundamental objectives of trend detection is to determine associations between a topic and periods of time during which the topic has significant meaning. Conceptually, the problem of topic-item association discussed in this paper can be thought of as a transformation of trend detection: items in our problem corresponds to time in trend detection.

In a previous study by Swan et al. on trend detection, a statistical measure is used to find a time period that has a statistically significant association with the occurrence of a given term [1]. Although a term can be one of the elements for explaining a topic, a single term is not sufficient for understanding the entire topic. For handling the complicated concept (and data) of a topic, latent Dirichlet allocation (LDA) has recently been introduced [2]. LDA is an effective unsupervised learning methodology for finding topics in document collections, which is also used in trend detection for identifying topics in a document set.

For example, Wang et al., proposed "Topics over Time", an LDA-style learning method for detecting trends [3]. It captures not only topics but also how the topics change over time. We also use LDA for identifying topics in a collection of documents. We then associate items with each identified topic.

2.2 Information Extraction/Semantic Relation Extraction

Our approach can also be thought of as falling under the category of information extraction or semantic relation extraction because it can be regarded as a method for discovering meaningful relationships between scenes and items. There have been many studies for extracting semantic relations between entities from a corpus. The relations include basic ones, such as *is-a* and *part-of*, as well as more complicated ones such as *reaction-of* and *trouble-of* [4,5,6]. In these studies, prior to extracting the target relation from documents, methods of identifying the target relation in the documents are formulated. Identification usually requires knowledge about the target relation such as particular patterns between entities, taxonomy, and positive examples of entities that will be used for supervised machine learning. Unlike the studies mentioned above, our approach does not rely on such knowledge prepared for a particular target because we would like to discover a wider variety of relationships between items and scenes, including ones that are difficult to formulate.

2.3 Cluster Ensembles

One of the contributions of this paper is presenting the technique for improving the quality of clustering results called PACA. The concept of the technique is to produce a set of clusters of higher quality by combining results of two different types of clustering algorithms.

There is a widely used technique called cluster ensembles that combines multiple clustering results generated by different algorithms into a single clustering result [7,8]. It produces a consolidated result without accessing the features or algorithms that determined the original partitions. PACA differs from cluster ensembles on two points. First, it improves the clustering quality by removing misplaced elements, not by relocating them. Second, it uses the characteristics of original clustering algorithms in producing a new partition, while cluster ensembles do not use such information.

3 Proposed Approach

Our approach for discovering scene-item associations is outlined as follows:

Step 1: Prepare a set of documents D and a list of items L.
Step 2: Identify scenes in D using LDA.

Step 3: Partition D into clusters, i.e., $D = \bigcup_i C_i$ so that C_i respectively correspond to identified scenes.

Step 4: Associate $x \in L$ with C_i if occurrence of x in C_i is significant.

In step 1, documents in D are selected so that D becomes topically consistent to some degree. This is because it is empirically known that LDA works poorly in identifying scenes in step 2 when D covers a broad range of topics. To meet this requirement, the condition to contain a predefined keyword q is set on documents in D. As a result, what are identified in step 2 are scenes naturally related to q. For items, any term can be chosen for elements of L as desired (which can even be all the terms appearing in D).

In step 3, we used the relationships between documents and topics estimated using LDA to partition D into clusters. LDA takes a set of documents and the number k of topics as inputs, and produces the distribution (θ_{ij}) $(j = 1, 2, .., k)$ of topics in a document d_i (as well as the distribution of words in each topic), where θ_{ij} is the probability of the j-th latent topic occurring in d_i. Based on the distribution, the set of documents can be divided into clusters by grouping those documents, d_m and d_n, satisfying the following condition:

$$\operatorname*{argmax}_{j} \theta_{mj} = \operatorname*{argmax}_{j} \theta_{nj}.$$

Let this clustering method be denoted as LDA(k).

In step 4, the χ^2 test on the contingency table below was used to determine whether occurrence of x in C_i is significant, where each number in the cells indicates the number of documents satisfying each combination of conditions (e.g., c is the number of documents in which x occurs but does not belong to C_i). In the test, we set the significance level to 0.05. That is, we associated x with C_i if the p-value was less than or equal to 0.05.

	x	$\neg x$
C_i	a	b
$\neg C_i$	c	d

It is expected that C_is consist of positive instances of the identified scenes. That is, we expect that C_i mostly consists of documents writing about the same identified scene. However, we found that there is a non-negligible amount of irrelevant documents in each cluster, as shown in section 5.2. We devote the next section to discussing our two techniques for improving the quality of clustering.

4   Improving Quality of Clustering

In this section, we discuss ways to improve the quality of clustering results obtained by LDA(k). After introducing indices for measuring the quality of the clustering results, we describe our idea for improving clustering quality. Our proposed techniques are based on this idea.

4.1 Evaluation Indices

We evaluated the quality of a clustering result $\mathcal{C} = \{C_j\}$ based on widely used metrics: purity P, inverse purity IP, and F-value (harmonic mean of P and IP), which are defined as follows:

$$P(\mathcal{C}) = \frac{1}{N} \sum_j \max_i |C_j \cap S_i|, \; IP(\mathcal{C}) = \frac{1}{N} \sum_i \max_j |C_j \cap S_i|, \; F(\mathcal{C}) = \frac{2}{\frac{1}{P} + \frac{1}{IP}},$$

where $N = \sum_i |C_i|$[9]. Purity is used to evaluate the coherence of C_is, i.e., the degree to which each C_i consists of documents belonging to a single scene. On the other hand, IP measures the degree to which documents about a scene are concentrated in a single C_i.

4.2 Eliminating Misplaced Elements

Our idea for improving clustering quality is to simply remove misplaced elements, not relocate them. Let us explain the idea in more detail using an example.

Suppose that D consists of 18 documents, $d_1,.., d_{18}$, which should be classified into four groups $S = \{S_1, S_2, S_3, S_4\}$ from the viewpoint of the scenes they are about. The document symbols (e.g., d_1) are colored so that document-scene correspondence can be easily recognized:

$$S_1 = \{d_1, d_2, d_3, d_4, d_5\}$$
$$S_2 = \{d_6, d_7, d_8, d_9, d_{10}, d_{11}\}$$
$$S_3 = \{d_{12}, d_{13}, d_{14}, d_{15}, d_{16}, d_{17}\}$$
$$S_4 = \{d_{18}\}.$$

Assume also that we obtained the following partition $\mathcal{C} = \{C_i\}$ of D by using LDA(k):

$$C_1 = \{d_1, d_2, d_3, d_9, d_{10}, d_{15}\},$$
$$C_2 = \{d_4, d_6, d_7, d_8, d_{16}, d_{17}\},$$
$$C_3 = \{d_5, d_{11}, d_{12}, d_{13}, d_{14}, d_{18}\}.$$

Then, $P(\mathcal{C}) = 1/2$, $IP(\mathcal{C}) = 5/9$, and $F(\mathcal{C}) = 10/19$. We believe that finding the perfect solution S is ideal (and difficult at the same time) but not the only one to the problem of improving the quality of the clustering result. Our idea is to improve the quality of clustering by removing elements that compromise purity. We call this process purification of $\mathcal{C}$. Formally, purification can be achieved by

$$\hat{C}_i = C_i \cap S_k, \quad k = \underset{j}{\mathrm{argmax}} \, |C_i \cap S_j|. \tag{1}$$

Based on this formula, we obtain purified clusters:

$$\hat{C}_1 = \{d_1, d_2, d_3\}, \quad \hat{C}_2 = \{d_6, d_7, d_8\}, \quad \hat{C}_3 = \{d_{12}, d_{13}, d_{14}\}.$$

In fact, we need to find ways other than using Equation (1) because we do not exactly know S.

4.3   Use of Auxiliary Clustering Algorithms

One of our purification techniques uses clusters generated using another clustering algorithm. By replacing $\mathcal{S}$ in Equation (1) with a set of clusters $\mathcal{Q} = \{Q_i\}$ obtained using another clustering algorithm, which can be thought of as an approximation of $\mathcal{S}$, we have

$$\mathcal{I}(C_i, \mathcal{Q}) = C_i \cap Q_k, \quad k = \underset{j}{\operatorname{argmax}} |C_i \cap Q_j|.$$

We call this technique purification of $\mathcal{C}$ with $\mathcal{Q}$, or purification of $\mathcal{C}$ by using an auxiliary clustering algorithm. For convenience, let us use a simpler expression, $\mathcal{I}(\{C_i\}, \mathcal{Q}) = \mathcal{I}(\mathcal{C}, \mathcal{Q})$ instead of $\{\mathcal{I}(C_i, \mathcal{Q})\}$. Let us also write $\#\mathcal{I}(\mathcal{C}, \mathcal{Q})$ to indicate the number of documents retained after purification, i.e., documents contained in any $\mathcal{I}(C_i, \mathcal{Q})$s.

For a better understanding of this technique, we present an example where we have the following clusters generated using an auxiliary algorithm:

$$Q_1 = \{d_1, d_2, d_3, d_9\}, \qquad Q_2 = \{d_4, d_5\}, \qquad Q_3 = \{d_{10}, d_{12}, d_{13}, d_{14}\},$$
$$Q_4 = \{d_6, d_7, d_8, d_{16}, d_{18}\}, \quad Q_5 = \{d_{11}, d_{15}\}, \quad Q_6 = \{d_{17}\}$$

Then, each $\mathcal{I}(C_i, \mathcal{Q})$ can be easily calculated as follows:

$$\mathcal{I}(C_1, \mathcal{Q}) = C_1 \cap Q_1 = \{d_1, d_2, d_3, d_9\}$$
$$\mathcal{I}(C_2, \mathcal{Q}) = C_2 \cap Q_4 = \{d_6, d_7, d_8, d_{16}\}$$
$$\mathcal{I}(C_3, \mathcal{Q}) = C_3 \cap Q_3 = \{d_{12}, d_{13}, d_{14}\}$$

Now, we have $P(\mathcal{I}(\mathcal{C}, \mathcal{Q})) = IP(\mathcal{I}(\mathcal{C}, \mathcal{Q})) = F(\mathcal{I}(\mathcal{C}, \mathcal{Q})) = 9/11 \approx 0.818$, which is clearly larger than the original value $F(\mathcal{C}) = 10/19 \approx 0.526$. Note that this is also greater than the F-value of the auxiliary clustering algorithm $F(\mathcal{Q}) = 35/54 \approx 0.648$.

By definition, this purification may fail, i.e., $F(\mathcal{I}(\mathcal{C}, \mathcal{Q})) \leq F(\mathcal{C})$, if the performance of the auxiliary clustering algorithm is poor. Although we have not yet fully clarified the requirements for the auxiliary clustering algorithm to improve clustering quality, we verified that all those introduced below work effectively.

Agglomerative Clustering. Agglomerative hierarchical clustering is a common classical data clustering technique [10]. An agglomerative clustering algorithm initially lets each element of D form its own cluster. Then, two clusters are greedily merged into a larger one step by step. That is, at each step, two clusters with the shortest distance between them are selected to be merged into one larger cluster. The distance between clusters, which is called the linkage criterion, can be defined in multiple ways. Each criterion is calculated based on the distance among elements belonging to the clusters. Commonly used linkage criteria include single-linkage and average-linkage and are respectively defined as

$$L_{single}(C_1, C_2) = \min_{d_1 \in C_1, d_2 \in C_2} \operatorname{dist}(d_1, d_2) \, and$$

$$L_{average}(C_1, C_2) = \frac{1}{|C_1||C_2|} \sum_{d_1 \in C_1} \sum_{d_2 \in C_2} \text{dist}(d_1, d_2),$$

where $\text{dist}(d_1, d_2)$ denotes the distance between d_1 and d_2.

At each step k of agglomerative clustering, we obtain a set of clusters. Let $ACS(k)$ and $ACA(k)$ denote the methods for obtaining such document clusters respectively using the single-linkage and average-linkage criteria. For $\text{dist}(d_1, d_2)$, we use the standard distance function using the cosine measure [11].

K-means. The K-means algorithm divides data (i.e., documents) into k clusters so that each datum belongs to the cluster with the nearest mean [12]. Since this problem is NP-hard, heuristic algorithms are commonly used to obtain local optimal solutions. We refer to partitioning a document set into k clusters using a heuristic K-means algorithm as $KM(k)$.

Community Detection. A set of documents can also be partitioned into clusters by using an algorithm developed from research on complex networks [13] for solving the community detection problem. Community detection is a research theme aimed at finding communities in a network, where a community is defined as a group of nodes that are more densely connected to each other than to the rest of the network [14]. By using a community detection algorithm, the node set V of a network can be divided into groups of nodes, which yield a partition of V. Community detection algorithms can be used for clustering documents as follows. (1) Extract k feature words from each document. (2) Create a network of documents where a link is generated between two documents if they share one or more words. (3) Apply a community detection algorithm to the network for obtaining document clusters. It is worth noting that a document network becomes denser as k increases because the number of links in the network increases along with the increase in k.

There have been a number of algorithms developed for solving the community detection problem [14]. We tried some of these algorithms, and compared their performances. We then found that the walktrap algorithm [15] outperformed the other algorithms, which is consistent with the results of a preceding study [16]. Let $CD(k)$ denote the document clustering method using the walktrap community detection algorithm.

4.4 Use of Topic Distribution of LDA

Our other prospective purification technique uses the topic distribution of LDA as follows. For each C_i, let the documents in C_i be ranked in descending order according to the topic probability:

$$C_i = \{d_{j_1}, d_{j_2}, ..., d_{j_n}\}, \quad \theta_{j_1 i} \geq \theta_{j_2 i} \geq ... \geq \theta_{j_n i}$$

where $\theta_{j_k i}$ is the probability of the i-th latent topic occurring in document d_{j_k}. Then, intuitively, it can be said that d_{i_k} focuses more clearly on the topic than

d_{i_m} if $k < m$. Therefore, it is expected that we can increase the purity of C_i by removing documents in the latter part of this sequence. In practice, we tried improving the purity of LDA(k) by removing half the documents based on our idea. This is because its purity was around 0.5 as shown in the next section. Let LDA*(k) denote the above-described procedure of purifying clusters obtained using LDA(k).

5   Experiment

In this section, we describe and present the results of an experiment of associating items with scenes.

5.1   Data Set

We used data (pages) of social Q&A as the collection of documents for the experiment because they have the following advantageous characteristics for identifying scenes.

(a) Content. The content of each page is about a problem a questioner is facing in her/his life, and the page can further be associated with the scene in which the problem will likely occur.
(b) Granularity. A single scene is associated with each document.
(c) Quality. Few noise pages (e.g., spam pages, affiliate advertising pages) are mixed in the data.

We thought that a set of blog entries would be another candidate for the data. However, we found that there are more cases in blogs than in Q&A in which conditions (b) and (c) are not satisfied. People sometimes include various events that happened in a day into a single entry. There are also cases where shops introduce their products in their blogs.

For actually obtaining Q&A pages, we first searched the Web using a query of q (cf. section 3) together with the option for specifying the target of the search within the Q&A website. Then, we retrieved pages listed on the search results. For this experiment, we used `oshiete.goo.ne.jp` and "pumpkin" respectively for the Q&A site and q. We then obtained 728 relevant documents. It should be noted that all the documents used for the experiment were written in Japanese.

For items, we prepared a list of places made up of names for locations (e.g., park, crossing), buildings and institutions (theater, school), shops (bookstore, butcher), and types of rooms (living, kitchen). We gathered these items from relevant lists and tables on the Web.

5.2   Verifying Quality of Document Clusters

Manually Identified Scenes. For quantitatively evaluating the quality of the clustering result obtained in step 3 of Section 3, we manually associated each document in D with a scene that document was about. In the association task, two situations were distinguished as different scenes if we could find differences in characteristic attributes and elements of the situations, such as time, place, objects, people, actions of people, and the ways people interact with each other. As a result, we identified 43 different scenes associated with one or more documents in D. Ten of the 43 scenes are listed in Table 1. For convenience, we sequentially assigned numbers to the identified scenes. Let S_i denote the set of documents that were associated with the i-th scene.

Table 1. Examples of identified scenes related to "pumpkin"

Scene	Role/Characteristics of pumpkin in scene
Cinderella	Pumpkin is turned into coach in story
Barbecue	Ingredient often used in barbecue
Breeding small animals	Pumpkin seeds as feed for small animals
Growing pumpkins	Agricultural crops
Halloween	Carved into Jack O'Lanterns
Having trouble cutting pumpkins	Pumpkins are not easily cut with knives since they have very hard skin
Overcoming stage fright	"Imagine you are talking to pumpkins" is commonly used Japanese phrase to calm someone's nerves
Preparing baby foods	Common ingredient in baby food
Preparing health-conscious foods	High-fiber and highly nutritious ingredient
Removing stain	Pumpkins stain clothes

Quality of Clusters by Using LDA. We measured the quality of clustering results of LDA(k) as a function of parameter k. As shown in Figure 1, LDA constantly yielded an F-value around 0.55, independently of the choice of parameter k. It was also confirmed that LDA(k) successfully generated clusters that correspond to each of the scenes listed in Table 1. However, it should be noted that purity was not so high, i.e., positive instances (relevant documents) barely retained a majority in each C_i on average.

Evaluation of Purification Techniques. We verified the two cluster purification techniques explained in Section 4 for improving quality of document clusters obtained using LDA(k). As shown in Fig. 1, the value of parameter k has little effect on the performance of LDA(k); therefore, we arbitrarily chose $k = 400$. We denote LDA(400) as $\mathcal{L}$.

First, we compared two implementations of the technique using an auxiliary clustering algorithm: we tried purifying $\mathcal{L}$ by using the auxiliary clustering algorithms ACS(k) and ACA(k). The results are shown in Figure 2 (a). The graphs indicate changes in F-values of $\mathcal{I}(\mathcal{L}, \text{ACS}(k))$ and $\mathcal{I}(\mathcal{L}, \text{ACA}(k))$ (left Y axis) as well as the number of retained documents $\#\mathcal{I}(\mathcal{L}, \text{ACS}(k))$, and $\#\mathcal{I}(\mathcal{L}, \text{ACA}(k))$ (rightY axis) along with the increase in k. Note that we can determine the amount of documents lost by using each purification technique by calculating

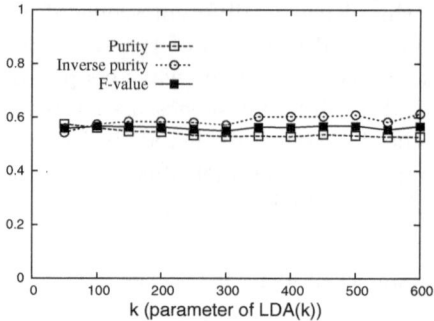

Fig. 1. Purity, inverse purity, and F-value of clustering results obtained with LDA(k)

the difference between $|D| = 728$ and the number of retained documents. The open and filled points (rectangles and circles) are respectively plots of ACS(k) and ACA(k). In both ACS(k) and ACA(k), considerably high F-values, even close to 0.9, could be achieved by purification for small k. On the other hand, we could obtain only a limited number of documents . This is because the size of each cluster is mostly small in the early stage of agglomerative clustering.

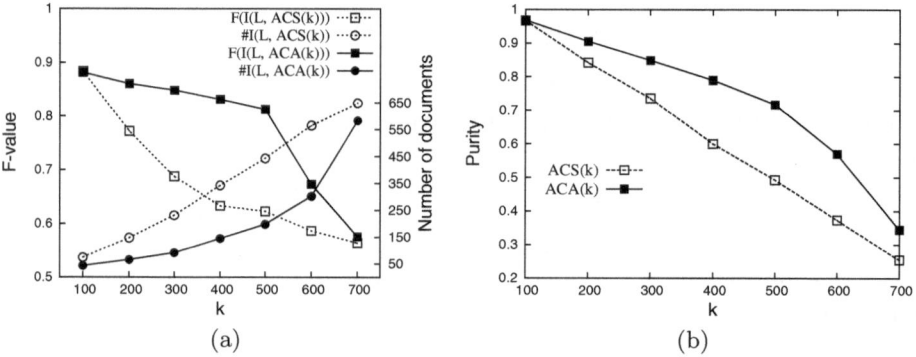

Fig. 2. (a) Effects of purification by using ACS(k) and ACA(k). (b) Changes in purity of ACS(k) and ACA(k) along with k.

In Figure 2 (a), although ACS(k) and ACA(k) basically share similar changing trends, they differ from each other in the rate of changes. For example, the F-value of ACS(k) decreases more quickly than that of ACA(k) as k increases. We found that this difference comes from the difference in the style of changes in the purity of the auxiliary clustering algorithms. Figure 2 (b) compares changes in purity of ACS(k) and ACA(k). It shows that the purity of ACA(k) decreases more slowly than that of ACS(k) until around $k = 500$. This explains the slower decline in the F-value of ACA(k) in (a).

We have seen in ACA(k) and ACS(k) that there is a trade-off between the resulting quality of purification and the number of retained documents, which is also true for other implementations of the purification techniques. This situation is clarified in Figure 3, where the number of retained documents are plotted against the F-values for each implementation. For example, each $\triangle$ in the figure corresponds to a pair $(F(\mathcal{I}(\mathcal{L}, \mathrm{CD}(k))), \#\mathcal{I}(\mathcal{L}, \mathrm{CD}(k))$ for some k. In ACS(k) and ACA(k), the F-value and number of retained documents monotonically change as k increases, as shown in Figure 2. For visualizing the overall trend in these changes, neighboring points are connected with lines in the plotting data of ACS(k) and ACA(k).

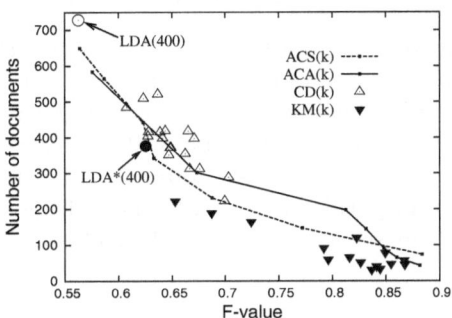

Fig. 3. Number of retained documents against F-value for purification techniques

The point corresponding to $\mathcal{L} = \mathrm{LDA}(400)$ is located at the upper-left corner of the figure. The ideal improvement in the quality of $\mathcal{L}$, which increases the F-value but does not lose the number of documents, corresponds to the positive horizontal translation of the point in the graph. In fact, points are distributed from upper-left to lower-right in the graph. This is a reasonable (and expected) result because our idea of cluster purification is to intentionally discard improperly classified documents to achieve better quality. This graph shows that we can successfully gain quality at the expense of losing the amount of data with the proposed approach.

Suppose that there are two implementations of the purification techniques that yield similar quality improvement. It can be said that one is superior to the other if one loses less documents than the other. More generally, an implementation is superior if plots in Figure 3 are distributed closer to the upper-right corner. From this viewpoint, LDA*(k) is inferior to CD(k) and ACA(k), and KM(k) is inferior to ACS(k) and ACA(k). This implies that we should use auxiliary clustering algorithms rather than topic distribution of LDA for cluster purification. It is also implies that we should use ACA(k) or ACS(k) instead of KM(k) for the auxiliary clustering algorithm.

5.3   Associating Items with Scenes

In this section, we present associations between scenes and items obtained in the experiment. We evaluated two ways of making associations in the experiment: one is to associate items with the original document clusters obtained using LDA. The other is to use purified clusters obtained with PACA using $CD(k)$ as the auxiliary clustering algorithm. In other words, the former involves conducting the χ^2 test on the contingency table introduced in 3. The latter involves using the table below:

	x	$\neg x$
$\mathcal{I}(C_i, CD(k))$	a	b
$\neg\mathcal{I}(C_i, CD(k))$	c	d

We tried associating items with scenes listed in Table 1. Remember that L is a list of places, and making associations in the experiment is equivalent to finding places related to each scene. Therefore, scenes that are location independent may fail to be associated with items. In fact, 4 out of 10 scenes could not be associated with any of the items in L. Table 2 lists some of associated items respectively using original clusters and purified clusters. In the rightmost column, items that were associated with the scenes by the purified clusters but not by original clusters are listed.

Table 2. Items associated with identified topics

Scene	Items associated based on	
	original clusters	purified clusters
Barbecue	park, butcher shop	100-yen shop
Growing pumpkins	field, garden	—
Halloween	church, toy store	parking lot
Preparing baby foods	bookstore	nursery
Preparing diet foods	hospital, pharmacy	stairway
Breeding small animals	toilet, hospital	—

The results indicate that the proposed approach is effective. The items in the tables are mostly of places where the problem (posed by questioners) will likely occur as well as places where we can find solutions to the problem. It should be noted that we could find interesting items — seemingly unrelated but actually related to the scenes — by using purified clusters. For example, although "barbecue" would not usually bring "100-yen shop" (one dollar shop in Japan) to our mind, it was associated from the proposed method. This is because some people would suggest a questioner to buy basic items for barbecue, which is usually available at 100-yen shops. Therefore, "100-yen shop" is practical information for those who are planning to have a barbecue. Associations with these interesting items were often difficult to find without the purification technique because the occurrence frequency of those items in the document set were fairly low: the number of documents related to "barbecue" that contained "100-yen shop" was less than 1/6 of those containing "park".

Additionally, we would like to briefly introduce the results showing that we can discover different types of information by selecting different types of items for L. We conducted another experiment using a list of items consisting of commodities. With the list, we could find necessary or useful commodities in each scene. For example, items that were associated with the scene of barbecue included match, lighter, hat, plastic bag, glove, uchiwa (Japanese fans), chopsticks, umbrella, and mobile phone.

6   Conclusion

We discussed the problem of discovering associations between scenes in our daily lives and their characteristic items. For solving this problem, we developed an approach that first prepares a list L of items and a set of Web documents D relevant to a keyword q. Then, D is divided into clusters using the distribution of topics in documents estimated using LDA. The obtained clusters can be regarded as corresponding to characteristic scenes. We found that consistency of the obtained clusters were not so high (the degree of consistency was around 60%). To improve consistency, we developed two simple techniques that purify original clusters (i.e., eliminates unwanted elements in each cluster). The most effective of the two, PACA, purifies original clusters created using a cluster algorithm by using the outcome from another clustering algorithm.

We conducted an experiment to evaluate the proposed approach. For the experiment, we used data (pages) of a social Q&A site in which scenes were identified. For items, we prepared a list of places. We confirmed that situations where people are facing typical problems were successfully identified as scenes. Furthermore, places where the problems are likely to occur, or places where we can often find the solutions to the problems were associated with the relevant scenes. We also verified that PACA could discover associations between scenes with items that were seemingly unrelated to but in fact practically useful in the scenes.

With the outcome of the proposed approach, scenes can be compared on the basis of their associated items. The similarity and difference of two scenes can be understood by analyzing the similarity and difference between their associated items. We believe that associated items would also be useful for trend analysis of scenes by tracking changes of the associated items.

References

1. Swan, R., Allan, J.: Extracting significant time varying features from text. In: Proceedings of the 8th International Conference on Information and Knowledge Management, pp. 38–45 (1999)
2. Blei, D.M., Ng, A.Y., Jordan, M.I.: Latent Dirichlet allocation. The Journal of Machine Learning Research 3, 993–1022 (2003)

3. Wang, X., Mccallum, A.: Topics over time: A non-Markov continuous-time model of topical trends. In: Proceedings of the 12th ACM SIGKDD International Conference on Knowledge Discovery and Data Mining, pp. 424–433 (2006)
4. Girju, R., Badulescu, A., Moldovan, D.: Automatic Discovery of Part-Whole Relations 32(1), 83–135 (2006)
5. Pantel, P., Pennacchiotti, M.: Espresso: leveraging generic patterns for automatically harvesting semantic relations. In: Proceedings of the 44th Annual Meeting of the Association for Computational Linguistics, pp. 113–120 (2006)
6. De Saeger, S., Torisawa, K., Kazama, J.: Looking for trouble. In: Proceedings of the 22nd International Conference on Computational Linguistics, pp. 185–192 (2008)
7. Strehl, A., Ghosh, J.: Cluster ensembles — a knowledge reuse framework for combining multiple partitions. The Journal of Machine Learning Research 3, 583–617 (2003)
8. Ghaemi, R., Sulaiman, M.N., Ibrahim, H., Mustapha, N.: A Survey: Clustering Ensembles Techniques. World Academy of Science, Engineering and Technology 38, 644–653 (2009)
9. Zhao, Y., Karypis, G.: Criterion functions for document clustering: Experiments and analysis. Technical Report CS 01-040, Department of Computer Science, University of Minnesota (2001)
10. Jain, A.K., Murty, M.N., Flynn, P.J.: Data clustering: a review. ACM Computing Surveys 31(3), 264–323 (1999)
11. Salton, G.: Automatic Information Organization and Retrieval. McGraw-Hill (1968)
12. McQueen, J.: Some methods for classification and analysis of multivariate observations. In: Proceedings of the 5th Berkeley Symposium on Mathematical Statistics and Probability, pp. 281–297 (1967)
13. Newman, M.E.J.: Networks: An Introduction. Oxford University Press (2010)
14. Fortunato, S.: Community detection in graphs. Physics Reports 486, 75–174 (2010)
15. Pons, P., Latapy, M.: Computing communities in large networks using random walks. Journal of Graph Algorithms and Applications 10(2), 191–218 (2006)
16. Orman, G.K., Labatut, V.: A Comparison of Community Detection Algorithms on Artificial Networks. In: Proceedings of the 12th International Conference on Discovery Science, pp. 242–256 (2009)

An Epistemic Equivalence
for Predictive Social Networks Analysis

Christophe Thovex and Francky Trichet

LINA (UMR-CNRS 6241), University of Nantes
2 rue de la Houssiniere, BP 92208
44322 Nantes cedex 03, France
{christophe.thovex,francky.trichet}@univ-nantes.fr

Abstract. The paper presents a semantic and dynamic model providing predictive recommendations and trend predictions, based on the analysis of social networks and social media. It introduces a significant completion of the interdisciplinary paradigm based on the analogy of information flows and current, commonly used in social networks analysis. It defines a social graph structure entailing knowledge engineering, which contributes to semantic clustering and classification of the Social Web and participates in a model of predictive recommendations based on natural laws in electro-physics. These multidisciplinary contributions are applied to team performance and social climate optimization, within collaborative organizations (patented model). Outcomes are presented in line with the SOCIO-PRISE project, funded by the French State Secretariat at the prospective and development of the digital economy.

1 Introduction

Human capital is a quality proper to individuals. Social capital is a
quality created between actors [3].

Sociology studies the behaviour of individuals and groups in society. Social Networks Analysis (SNA) is a an approach of graph-mining based on sociology and applied to social graphs, initially represented by sociograms [12]. It commonly uses dynamic models based on the analogy of information flows and electric current flows [13] [2], in order to improve topological measures such as the standard betweenness centrality defined in [6]. We observe the recent integration in SNA models, of the semantics intrinsic to the shared content related to social networks (*i.e.*, the endogenous content), resulting in semantic SNA models. Semantic SNA attempts are sustained since a few years, with contributions such as [11] and [7].

Dynamic SNA associates the elemental notions of graph theory, such as degree and geodesic distance[1], with the study of interdependent behaviours, so as to

---

[1] The degree of a vertex v is the quantity of edges connected to v, and the geodesic distance between a pair of vertices (x, y) is the number of edges in the shortest chain $[x, y]$.

A. Haller et al. (Eds.): WISE 2011 and 2012 Combined Workshops, LNCS 7652, pp. 201–214, 2013.

define and foresee the structures and flows evolutions within social networks. Semantic SNA focuses on the conceptual aspects of social graphs. It is based on the semantics contained in the media exchanged by the actors, within social networks.

The semantic, electrodynamic and predictive model we advocate improves the epistemic paradigm introduced with the standard centrality measures defined in [13] and [2]. These interdisciplinary SNA measures are based on the analogy between information flows in SNA and current flows in electro-physics. Their outcomes are recognized and currently implemented in most of the SNA software applications - *e.g.*, Pajek, NetMiner, Gephi.

First, we define an improvement of the current epistemic paradigm in SNA, based on KRIPKE's semantics[2] [10]. Then, we define (1) a semantic structure of social networks within which relationships between individuals are based on the endogenous information they share, (2) a socio-physical metric of tension of a social network, within a professional social network, and (3) a semantic and predictive model of dynamic SNA, dedicated to Enterprises Social Networks Analysis (ESNA - patented model).

Our work is applied to economic performance and social climate optimization, and is experimented and approved by a community of experts in human capital management, in line with the SOCIOPRISE project [1]. SOCIOPRISE was funded by the French State Secretariat at the prospective and development of the digital economy, and was realized in partnership with the *OpenPortal Software* company, which provides French leading software solutions for human capital management - `http://www.openportal.fr`.

In this paper, the section 2 presents the principles and methods related to our contributions, encountered in SNA. Section 3 defines our theoretical and epistemic contribution. Section 4 sums up the purpose of the SOCIOPRISE project and defines a skills network structure as a base for the experimentation of our theoretical contribution. Section 5 presents the predictive model of recommendations we define, based on knowledge engineering and on the natural balance of electronic flows. Section 6 sums up results obtained with a real dataset in the context of SOCIOPRISE. Finally, the section 7 presents the conclusions and perspectives of our work.

2   Dynamic and Semantic Models

The section sums up the main dynamic and/or semantic principles and methods encountered in SNA models, from the sociograms presented in [12], to the latest outcomes in SNA for the Semantic Web presented in [5].

2.1   Dynamics and Multidisciplinarity

In [12], social graphs named *sociograms* were introduced for the analysis of dynamic behaviour in people groups. Our approach of SNA is interested in works

---

[2] KRIPKE's semantics is a metaphysic approach using modal logic and intuitionist logic in restricted conceptual worlds, named *possible worlds*, for reasoning on assertions and ontological contexts.

involving connate disciplines in dynamics and networks study. For instance, KIRCHHOFF's point rule and graph theory evoke an epistemic equivalence, and their multidisciplinary study leads to results such as the demonstration of current flows unity and continuity in large graphs, established in [18].

The betweenness centrality defined in [6] is a "standard" SNA measure. [6] has probably inspired the definition of a set of centrality measures based on the comparison of information flows with electronic flows, in [13] and [2], so as to cope with the need of more relevant SNA metrics. Looking for a way to take real information flows into account, NEWMAN introduced a significant contribution in SNA, comparing arcs to electric resistances in a dynamic model based on the KIRCHHOFF's point rule [13].

Metaphysic bridge-building between SNA and electromagnetic force[3] is useful to SNA models, but also to physics. For instance, in [20] physicians have recourse to centrality measures so as to prevent defaults in electric power grids. Our work defines and experiments a semantic completion of the electro-physic metaphor commonly employed in SNA.

2.2   Semantics in Social Networks Analysis

An ontology is an explicit specification of a shared formalisation. It represents the concepts, objects and other entities supposed to exist in an interest area, with their relationships [8]. Semantic SNA studies the conceptual aspects of social networks. It is based on knowledge and ontology engineering coupled with SNA principles, and concerns the Social and Semantic Web [7]. Since [11], more and more works are published in the domain and various markets are waiting for innovative applications resorting to semantic SNA - *e.g.*, collaborative enterprises, marketing, mobile and social apps.

Basically, text analysis and mining produce a set of statistical models which provide a gateway between syntactic and semantic levels in natural language processing. The JACCARD index refinement defined in [15], improves the JACCARD's measure of semantic similarity between terms and corpora. The standard Term Frequency (TF) measure introduced in [16], and the Inverse Document Frequency (IDF) defined in [17] are frequently improved as $TF.IDF$ measures, such as in [14]. Semantic SNA and ontology-building benefit each other, as defined in [9] with a three-dimensional model crossing social graphs, annotations (tags) and consensual ontologies. The latest works in semantic SNA aim at making operational the outlines of SNA using ontologies and Semantic Web languages, and pave the way for statistical and semantic analysis of the Social Web [5] [19].

---

[3] In modern physics, the electromagnetic force is one of the 4 forces enabling the creation of Universe. The three other forces are: gravitation, weak nuclear force and strong nuclear force.

3   An Epistemic Equivalence in Dynamic SNA

Based on the KRIPKE's semantic [10], we have detected and resolved a semantic incoherence due to an incompleteness of the electro-physic metaphor introduced in SNA to represent the information behaviour in social networks [13] [2][4].

Definition 1. *In a* KRIPKE*'s model* $\mathcal{M} = \{W, R, h\}$*,* W *is a set of worlds* w*,* R *is a binary relationship of accessibility in* W*, and* h *is a function representing all the propositions* p *such as* $h(p) = true$ *in all worlds* w*. The formalism* $\mathcal{M}, w \vDash p$ *means* p *is true in the world* w *of the model* $\mathcal{M}$*.*

Definition 2. *Some propositions are Kripke-satisfiable if it exists at least one world of a model in which they are true. A proposition is Kripke-valid if it is true in every world of every* KRIPKE*'s model, or in every world of a unique* KRIPKE*'s model.*

Definition 3. *A proposition* p *is Kripke-equivalent to* p' *if* $p \Leftrightarrow p'$ *in every world of every* KRIPKE*'s model, or in every world of a unique* KRIPKE*'s model.*

Let a proposition Ω : OHM's law $= true$ and a proposition κ : KIRCHHOFF's laws $= true \wedge$ KIRCHHOFF's laws $\Rightarrow \Omega$. Ω means OHM's law is respected and κ means KIRCHHOFF's laws are respected and imply OHM's law is respected. In the common world as in electro-physics, κ is true because KIRCHHOFF's laws are based on the OHM's law[5].

Let e, possible world restricted to electro-physics and g, possible world restricted to graph theory. We define the KRIPKE's model based on the electro-physic metaphor employed in SNA [13], such as $\mathcal{M} = W, R, h$ with $W = \{e, g\}$, $R = \{(e; g), (g; e)\}$. In this model, the propositions Ω and κ are true in the world e, because OHM's law and KIRCHHOFF's laws are respected and dependant in electro-physics.

In the world g, Ω is false because OHM's law is not respected in graph theory with $\mathcal{M}$ based on [13], and κ is false because it implies Ω.
We obtain the equations (1) and (2):

$$\mathcal{M}, e \vDash \Omega, \kappa. \tag{1}$$

$$\mathcal{M}, g \nvDash \Omega, \kappa. \tag{2}$$

Rightfully observing the KRIPKE's semantic, Ω and κ are Kripke-satisfiable thanks to (1), but they are not Kripke-valid because of (2). Furthermore, releasing constraints by replacing κ with κ : KIRCHHOFF's laws $= true$ is not sufficient to make it Kripke-valid, as long as Ω remains false in g ($\mathcal{M}, g \nvDash \Omega$).

---

[4] Our findings concern all graph analysis models entailing the KIRCHHOFF's laws (not only SNA models).

[5] OHM's law defines the tension U as the product of electric intensity and resistance. KIRCHHOFF's laws define the sum of outgoing intensities as a value equal to the sum of incoming intensities in a point of a solid circuit, and the sum of potential differences (*i.e.*, the sum of tensions) along a mesh as a null constant.

Therefore, according to KRIPKE's semantic, the demonstration of a semantic incoherence within the electro-physic metaphor commonly employed in SNA is established.

Finally, we introduce an epistemic completion resolving the semantic incoherence we have demonstrated. Making κ Kripke-valid comes to modify $h = \oslash$ in $\mathcal{M} = W, R, h$ such as $h = \{h(\Omega), h(\kappa)\}$. Pragmatically, making κ Kripke-valid only needs Ω to be true in g. Once we have $h = \{h(\Omega), h(\kappa)\}$ in our KRIPKE's model, we have a Kripke-equivalence with $\Omega \Leftrightarrow \kappa$, when $\Omega \Rightarrow \kappa$ is true[6].

3.1   Stress at Work and Shared Knowledge

The linguistic Term Frequency metric TF is defined in [16] as the number of occurrences of a term divided by the number of terms, for a document D.

$$TF = |term| \subset D/|Terms| \subset D$$

Regarding people within a social network platform/database, TF quantifies the relative pregnancy of consumed and/or produced knowledge, based on the written and/or read terms denoting knowledge in documents[7]. It provides a knowledge use intensity weighing the relationship(s) of people with knowledge. We name it *semantic intensity*.

The rarity of a term denoting knowledge is usually given by the linguistic Inverse Document Frequency metric IDF. TF and IDF are generally combined as a standard $TF.IDF$ metric. IDF is defined in [17] as the number of documents divided by the number of documents containing a given term, in a corpus[8].

$$IDF = |Doc| \subset Corpus/|Doc \supset term| \subset Corpus$$

We use IDF to calculate the rarity of the terms denoting knowledge within the endogenous content of social networks. In our model of SNA, IDF is coupled to TF in order to weigh the connections between individuals and knowledge within social networks.

In the context of the SOCIOPRISE project, we are interested in Enterprises Social Networks Analysis (ESNA). For a person in an ESN, we state stress at work increases in proportion to the knowledge use intensity and to the knowledge rarity. In other words, the semantic intensity metric combined to the rarity metric IDF in a $TF.IDF$ factor provides an indicator of stress at work for ESNA. We are conscious of the triviality of such a simple factor compared to all the quantifiable or undefined causes of stress at work. However, it represents a plausible stress metric in the context of models respecting the epistemic equivalence that we have defined in section 3. Therefore, our model integrates an implementation of the BM25 formula defined in [14], considered as a state-of-the-art refinement of the $TF.IDF$ based formulas according to [21].

---

[6] The KRIPKE's model entails $\Omega \Rightarrow \kappa$ when Ω is true in g.

[7] We compare a digital resource within the social network to a document.

[8] We don't consider noise words and stop words as terms denoting knowledge.

3.2   Interdisciplinary Analogy as Epistemic Equivalence

Respecting our epistemic equivalence, we introduce the OHM's law[9] in $\mathcal{M} = W, R, h$. The semantic intensity we have defined in the world g, based on TF and representing information/knowledge flows intensity, is equivalent to electronic flows intensity in the world e.

Theorem 1. *Let $\Omega \Rightarrow \kappa$, $\Omega : U = I.R \vDash e$, stress $\equiv TF.IDF \vDash g$, then*

$$TF \equiv I \Leftrightarrow (IDF \equiv R)^{\wedge}(U \equiv TF.R \equiv I.IDF) \vDash g$$

and

$$TF.IDF \equiv I.R \Leftrightarrow stress \equiv U \tag{3}$$

Based on modal logic and on KRIPKE's semantics, theorem 1 defines knowledge use intensity (*i.e.*, the *semantic intensity*) coupled to knowledge rarity as a factor of stress in g (possible world restricted to graph theory), and defines tension in e (possible world restricted to electro-physics) as an equivalence to stress in g. Equation (3) represents the semantic equivalence between stress and electro-physic tension for interdisciplinary graph analysis.

Proof. Introducing equation 3 in $\mathcal{M} = W, R, h$ as a new proposition $\Upsilon : TF.IDF \equiv I.R \Leftrightarrow stress \equiv U$, with $\mathcal{M}, g \vDash \Upsilon$, we implicitly define $\mathcal{M}, g \vDash \Omega$ (*i.e.*, Ω becomes true in g).
$h = \{h(\Upsilon), h(\Omega), h(\kappa)\}$ and equations (1) and (2) respectively turn into (4) and (5)[10].

$$\mathcal{M}, e \vDash \Upsilon, \Omega, \kappa \tag{4}$$

$$\mathcal{M}, g \vDash \Upsilon, \Omega, \kappa \tag{5}$$

Υ, Ω and κ are Kripke-satisfiable and Kripke-valid in $\mathcal{M}$ thanks to h. Ω and κ are Kripke-equivalent.

Introducing the OHM's law in the world g thanks to Υ, our theoretical approach defines (1) an epistemic equivalence resolving the semantic incoherence found in the electro-physic metaphor commonly employed in SNA; (2) a demonstration based on modal logic, of the semantic equivalence between tension/intensity as electro-physic flows metrics, and stress related to information/knowledge flows in ESNs. Introducing the OHM's law prior to the KIRCHHOFF's point law in SNA, our epistemic paradigm improves the interdisciplinary bridge-building between information flows behaviour and electronic flows behaviour previously introduced in [13] and [2]. This theoretical contribution paves the way for graph-based applications taking advantage of electronic flow-based rules in graph theory, so as to discover further heuristics merging knowledge engineering and electro-physics in SNA.

---

[9] $U = I.R$, meaning *tension = Intensity.Resistance*.
[10] We don't study the possible uses of equation (4), which semantics is coherent in e.

4   Skills Network, Socio-Semantic Graph Structure

Our theoretical approach is applicable to various social graph structures. We apply it to a socio-semantic structure of social network named skills network, which facilitates the applicative approach and makes our theoretical contributions easier to perceive. It is defined below, after the presentation of our applicative purpose.

Our applicative objective is to define an innovative and decisional model based on dynamic and semantic SNA, which purpose is to foster social performance and social climate within workgroups. This model produces some visual indications and predictive recommendations for the performance optimization and the reduction of psychosocial risk, fostering the agile and skills-based work organization defined in [4].

Our experimentations aim at the discovery of multidimensional synergies between knowledge engineering and physical aspects of the analysis of Collaborative Enterprises Social Networks (CESN). We define a CESN as a heterogeneous graphic structure, based on the numerical marks and contents available in enterprises information systems and representing various collaborative relationships within enterprises.

We define a hybrid graph structure, named *skills network*, which represents the studied CESN and the semantic network induced by its endogenous dataset. The nodes of the semantic network represent the meaningful terms found in the endogenous content and indexed. The arcs of the semantic network represent some semantic relationships, such as synonymy and/or hyponymy, defined in a thesaurus or an ontology. In the context of the SOCIOPRISE project, we solely represent synonyms found in a predefined thesaurus, for each term denoting knowledge found in the endogenous content. Our theoretical findings provides a new heuristic for defining coherent semantic weighs within skills networks, which is going to be presented in section 5.

We have developed a skills network research system enabling the automatic building of skills networks based on keywords researches. These restricted networks are named skills networks *on demand*. They make knowledge communities and noticeable individuals accurately identifiable regarding seized keywords, with the help of SNA metrics. The figure 1 shows an example of skills network on demand.

In our current experimentations as in figure 1, a skills graph on demand comprises three types of nodes and four types of relationships[11]. The nodes are either of a type (1) keyword - *Key knowledge (Kk)*, (2) term associated to keyword - *Relative knowledge (Rk)* or (3) *People (Pkr)*. The relationships are either of a type (1) *Key knowledge/Key knowledge*, (2) *Key knowledge/Relative knowledge*, (3) *People/Key knowledge* or (4) *People/Relative knowledge*. The set of arcs $akk(Kk_i, Kk_j)$, making a unique circuit is named aKk. The relative knowledge makes a star-network around the nodes Kk_i (Key knowledge), by

---

[11] Types and number of relationships can change depending on the semantic referential used for skills graph automatic building.

the set of arcs $aRk \quad \forall \; ark\,(Rk_i, \; Kk_j) \in aRk, \quad Rk_i \in \; Rk, \quad Kk_j \in Kk$. An individual is associated to the key and/or relative knowledge present in the endogenous content with which he interacts - *i.e.*, any predefined role, such as "client" or "consultant". These interactions are classified accordingly with the role types, role of a type *production* (P) and of a type *consumption* (C). An individual Pkr is associated to a knowledge $Kk_i \vee Rk_j$ if he produces or consumes a resource indexed by $Kk_i \vee Rk_j$ - *e.g.*, document, annotation. Figure 1

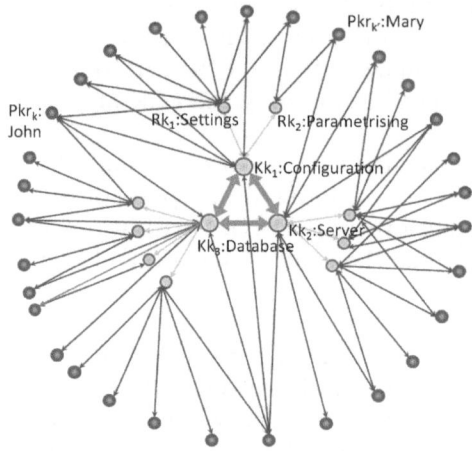

Fig. 1. Basic structure of skills network on demand

illustrates the result of a query on "Configuration, Server, Database". It returns a skills networks entailing 31 individuals producing and/or consuming the 3 keywords and/or their respective related knowledge. For instance, John produces and/or consumes content in the domain "database configuration and settings", while Mary produces and/or consumes content in the domain of "parametrizing server". The graphic structure illustrated represents the part of the competencies of John and Mary related to the knowledge domain denoted by the keywords.

5 Dynamic Weighing of Skills Network on Demand

Our keywords-based and automatic process of skills networks building stands on a lexico-semantic indexation of the enterprise content associated to people - *i.e.*, the digital resources produced and/or consumed within enterprise software. First, we apply our theoretical approach thanks to a process propagating initial values calculated by the BM25 formula, in respect to the OHM's law and KIRCH-HOFF's laws. Our process enables to calculate values of semantic tension/stress, semantic intensity and resistance for each element of a skills network, respecting the epistemic paradigm we introduce. Then, we define a predictive recommendation model based on a retro-propagation process and on the natural balance of electronic flows.

5.1   Lexico-Semantic Research of Skills Networks

Skills networks on demand are automatically built depending on keywords. They use a lexico-semantic research service requiring indexation of the endogenous content. The enterprises corpora being usually consequent, for their semantic indexing, we momentarily reject the hypothesis of rich (and time-consuming) ontologies built by the experts. The software architecture deployed in the context of SOCIOPRISE comprises (1) lexical analysers for French and English language, (2) noise words lists, (3) French and/or English thesaurus, and (4) some multilingual full-text indexing-search services. It supports about 50 languages and remains quickly adaptable to various datasets.

In our main experimentation with a French corpus, the noise words list and the thesaurus are simply built using French lexicons and synonyms dictionaries. This part of our indexation and research process could be improved with specialized dictionaries, ontologies and/or domain-specific thesauri such as SKOS thesauri[12], in order to provide more relevant skills networks on demand.

Thanks to the BM25 formula integrated in our experimental architecture, we are in measure to calculate values of production Pv and/or consumption Cv for each triplet (term; document; individual) within a skills network. Each pair (term; individual) is represented by a pair of symmetric arcs within skills networks, which are respectively weighed with the sum of Pv and the sum or Cv calculated for all the documents produced and/or consumed by the individual. Costs and execution times are distributed between indexation phase and research phase, in order to meet the general performance and scalability requirements imposed in the context of the SOCIOPRISE project.

5.2   Integrated Rules of Electronic Flows Distribution

The KRIPKE's model defined in section 3 provides an epistemic equivalence between graph theory and electro-physics, based on the coherence of OHM's law and the KIRCHHOFF's point law and represented in equation 3. It introduces a rational improvement of the interdisciplinary paradigm defined in [13] [2] and commonly accepted in SNA.

Within skills networks, a person Pkr_k produces and/or consumes knowledge in several documents. For a same term, each relation of the person to a single document owns a $TF.IDF$ value, calculated with the BM25 formula [14] [21]. The sum of all the $TF.IDF$ values gives the production and consumption weighs respectively to the pair of symmetric arcs connecting an individual to a term, denoting produced/consumed knowledge. According to our epistemic paradigm, this weigh is named *semantic tension Ts*.

For a same term denoting knowledge, the IDF part of the semantic tension Ts is constant in the endogenous content, due to the definition of Inverse Document Frequency. According to equation 3, IDF is equivalent to R that we name

---

[12] Simple Knowledge Organization System (SKOS) is a request for comments (W3C) specific to multilingual thesaurus management.

semantic resistance, and TF is equivalent to the semantic intensity Is introduced in section 3.1.

Based on the values $Ts \equiv U$, $Is \equiv I$ and $IDF \equiv R$ weighing the arcs connecting individuals to knowledge, we define a propagation process respecting the OHM's law and the KIRCHHOFF's point law. Pragmatically, our process respects the rules corresponding to current distribution in parallel and/or serial group of resistances.

For instance, the total resistance TR of a group of n parallel resistances R is formalised by the equation (6). Equation (7) formalises the generic equation (6) applied to the arcs connecting individuals $(Pkr_k \ldots Pkr_n)$ to a single term denoting relative knowledge Rk_j (*cf.* section 4):

$$TRpa = \frac{1}{\sum_{k=1}^{n} 1/R_k} \tag{6}$$

$$TR\,(Pkr_k, Rk_j) = \frac{1}{\sum_{k=1}^{n} 1/Rapkr_k\,(Pkr_k, Rk_j)} \tag{7}$$

Let $Iapkr_k\,(Pkr_k, Rk_j)$, semantic intensity in an arc connecting an individual Pkr_k to a term denoting relative knowledge Rk_j. Resistances in serial groups add each others, hence the local tension $LT(akr(Rk_j, Kk_i))$ in an arc connecting Rk_j to a keyword Kk_i is defined by the formula (8).

$$\sum_{k=1}^{n} Iapkr_k\,(Pkr_k, Rk_j).\,(TR\,(Pkr_k, Rk_j) + Rakr(Rk_j, Kk_i)) \tag{8}$$

Formula (8) defines how the intensity produced and/or consumed in a group of parallel resistances is collected and converted to tension in a serial-parallel component.

$I(Kk_i)$, total intensity collected by a keyword, depends on the distribution of intensity in a hierarchy of groups of serial-parallel resistances which hierarchical depth is basically 1 or 2. For m arcs $ark(Rk_{j...m}, Kk_i)$, $I(Kk_i)$ is defined by the formula 9. Let $TI = \sum_{k=1}^{n} Iapkr_k\,(Pkr_k, Kk_i)$:

$$\sum_{j=1}^{m} \left(LT(Rk_j, Kk_i) / \frac{TR(Rk_{j...m}, Kk_i)}{R(Rk_j, Kk_i)} \right) + TI \tag{9}$$

In the formula 9, all the arcs connected to a node Kk_i are processed as a group of parallel resistances, in which the intensity of an arc depends on the inverse proportion of its resistance compared to the total resistance of the group. Formulas 8 and 9 enable to weigh skills networks on demand based on semantic hierarchies endowed with varying depth.

5.3   Balanced Tension Core Values

In an electric mesh defined by the points (A, B, C), $U(AB)+U(BC)+U(CA) = 0$ - KIRCHHOFF's mesh law.

In skills networks on demand resulting from multiple keywords queries, we note $GC(Kk_i, aKk)$ the unique mesh at the core of skills networks on demand, where aKk represents the set of arcs connecting the nodes Kk_i by symmetric pairs - *cf.* figures 1 and 2. In order to verify the mathematical validity of our model, we calculate the semantic tension within a mesh $GC(Kk_i, aKk)$ of n nodes. With $n > 1$, based on the produced and consumed tension of each node, $U(AB)$ is the difference of potential between the produced tension Tps and the consumed tension Tcs of the pair of nodes (A, B). We state the arcs in aKk have the same resistance so as to simplify the calculus and to define $U(AB) = (Tps(A)/n-1) - (Tcs(B)/n-1) + (Tps(B)/n-1) - (Tcs(A)/n-1)$. During our experimentations, we have noticed that all the experimented meshes, comprising up to 5 nodes, rightfully observe the KIRCHHOFF's mesh law and confirm the accuracy of our propagation process.

5.4 Balanced Intensity Core Values

The sum of incoming current intensities in a node is equal to the sum of outgoing current intensities - KIRCHHOFF's point law.

Each node Kk_i of the mesh $GC(Kk_i, aKk)$ verifies the KIRCHHOFF's laws by the produced flow or the consumed flow it collects. However, produced flows are not necessarily equal to consumed flows. Regarding our epistemic paradigm, the balance between produced and consumed tension within skills networks represents an optimal state, in terms of performance and stress at work. Smoothing the difference between produced and consumed tension without changing the total intensity (*i.e.*, the collaborative activity) within skills networks comes to optimize the stress and activity distribution, in respect to individual knowledge and knowledge communities. In the paradigm we define, tension value depends on knowledge use intensity (*i.e.*, semantic intensity) and on knowledge rarity (*i.e.*, semantic resistance). Respecting our paradigm, we don't act on semantic resistance, because it represents a factor difficult to control and to manage in an enterprise. Semantic resistance is constant for each arc of a given skills network. Therefore, semantic intensity enables to indirectly act on tension (*i.e.*, on stress at work, in respect to our paradigm). Semantic intensity represents a factor easy to manage in an enterprise.

Thus, we define a principle for smoothing tension, based on the average intensity value within $GC(Kk, aKk)$. It produces $Ip = Tc$, where Ip is the total produced intensity and Ic is the total consumed tension. In figure 2, we present a simulation of produced/consumed tension balanced in a three-keywords mesh. Figure 2 shows how balanced values Ip and Ic are calculated, thanks to the average intensity deduced from the initial values Ips_i and Ics_i. The balance conserves the total intensity within the mesh (*i.e.*, $1450 + 1380 = 1415 * 2$) and still respects the KIRCHHOFF's mesh law. We use the balanced values to initialize a process of retro-propagation rightfully respecting the theoretical model implemented for propagation.

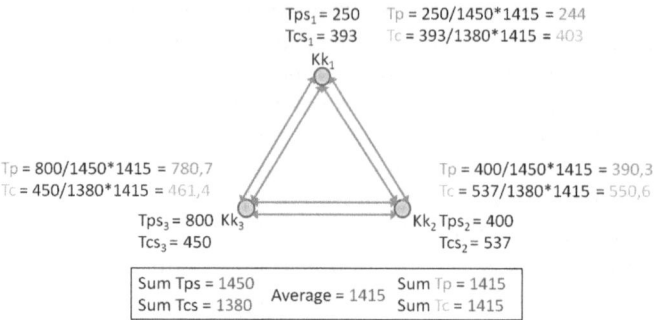

Fig. 2. Balance of a three-keywords mesh in a skills network on demand

5.5   A Natural Load-Balancing Recommendation

Semantic values are collected by the keywords at the core of skills networks and production/consumption flows are balanced depending on a simple average value, in respect to our epistemic constraints. The retro-propagation of the balanced values from the core of the the skills network to its extremities produces balanced semantic values to be compared to the initial values, for each element of the network. Theoretically, the balanced values represent the optimal stress and activity distribution, in respect to individual knowledge and knowledge communities - *i.e.*, a natural and predictive recommendation. We use a retro-propagation process respecting the model defined in section 5.2.

6   Predictive Recommendation with a Real Dataset

The corpus we use in line with the SOCIOPRISE project stems from a real business process, supported by a workflow application running over a relational database. The studied dataset comprises about 250 000 commented steps, regarding 33 000 entries - *i.e.*, some initial actions of entry in the workflow. These entries comes from about 540 persons and are carried through issues by 80 collaborators. Entries and steps comprise various textual metadata and comments.

We generate a skills network on demand, automatically built from a research with the French keywords "validation, message, erreur". Then, we apply our propagation / retro-propagation model and compare the initial and the predictive recommended values. We observe the following results: (a) intensities are stronger than tensions, for both the initial and the balanced values; (b) the initial values are more sparse than the balanced ones, regarding mean values; (c) the balanced tension values make *hot spots* absorbed by new small groups appearing in the nearest knowledge domain; (e) average values are lower for the balanced values than for the initial values - about 2,5 times lesser for the intensity values and 14 times lesser for the tension values, according to our observations. We also observe an unexpected particularity, concerning the main hot spot of semantic tension, which corresponds with a generic account shared by various collaborators. Appearing with initial values, this hot spot disappears in the predictive recommendation and a new group appears. This new group partially comprises the individuals who use the shared account, although there is no explicit information associating these individuals to the shared account in the studied database. We have not yet determined if it results from coincidence or from the regular behaviour of the model. We also have noticed terms exhibited as recommendations for social capital management, within the studied networks.

The experts in human capital management involved in the SOCIOPRISE project are specialised consultants of the company *OpenPortal Software*, managing large client relationships. According to the experts, the dynamic model of predictive SNA we have defined produces (1) a principle of reduction of the tension/stress at work, (2) some precise indications for training management (training plan), and (3) some recommendations for social capital management - *e.g.*, employs and careers management and development. Results of the SOCIOPRISE project have been presented in a large professional event (Human Resources 2012, Paris - France) and a new project is currently funded by the *OpenPortal Software*, aiming at the production of innovative end-user software and services.

7   Conclusion and Perspectives

Based on the KRIPKE's semantic [10], we have detected and resolved a semantic incoherence due to an incompleteness of the electro-physic metaphor commonly employed in Social Network Analysis. Our theoretical approach defines an epistemic paradigm applicable to various social graph structures. We apply it to a socio-semantic structure of social network named skills network on demand, which represents a contribution to semantic and dynamic SNA and makes our theoretical contributions easier to perceive.

From an applicative standpoint, our epistemic paradigm produces first a metric of stress at work based on individual/collective knowledge and on spontaneous numerical activity, which bias might be lower than questions-based surveys. Comparison remains to experiment. Still based on our epistemic paradigm, we define a semantic and predictive model of dynamic SNA (patented model), dedicated to Enterprises Social Networks and experimented with a professional dataset. Its application to social capital management helps in fostering psychosocial risk reduction, professional well-being and overall, peaceful social climate.

From a theoretical standpoint, the metaphysic approach we develop paves the way for electrodynamics and thermodynamics metaphors in graph models, possibly resulting in future dynamic SNA measures. The applicative perspective of such attempts, in case of success, might be wider than the domain of SNA, extending to various applications based on graph theory.

References

1. The socioprise project (2009-2012),
   http://www.cxp.fr/gespointsed/imgbreves/CP_OpenPortal-_Socioprise-_
   Innovation-_Reseaux_sociaux_dentreprise.pdf

2. Brandes, U., Fleischer, D.: Centrality measures based on current flow. In: Diekert, V., Durand, B. (eds.) STACS 2005. LNCS, vol. 3404, pp. 533–544. Springer, Heidelberg (2005)
3. Burt, R.: The Social Capital of Structural Holes, ch. 7, pp. 148–190. Russell Sage Foundation, New York (2002)
4. Demailly, L.: Politiques de la relation: approche sociologique des métiers et activités. Presses Universitaires Septentrion (2008)
5. Erétéo, G.: Semantic Social Network Analysis. Ph.D. thesis, Université de Nice Sophia-Antipolis, Laboratoire d'Informatique, Signaux et Systèmes de Sophia-Antipolis (L3S, UMR6070 CNRS) (2011)
6. Freeman, L.: A set of measures of centrality based on betweenness. Sociometry 40, 35–41 (1977)
7. Gruber, R.T.: Collective knowledge systems: Where the social web meets the semantic web. Web Semantics: Science, Services and Agents on the World Wide Web 6(1), 4–13 (2008)
8. Gruber, T.: Toward principles for the design of ontologies used for knowledge sharing. International Journal of Human Computer Studies 43(5/6), 907–928 (1995)
9. Jung, J.J., Euzenat, J.: Towards semantic social networks. In: Franconi, E., Kifer, M., May, W. (eds.) ESWC 2007. LNCS, vol. 4519, pp. 267–280. Springer, Heidelberg (2007)
10. Kripke, S., Kripke, S.A.: Philosophical Troubles: Collected Papers, vol. 1. Oxford University Press (2011)
11. Mika, P.: Ontologies are us: A unified model of social networks and semantics. In: Gil, Y., Motta, E., Benjamins, V.R., Musen, M.A. (eds.) ISWC 2005. LNCS, vol. 3729, pp. 522–536. Springer, Heidelberg (2005), http://citeseerx.ist.psu.edu/viewdoc/summary?doi=10.1.1.60.2861
12. Moreno, J.: Who shall survive? - (Trad. fr) Fondements de la sociométrie. PUF (1934)
13. Newman, M.: A measure of betweenness centrality based on random walks. Social Networks 27(1), 39–54 (2005)
14. Robertson, S.E., Sparck Jones, K.: Relevance weighting of search terms. Journal of the American Society for Information Science 27(3), 129–146 (1976)
15. Rogers, D., Tanimoto, T.: A computer program for classifying plants. Science 132, 1115–1118 (1960)
16. Salton, G., MacGill, M.: Introduction to Modern Information Retrieval. In: Retrieval Refinements, ch. 6, pp. 201–215. McGraw-Hill Book Company (1986)
17. Sparck Jones, K.: A statistical interpretation of term specificity and its application in retrieval. Journal of Documentation 28, 11–21 (1972)
18. Thomassen, C.: Resistances and currents in infinite electrical networks. J. Comb. Theory Ser. B 49(1), 87–102 (1990)
19. Thovex, C., Trichet, F.: Static, Dynamic and Semantic Dimensions: From Social Networks Analysis to Predictive Knowledge Networks. In: Dynamic Analysis for Social Network. iConcept Press Ltd. (2012)
20. Wang, Z., Scaglione, A., Thomas, R.: Electrical centrality measures for electric power grid vulnerability analysis. In: Proceedings of the 49th IEEE Conference on Decision and Control, CDC 2010, Atlanta, Georgia, USA, December 15-17, pp. 5792–5797. IEEE (2010)
21. Zaragoza, H., Craswell, N., Taylor, M., Saria, S., Robertson, S.E.: Microsoft cambridge at trec-13: Web and hard tracks. In: Proceedings of Text REtrieval Conference, TREC 2004 (2004)

Dynamic Clustering Process to Calculate Affinity Degree of Users as Basis of a Social Network Recommender

Andrea Zanda, Ernestina Menasalvas*, and Santiago Eibe

Universidad Politecnica de Madrid, Facultad de Informatica, Spain
andrea.zanda@alumnos.upm.es, {emenasalvas,seibe}@fi.upm.es

Abstract. Social networking has become a reality: links, activities, and recommendations are proposed by networked friends every moment. There is a need to filter such information to make user enjoy such an experience. In the same way recommender systems where proposed to ease the browsing experience of navigators, nowadays recommenders are required to help users in sharing and obtaining the appropriate information on the social networks. The challenges behind are not only related to the continuous evolution of information being shared, but also by the fact that ubiquity is today a reality. Consequently recommender should take into account the context of the user to whom recommendations are being done. In this paper we present a recommender for social networks that is able to update recommendations as information evolves. The recommender is based on a graph build on basis of a data mining component that extract knowledge on relations and information exchanged by users. The mining component can run autonomously so recommendations can be updated if required. The paper also presents preliminary analysis on the performance of the proposed recommender.

1 Introduction

In recent years, people have increasingly been using *social networks* to share experiences and content, interact with others, learn and disseminate knowledge. Social networking is now supported by mobile devices. Another aspect of mobile devices is that they can access *sensor information* to provide reliable and relevant services to people.

The analysis of on-the-move interactions in social networks is now gaining further attention from researchers. In particular, this topic poses new challenges for the data mining community. As such, there is a need for flexible and robust community discovery technologies to characterize patterns in the community and develop personalized recommendations for users that can be updated as their behavior changes.

By endowing mobile/ubiquitous devices with the capability to generate data mining models, it may be possible to create a new smart world in which computational intelligence is distributed throughout the physical environment. In particular, these data-mining models could provide the mobile users of social networks with recommendation services. One important issue to take into account though is the way in which models will have to be recalculated to keep them updated as user behavior changes.

Social network users generate and share incredibly high quantities of information regarding news, groups, events, and social interactions that are turning difficult to be

---

* Project partially financed by Project TIN2008-05924.

A. Haller et al. (Eds.): WISE 2011 and 2012 Combined Workshops, LNCS 7652, pp. 215–225, 2013.

managed. Thus, social network data focus on actors and relationships, which means that the features of an actor are represented by relationships with other actors in that network. A graph is a natural way to represent actors and relationships among them; the nodes represent the actors, and the edges represent the relationships [5].

In [8], a social mobile system for recommending activities is presented. The recommendations made are based on a static graph, this is to say, once the graph is calculated it is not updated to reflect changes on users and their behaviors. Consequently, this paper proposes a hybrid dynamic recommender which is able to dynamically update the engine for the recommendations. In [7], the EE-Model dynamically update a data mining model, so here we propose to integrate SOMAR with the EE-Model. In our approach we propose to use the algorithm presented in [7] to update the clustering in which the graph computation is made.

The rest of the paper has been organized as follows. In Section 2, we firstly present the basic ideas behind the EE-Model computation and later we present the architecture of SOMAR. Section 3 presents our approach. Thus the process concerning the re-computation of the social graph is presented. In section 4 experimental results of the computation of the EE-Model for a clustering algorithm with synthetic social network data data are presented. To end with, section 5 summarizes the main conclusions of the research conducted so far together with the discussion of the future work.

1.1   Related Work

SocialFusion, which is described in [2] , combines information from three sources (SN, sensor data and mobile data) to provide recommendations, whereas WhozThat, which is described in [1], provides context-aware audio by combining Facebook and mobile phones. However, both of these tools generate recommendations by sending sensitive information to a third party. Besides user personalization and assistance, privacy preservation is also necessary in social networking.

In [8] SOMAR, a social recommender was presented. SOMAR uses social network information and local data from a user of a mobile device to recommend activities on the basis of the users social relationships, which are made explicit by a *social graph* that is built with data mining clustering techniques. The social graph is a representation of the users social relationships with two key characteristics: (i) the nodes of the graph can be friends or groups of friends (depending on the clustering process), and (ii) the graph is based on mutual friendship and the quantity of relationships among users. An analysis of the social graph used as basis for the social recommender can be found in [6]. Further, several types of social graphs can be found in literature, differentiating in the structure as well in the representing information, in [4] a review of social graphs for Web 2.0 with focus on social tagging can be found as an example.

In order to make SOMAR dynamic, we integrate a mechanism presented in [7]. The mechanism makes it possible to automatically decide the proper configuration of a mining algorithm. The mentioned method is based on a cost model (EE-Model) which is able to predict the resource consumption of the mining algorithm execution (efficiency) and the quality of the mining model (efficacy). Based on this model the most appropriate configuration of a data mining algorithm can be chosen depending on the available resources and performance required. It is important to note that the process to generate the EE-model is calculated off-line as it requires to analyze historical executions of the algorithm under different scenarios for which the best configuration wants to be predicted. Once the model is calculated, it can be efficiency applied on a mobile device.

2   Preliminaries

2.1   Automatic Configuration Based on the EE-Model

The EE-model was firstly proposed in [7] to face the problem of autonomous configuration of data mining algorithms. In fact the EE-model makes it possible to estimate the efficiency and efficacy of a certain configuration of an algorithm what is later used to be able to establish a suitable configuration of an algorithm under certain context constrains facilitating the autonomous execution of algorithms in mobile devices.

The EE-Model is a predictor of algorithm's behavior. The inputs of this modules are: (1) an algorithm configuration, (2) dataset information. The outputs will be the quantity of memory, CPU and other measures associated to the resource consumption as well as the accuracy on the mining model.

The steps to build the EE-Model are the following:

1. Selecting the data mining algorithm. Any data mining algorithm is possible, also a measure associated to the quality of the results has to be chosen.
2. Selecting the datasets. A number of dataset are needed to test the data mining algorithm performances with different datasets.
3. Selecting features. The feature are the one associated to the problem description, as for example dataset information and algorithm's configuration, and also the information we want to predict (i.e. CPU, memory, battery of the device).
4. Generating a set of configurations. Here, a representative set of algorithm's configurations is chosen to test the data mining algorithm.
5. Collect data. In this step the data mining algorithm is executed with different datasets and different configuration and the results are stored.
6. Applying machine learning techniques. The data of the previous step are analyzed with machine learning techniques (i.e. linear regression) to discover the relations between the features describing the problem and the results.

Once the cost model is built, the author proposes another component called *configurator*, whose aim is to find the optimum configuration. For this aim the algorithm is able to select an algorithm configuration from a fixed set of configurations on the basis of the predictions. It is a multi-objective decision problem, in which the constraining functions are the resource consumption and the quality of the model.

It is worth mentioning that the process behind computing the EE-model of an algorithm and the configurator are time and resource consuming but these processes are not required to be executed in the mobile device. What will be executed in the mobile is the result of the configurator what as we will see is a light process.

2.2   SOMAR Recommender

SOMAR is a social network-based recommender for mobile devices ([8]). SOMAR accesses information from three sources: (1) mobile data in terms of call history or contacts, (2) sensor data in terms of sensor location and (3) Facebook data. In Figure 1, it is shown how these inputs are used to first generate the recommender and the modules required for its later application.

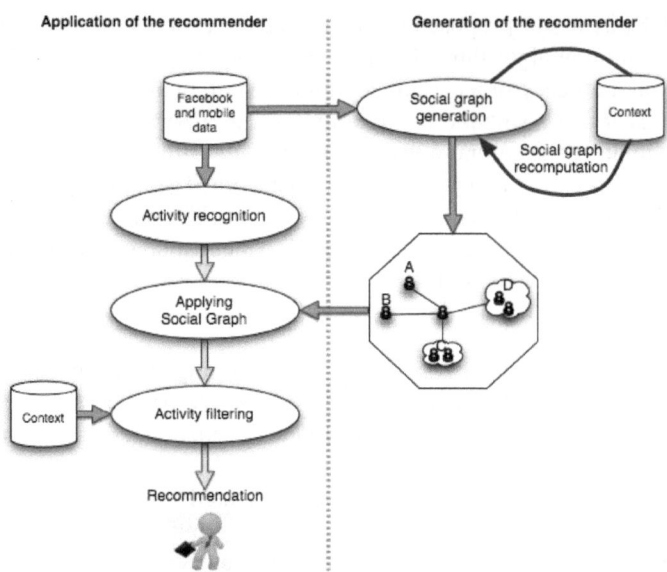

Fig. 1. Recommendation input, process and output. The process can be divided into two phases: the generation of the recommender and the application of the recommender.

Once the recommender is generated to provide the user with a correct recommendation, the following process must be followed in SOMAR:

- Gather information from the available sources and manipulate it so that it can be analyzed in the following steps.
- Detect the activities that are shared on the social network. Then, information can be processed and analyzed.
- Detect the degree of affinity of each activity for a certain user to choose the most appropriate activity from those identified in the above steps.
- Gather information on user context to refine the recommendations.

The generation of the recommender is based on a social graph that is built with clustering techniques and it is composed by two modules:

- Social graph generator: the graph is built based on the the degree of affinity of the users that is calculated based on clustering techniques.
- Change detector and updater (CDU)

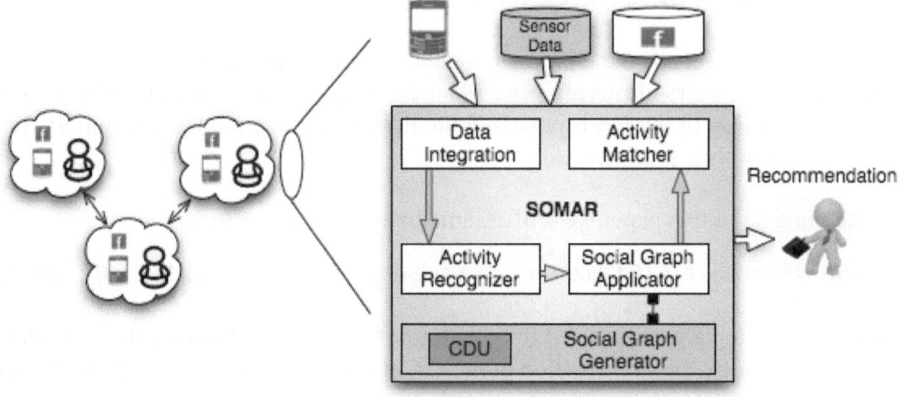

Fig. 2. A social network of users with a mobile phone and Facebook access. On the right, the zoom shows the SOMAR architecture.

Figure 2 shows the architecture with the two modules (in blue) described above that generate the recommender; the rest of the modules are responsible for providing an appropriate recommendation when solicited to do so (the reader can find a deeper description of SOMAR modules in [8])

3   Change Detector and Updater

The users change their interests as well their social relations during time, so a static social graph is not suitable to represent the evolving nature of user relations. Therefore, the social graph should evolve as the user social interactions evolve over time. In SOMAR, the CDU module is the responsible of detecting the changes and then updating the graph. Nevertheless it is not presented the way in which the update process is carried out.

In this paper we focus on the latter process and propose to use the mechanism presented in [7] to calculate the best configuration required for the clustering algorithm in charge of calculating the degree of affinity of users in which the social graph is based on.

There are two variables affecting the definition of a social graph, on one hand the interests of the users that might converge even if two users are not close friends and on the other hand the degree of friendship among friends. Consequently, the social graph needs to be updated whenever either there is a substantial change of degree of friendship with friends or the user for some reason changes interests. Each time this happens, the clustering of users has to be recomputed and to be able to known. Here is where the EE-model based mechanism plays its role to predict the most appropriate configuration of the clustering process that is suitable and feasible at each point.

4   Proposed Approach for Dynamically Updating the Social Graph

In this section we focus on the process to generate a EE-Model for the clustering algorithm supporting users clustering prior to social graph computation in SOMAR. Note that the generation of the EE-Model is done offline, then the models are imported and applied in the mobile platform.

For the purpose of this paper we will assume the following scenario:

- The system has a 2.16 GHz Core 2 processor and 2.5GB 667 MHz DDR2 SDRAM memory.
- The clustering algorithm used for generating the social graph is the the K-medoids, in particular the implementation of the C Clustering Library [3]. Given the k (number of clusters) parameter and a similarity matrix $n * n$, where n is the number of items in the dataset, the K-Medoids algorithm outputs a clustering model. K-Medoids has complexity $O((K(N - K))^2)$.
- We have data of a previous clustering model. In particular, the data we have used to run the experiments shown in the paper have been synthetic data that have been generated by a random function. The similarity matrix of each dataset contains the distances among the items, where the distance in this specific case is the number of mutual friends between two users. So, realistically means that we have generated dataset according to Facebook constraints on user's friendship. In particular, the distance between the user i and j is defined by:

$$Distance_{i,j} = Random(0, Max), where$$
$$Max = Min(Nfriends_i, Nfriends_j) * 0.3 \tag{1}$$

$Random(MIN, MAX)$ is a system function which returns a random value between MIN and MAX, while $Nfriends_x$ is the number of friends of the user x.

In this scenario if we want to generate the EE-Model according to the steps reviewed in 3, the steps would be as follows:

(1) Selecting the Data Mining Algorithm. As we have said the algorithm for which the EE-Model is being calculated is the K-Medoids.
(2) Selecting the Datasets. We have generated 8 different datasets according to the random function for the similarity matrix defined below.
(3)Selecting Features. We distinguish between:

- *Input features*: related to the two input elements: 1) dataset 2) algorithm parameters.
- *Output features* describe the output of the executions, they regard to two main aspects: 1) the efficiency of the algorithm in terms of consumed resources, and 2) the efficacy of the mining model as quality of the results. The output variables can be several, here we focus on persistent memory and CPU cycles for efficiency and sum of error of each item to its centroid for the efficacy.

Table 1 shows a description of the features used for the experiments.

Table 1. Input and output features

Items	Number of items	Integer
k	Number of clusters	Integer
Memory	Memory used	Integer
CPU	Number of CPU cycles	Integer
Error	Error of the clustering model	Real

(4) Generating a Set of Configurations. This step consists on defining the set of configuration for executing the algorithm. As the algorithm parameter is only one, the number of elements k, we define a number of configuration n equal to the number of elements between 2 and number of items divided by 3 ($2 < k > number\_items/0.33$).

(5) Collecting Data. The system used for the executions has a 2.16 GHz Core 2 processor and 2.5GB 667 MHz DDR2 SDRAM memory. In order to collect data about the performance of the execution a script running in parallel and controlling the thread has been developed. Even if the complete capabilities of the machine were available to execute the algorithm, some library has been used to discard all the overhead charge the operative system could generate.

(6) Applying Machine Learning Techniques. The collected data on executions is the input of a data mining algorithm, the objective is to build a model for each of the output variables. We used Linear Regression and REPTree as the techniques to build the models.

4.1 Analysis of the Results

First, we explore the collected data and in the following figures we see how the single output variables are correlated with the input variables.

In Figure 3, on the left we can see how the CPU cycles variable changes according the number of clusters, for lower number of clusters the computational cost of the algorithm decreases. In the same figure, on the right, we can see that the number of clusters does not affect the memory as the CPU cycles.

In Figure 3, on the left figure we see how CPU cycles and memory are correlated, when memory grows, the CPU cycles also grows, but faster. In the same figure, on the right figure, the sum of error of the clustering model decreases when the number of clusters increases. In fact, more clusters means to have more centroids and consequently the distance of every single item to a centroid is supposed to be smaller.

In Figure 5, we can notice the correlation between the number of items of the dataset and the output variables: CPU cycles and memory. On one hand (right) we can see how CPU cycles grows when the number of items grows as it is required more computation for the algorithm, on the other hand the memory also grows because the algorithm has to allocate more memory for the items.

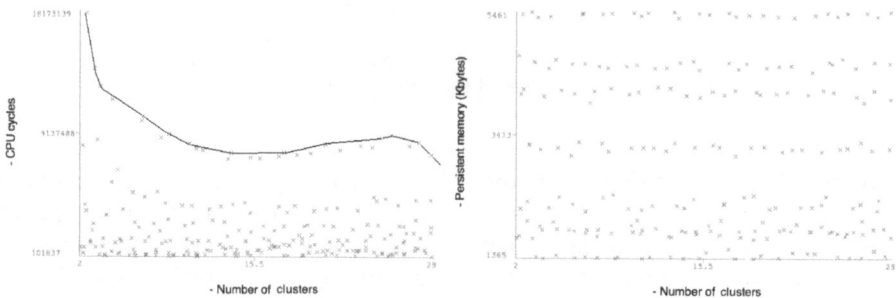

Fig. 3. The correlation between the number of clusters and CPU cycles (left), and memory (right)

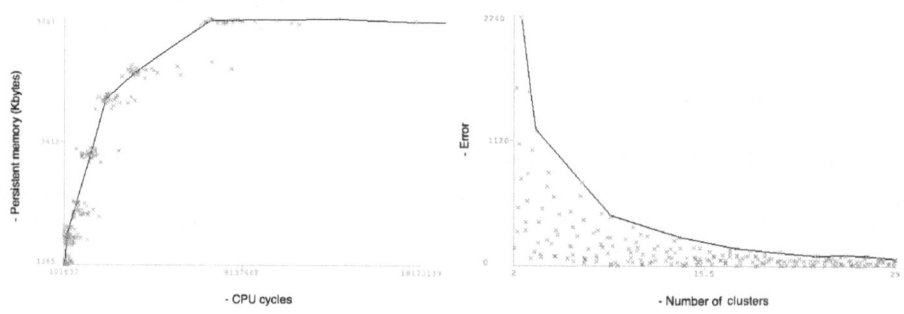

Fig. 4. The correlation between CPU cycles and memory on the the left, and the correlation between the number of clusters and the error of the model

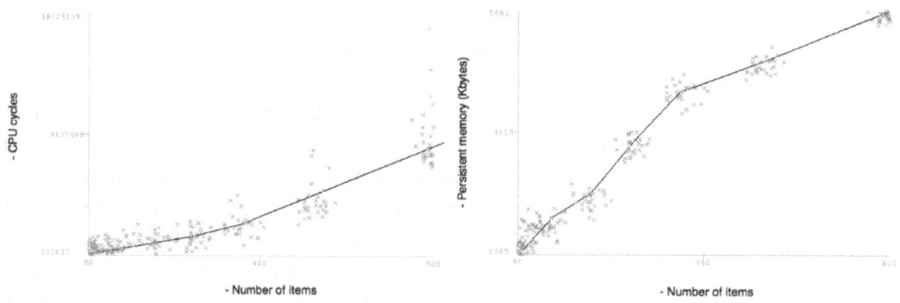

Fig. 5. The correlation between the number of items and CPU cycles (left), and memory (right)

After the exploration, we apply data mining algorithms to that dataset; we used Linear Regression and REPTree as the techniques to build the models. The result is a mining model for each of the output variables: memory, CPU cycles, error. Table 2 shows the information on the modes.

Table 2. Mining models obtained

	Error (REPTree)	CPU (REPTree)	Memory (LR)
Correlation coefficient	0.9331	0.9795	1
Mean absolute error	40.0489	235026.7556	0.1553
Root mean squared error	115.6515	618605.0428	0.2062
Relative absolute error	25.3224 %	10.6392 %	0.0124 %
Root relative squared error	42.022 %	20.5169 %	0.0147 %

We analyze the data mining models obtained for the tree output variables. The model for predicting the CPU cycles has been obtained from the REPTree algorithm, its correlation coefficient is close to 98%, while the relative absolute error is about 10.5%. By looking at Figure 4.1, we can see that the error (bigger Xs) is focused on small values of k, in particular when the number of items is big and $2 < k > 5$. Regarding to this error, we can underline that the generation of the social graph for a standard user will rarely use small values of k.

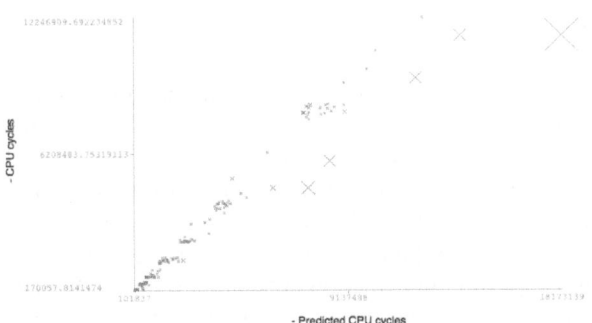

The model predicting the error (quality of the clustering model) has correlation coefficient equal to 93%, while the relative absolute error is equal to 25%. In Figure 4.1, we can see where the error grows. As the previous case, the models has lower accuracy when predicting low vales of k - the assumptions regarding the standard SOMAR users here are also valid.

The linear regression model for predicting the persistent memory is able to predict accurately, in fact the correlation coefficient is equal to 1 and the relative absolute error is equal to 0.01. From a close look to Figure 4.1, we can see that the error is concentrate for number of items equal to 300, nevertheless we can discharge is as the error is statistically non significant.

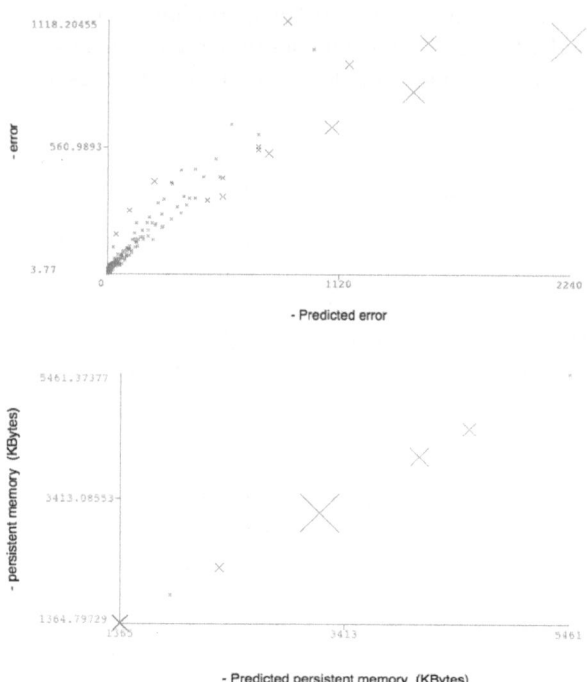

- Predicted error

- Predicted persistent memory (KBytes)

5 Conclusion

The use of social networks is widespread and the amount of both users and information being shared is growing at a dizzying pace. At the same time technological advances have made ubiquity a reality and users connect to share experiences from any mobile device. In order to to make the social network experience pleasant some mechanisms are required to help the user to manage, filter, and find the most appropriate information in each particular situation. In this paper, we have focused on the problem of recommending the user the most relevant information depending on his behavior and his interests in each moment. Based on previous results for autonomous configuration of a data mining algorithm, we have proposed to enhance the recommender SOMAR to provide the user with recommendations based on a dynamically updated social graph. In particular we have proposed to apply the EE-Model to obtain a model to be able to configure the K-medoids algorithm in which the social graph computation is based in SOMAR. Without any loss of generality, the process can be used for any other algorithm. It is worth noting that the EE-Model uses past configurations of the algorithm and data regarding its executions to be able to estimate the most appropriate configuration. As the process is resource and time consuming, we have shown how the EE-Model can be computed offline. Once the model to estimate the most appropriate configuration is obtained, this model can be executed on the mobile device and we have shown how to integrate it in SOMAR. Nevertheless, in this work we focus on the re-generation of the data mining model, a mechanism to detect changes and check the degradation of the mining model is left to future work. Promising results have encourage to continue working on the integration on SOMAR of other data mining algorithms to analyze the best option to help recommendations depending on the user, context and device being used.

Acknowledgements. This work was partially funded by the Spanish Ministry of Science and Innovation, Project TIN2008-05924.

References

1. Beach, A., Gartrell, M., Akkala, S., Elston, J., Kelley, J., Nishimoto, K., Ray, B., Razgulin, S., Sundaresan, K., Surendar, B., Terada, M., Han, R.: WhozThat? evolving an ecosystem for context-aware mobile social networks. IEEE Network 22(4), 50–55 (2008)
2. Beach, A., Gartrell, M., Xing, X., Han, R., Lv, Q., Mishra, S., Seada, K.: Fusing mobile, sensor, and social data to fully enable context-aware computing. In: Proceedings of the Eleventh Workshop on Mobile Computing Systems and Applications, HotMobile 2010, pp. 60–65. ACM, New York (2010)
3. de Hoon, M.J.L., Imoto, S., Nolan, J., Miyano, S.: Open source clustering software. Bioinformatics 20(9), 1453–1454 (2004)
4. Giatsoglou, M., Papadopoulos, S., Vakali, A.: Massive graph management for the web and web 2.0. In: Vakali, A., Jain, L. (eds.) New Directions in Web Data Management 1. SCI, vol. 331, pp. 19–58. Springer, Heidelberg (2011)
5. Harary, F., Norman, R.Z.: Graph theory as a mathematical model in social science (1953)
6. Ugander, J., Karrer, B., Backstrom, L., Marlow, C.: The anatomy of the facebook social graph. CoRR (2011)
7. Zanda, A., Eibe, S., Menasalvas, E.: Adapting batch learning algorithms execution in ubiquitous devices. In: Proceedings of the 2010 Tenth International Conference on Mobile Data Management: Systems, Services and Middleware, MDM 2010, Kansas city, USA. IEEE Computer Society (2010)
8. Zanda, A., Eibe, S., Menasalvas, E.: SOMAR: A SOcial Mobile Activity Recommender. Expert Systems with Applications (February 2012)

Data Leak Aware Crowdsourcing in Social Network

Iheb Ben Amor[1], Salima Benbernou[1], Mourad Ouziri[1],
Mohamed Nadif[1], and Athman Bouguettaya[2]

[1] Université Paris Sorbone Cité, Paris Descartes, France
[2] RMIT University, Australia
firstname.lastname@parisdescartes.fr,
iheb.ben-amor@etu.parisdescartes.fr,
athman.bouguettaya@rmit.edu.au

Abstract. Harnessing human computation for solving complex problems call spawns the issue of finding the unknown competitive group of solvers. In this paper, we propose an approach called *Friendlysourcing* to build up teams from social network answering a business call, all the while avoiding partial solution disclosure to competitive groups. The contributions of this paper include (i) a clustering based approach for discovering collaborative and competitive team in social network (ii) a Markov-chain based algorithm for discovering implicit interactions in the social network.

Keywords: Social network, outsourcing human-computation,privacy.

1 Introduction

A new tend of teamwork has been emerged unconstrained by local geography, available skill set, networking and deep relationships the *crowdsouring*. It is the action of outsourcing tasks, traditionally performed by an employee or contractor, to an undefined group of people through an open call [1]. Crowdsourcing applications should be enable to seek for people crowd on demand to perform a wide range of complex and difficult tasks. Thousands human actors will provide their skills and capabilities in response to the call. We introduce a type of crowdsourcing called *Friendlysouring* based on the efficiency of social network to outsource a task to be performed by people on demand instead of an open world as Mechanical Turk is doing. In fact, the interactions between people involved to answer a query become complex more and more and the collaboration leads to the emergence of social relations and a social network can be weaved for human-task environment.

Challenges. The goal of Friendlysouring system is to let people collaborating on a joint task in the crowd environment where they may seek for other members towards social crowd relationships for achieving a business goal. Thus, many competitive teams can provide a set of answers to the call. However, as the crowd

A. Haller et al. (Eds.): WISE 2011 and 2012 Combined Workshops, LNCS 7652, pp. 226–236, 2013.
© Springer-Verlag Berlin Heidelberg 2013

task is competitive between teams, it is important to group people in a manner there is no inter-teams leaking. Such mechanism will avoid the information leak between crowd people in different teams. Hence, the issues and challenges considered in our system, include, (i) how to build up and discover teams answering the query towards the social relationships, (ii) how to avoid the solution disclosure of the problem during the teams construction between competitive groups. In fact, people on demand collaborating to a task may share sensitive information (part of the problem solution)that may be propagated or forwarded to other crowd members in the social network.

Few works dealing with crowdousourcing are provided. In [2], the Trivia Masster system generates a very large Database of facts in a variety of topics, cleans it towards a game mechanism and uses it for question answering. In [7] is proposed a novel approach for integrating human capabilities in crowd process flows. In [6], the CrowdDB system uses human input via crowdsourcing to process queries that neither database systems nor search engines can adequately answer. Privacy and data leaking are not at all discussed in these works. Moreover, privacy have been introduced in social network as in [3] to design a wizard that may automatically configure a user's privacy settings with minimal effort from the user to aim policy preferences learning. [8], the authors introduced privacy protection tool that measures the amount of sensitive information leakage in a user profile and suggest self-sanitizing action to regulate the amount of leakage. The primovoter tools is unable to estimate the leakage based on a private data propagation, so it settle for a direct user connections and an installed applications on friend profiles.

Contributions. We address the aforementioned challenges by proposing the Friendlysourcing system to discover data leaking aware competitive teams answering the query through a social network. The discovering method is based on a k-means like algorithm to cluster the potential crowd people from the social network that are *close* to collaborate in the same team. The system will not group them in the clusters that are competitive, thus, avoid inter-teams data leaking. To handle such leak, a Markov chain-based algorithm is proposed to discover the implicit social relationships between crowd people. It knows exactly to whom the user data can be propagated in the social network and hence avoid to let crowd people grouped in different clusters. The approach is based on dynamic model that deals with effective rates of shared data and not only on static friend relationship between crowd members.

The rest of the paper is organized as follows: section 2 provides an overview of our Friendlysourcing system. The propagation process is described in section 3. In section 4 is discussed the clustering based approach to discover competitive clusters in the social network answering a business call. Finally, conclusion and future works are given in the section 5.

2   Overview of Friendlysourcing Framework

We devise a crowdsourcing architecture for discovering data leak aware collaborative and competitive teams. It incorporates two main components discussed in this paper, they are depicted in Figure1 and are namely *Data propagation process, clustering process*. Beforehand, the person responsible of company call may register to the friendlysourcing platform using the user interface. He describes the company activities and submits a query. Once the call is launched, it will be visible in our platform. Every social network member is authorized to register for a call using the user interface. The registered member can examine the details of the call and make comments.

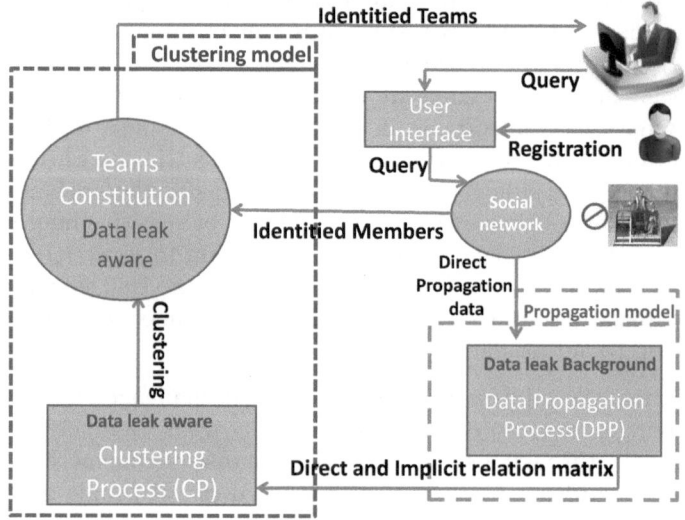

Fig. 1. Friendlysourcing architecture

- *Data Propagation Process:* When the registration is closed, the friendlysourcing system computes the data based on the information collected from the different social networks in order to data leak aware while discovering the teams. In the first step, the request is achieved by the propagation process, thus identifying the direct relationships between the social network members. After that, the process will discover implicit interactions in the social network and the maximum of data propagation between members. The approach is based on Markov chain model. The details are provided in the next section.
- *Clustering Process*
  In a second phase, starting from the whole data propagation calculated in the propagation process DPP, the Clustering process CP will group the crowd

users in the same clusters having strong propagation. That means, more the relationship probability is higher more the users need to be in the same cluster for the collaboration and not in competitive clusters.

- *Team Constitution*: The module will constitute the different team based on the provided result from the clustering process CP. It will use the user profile information provided from the social network. It will notify the users about the team discovering results.

3   A Markov Chain-Based Approach for Data Propagation

The social networks is a set of direct relationships between members. These direct relationships allow to compute the probability of data propagation between only direct friends. However, discovering competitive teams aware of data leak, needs to know all possible interactions. For handling the implicit/indirect relations between members we propose a Markov chain-based approach.

We present in this section a model and an algorithm of data propagation across the entire social network. This allows to compute all indirect interactions between all members and to know to whom the user data can be propagated to.

3.1   A Graph-Based Model of Data Sharing Relationships

Our model of the social network is a labeled directed graph $G \langle M, A, P \rangle$ where,

- $M = \{m_i\}$: set of nodes where each node represents a member of social network.
- $A = \{a_{ij} = (m_i, m_j)/(m_i, m_j) \in M\}$: set of edges where each edge represents a direct friend relationship between two members.
- $P = \{p_{ij}/\forall i, j \ p_{ij} \in [0,1]\}$: is set of labels where each label p_{ij} of the edge a_{ij} represents the rate/probability of data shared by the member $m_i \in M$ with his friend member $m_j \in M$.

In the given graph model, *friend* relationship is represented using edge A labeled with the real probability of shared data P.

The probability p_{ij} that the member m_i shares owned data with member m_j is computed in real time using the following formula:

$$p_{ij} = \frac{quantity \ of \ data \ that \ m_i \ shared \ with \ m_j}{quantity \ of \ data \ held \ by \ m_i}$$

Let's consider the example of a social network depicted in Figure 2:

- the arc (m_1, m_3) indicates that m_1 has friend relationship with m_3, and shares with him 90% of his data.
- the arc (m_1, m_2) indicates that m_1 has friend relationship with m_2 but he never shares with him any data.

The presented graph-based model is a set of direct relationships between members. These direct relationships provide the probability of data propagation between only direct friends. We present in the following the Markov propagation model to compute the probability of data propagation between indirect friends (such as propagation rate from m_1 to m_6 in Figure 2).

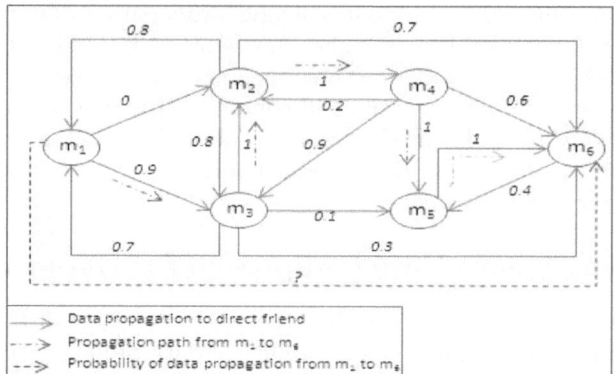

Fig. 2. Example of interactions between crowd members in the social networks

3.2   Markov Chain-Based Model for Data Propagation

Given an owned data of a member, we propose a Markov chain-based model to compute the propagation probability of this data in the entire social newtork.

In social networks, data is propagated from friend to friend following a Markov chain model [4]. That is, a social network member shares owned data only with his friends and, then, each friend shares the data with only their friends and so on.

Definition 1. *A Markov chain is a sequence of random variables $X_1, ..., X_n$ with the Markov property, namely that, the future state depends only on the the present state, and not on the past states. Formally,*

$$P(X_{n+1} = x | X_1 = x_1, X_2 = x_2, \dots, X_n = x_n) = P(X_{n+1} = x | X_n = x_n)$$

From this formal definition, the probability that a given member get a data depends on probability to get it from only his direct friends (and not from indirect friends).

The probability of data propagation between direct friends may be represented with Propagation Matrix defined as follows:

Definition 2. *Propagation Matrix of a social network is matrix that gives probability p_{ij} of propagating data between each couple of members (m_i, m_j):*

$$
\begin{array}{c}
 \\
m_1 \\
m_2 \\
m_3 \\
\dots \\
m_n
\end{array}
\begin{array}{cccccc}
m_1 & m_2 & m_3 & \dots & m_n \\
\left(p_{11} \right. & p_{12} & p_{13} & \dots & p_{1n} \\
p_{21} & p_{22} & p_{23} & \dots & p_{2n} \\
p_{31} & p_{32} & p_{33} & \dots & p_{3n} \\
\dots & \dots & \dots & \dots & \dots \\
p_{n1} & p_{n2} & p_{n3} & \dots & \left. p_{nn} \right)
\end{array}
$$

where

$$p_{ij} = \begin{cases} \dfrac{quantity\ of\ data\ that\ m_i\ shared\ with\ m_j}{quantity\ of\ data\ held\ by\ m_i} & if\ (m_i, m_j) \in A \\ 1 & for\ i = j \\ 0 & else \end{cases}$$

This propagation matrix has the following properties:

- $p_{ii} = 1$, which means that member m_i does not lost owned data when he shares it.
- $\sum_{k \in [1,n]} (p_{ik}) \neq 1$, because data may be propagated to several members at the same time.
- $p_{ij} \neq p_{ji}$, which means that a member m_i may share with a friend m_j a quantity of data different his friend m_j may share with him.
- $\exists (i,j)|(m_i, m_j) \in A \wedge p_{ij} = 0$, which means that members do not share necessarily data with their friends.
- if m_i and m_j are not direct friends then $p_{ij} = 0$.

The propagation matrix of the social network of figure 2 is given as follows:

$$\begin{array}{c} \\ m_1 \\ m_2 \\ m_3 \\ m_4 \\ m_5 \\ m_6 \end{array} \begin{array}{cccccc} m_1 & m_2 & m_3 & m_4 & m_5 & m_6 \\ \left(\begin{array}{cccccc} 1 & 0 & 0.9 & 0 & 0 & 0 \\ 0.8 & 1 & 0.8 & 1 & 0 & 0.7 \\ 0.7 & 1 & 1 & 0 & 0.1 & 0.3 \\ 0 & 0.2 & 0.9 & 1 & 1 & 0.6 \\ 0 & 0 & 0 & 0 & 1 & 1 \\ 0 & 0 & 0 & 0 & 0.4 & 1 \end{array} \right) \end{array}$$

3.3 A Markov Chain-Based Algorithm of Data Propagation

The propagation matrix of section 3.1 gives only probability of data propagation between direct friends.

But it is not sufficient to compute the probability that data is propagated from member to indirect-friend because:

1. Propagation matrix defined in Definition 2 does not give the real propagation probabilities between members. In Figure 2, the direct propagation probability from m_1 to m_2 is zero ($p_{12} = 0$). However, through m_3, data of m_1 may be propagated to m_2 with probability $0.9 \times 1 = 0.9$.
2. The Propagation matrix does not provide the data propagation to indirect-friends. It's indicate a zero value of sharing data with indirect friends, because members share their data only with direct friends. In the propagation matrix of figure 2, probability that data of member m_1 may be propagated to his indirect-friends m_5 and m_6 is zero because m_1 is no direct friend relationship with them. However, data may be propagated from m_1 to m_5 through m_3 with probability $0.9 \times 0.1 = 0.09$.
3. Propagation to indirect-friends is hard to calculate: as example, what is the probability that data of m_1 may be propagated to m_6 (probability of dotted

red arrow in figure 2)? To calculate this probability, we have to explore all
the paths allowing propagation of data from m_1 to m_6. Each one allows to
calculate a propagation probability. The final propagation probability is the
maximum of propagation probability of all the possible paths, wihich corresponds to the propagation risk. The path $(m_1, m_3, m_2, m_4, m_5, m_6)$ indicated
with dotted green arrows in figure 2 allows propagation of data from m_1 to
m_6 with the maximum probability $0.9 \times 1 \times 1 \times 1 \times 1 = 0.9$. However, it is
hard to calculate this probability because real social networks are complex.

For this reasons, we need to design an efficient algorithm that calculates the
optimal data propagation probability from the owner to all the members of the
social network. This algorithm is based on energy function we define as follows:

Definition 3. *The energy function p_i of member m_i is the probability that data
is propagated to member m_i. In our model, data is propagated following Markov
chain. That is:*

$$p_i = \underset{m_k \in N_{m_i}}{Max} (p_k \times p_{ki}) \tag{1}$$

where,

- N_{m_i} is a set of direct friends of m_i,
- p_k is the energy function of m_k,
- p_{ki} is the probability of propagating data from m_k to m_i.

To compute the energy function p_i for all members m_i of the social network, we
have to use an iterative algorithm [9]. We design our simple algorithm Algorithm
1..
  The algorithm processes as follows:

1. Initialisations: $p_{ow} = 1$, $\forall i \neq ow$ $p_i = 0$. That is, only the owner m_{ow} has
   the data.
2. Iterations: At each iteration, the algorithm computes p_i for $m_i \in N_{m_i}$ using
   formula (1).
3. Stop: The algorithm stops when the probability maximum of each member
   is reached.

Applying this algorithm on the propagation matrix of section 3.2, we get the
following matrix completed with indirect-friend propagation probabilities:

$$
\begin{array}{c c c c c c c}
 & m_1 & m_2 & m_3 & m_4 & m_5 & m_6 \\
m_1 & 1 & 0.9 & 0.9 & 0.9 & 0.9 & 0.9 \\
m_2 & 0.8 & 1 & 0.8 & 1 & 1 & 1 \\
m_3 & 0.8 & 1 & 1 & 1 & 1 & 1 \\
m_4 & 0.7 & 0.9 & 0.9 & 1 & 1 & 1 \\
m_5 & 0 & 0 & 0 & 0 & 1 & 1 \\
m_6 & 0 & 0 & 0 & 0 & 0.4 & 1 \\
\end{array}
$$

Algorithm 1. PROPAGATION PROBABILITY COMPUTING ALGORITHM

Require: $G\langle M, A, PM\rangle$ – labeled directed graph of the social network where PM in the propagation matrix

m_{ow} – owner of the data

m_r – recipient member of data that we want calculate the propagation probability

Ensure: p_r – probability that owned data is propagated to m_r

1: **print** $\mathcal{P} = (p_1, \ldots, p_{ow}, \ldots, p_r, \ldots, p_n)$: Energy function at the previous step.
2: **print** $\mathcal{PS} = (ps_1, \ldots, ps_{ow}, \ldots, ps_r, \ldots, ps_n)$: Energy function at the current step.
3: **print** *continue*: boolean value indicating if the optimum values of all members are reached.
   {I}nitializations
4: $p_{ow} = 1$ and $\forall m_i \neq m_{ow}, p_i = 0$
5: *continue* $\leftarrow$ *true*
   {I}terations
6: **while** *continue* **do**
7:    **for** each members $m_i \neq m_{ow}$ **do**
8:       $ps_i = Max_{m_k \in N_{m_i}}(p_k \times p_{ki}))$
9:    **end for**
10:   **if** $\mathcal{P} \neq \mathcal{PS}$ **then**
11:      $\mathcal{P} \leftarrow \mathcal{PS}$
12:   **else**
13:      *continue* $\leftarrow$ *false*
14:   **end if**
15: **end while**
16: **return** p_r

4 Data Disclosure Aware Clustering Process

Based on data propagation calculated in the previous section, the clustering process groups of crowd members into free data leak clusters.

Definition 4. *A cluster C is set of crowd members having or no strong propagation:*

$$C = \{m_i\} \text{ such that } \forall m_i, m_j \in C, p_{ij} \in [0, 1], p_{ji} \in [0, 1]$$

Definition 5. *Two clusters C_k and C_s are free data leak iff:*

$$\forall m_i \in C_k, \forall m_j \in C_s, p_{ij} \leq \eta \wedge p_{ji} \leq \eta$$

From the definition 5, we consider there is a risk of data leak between two clusters C_k and C_s if the propagation rate between all the members of the two clusters is less than a threshold η. The later is proposed as a value for which the data propagation of a crowd member is acceptable in a social network. The dynamacity of social member profil impactes the value of η, but it is out of the scope of the paper.

The propagation matrix calculated by the algorithm 1. is be updated as folllows:

$$\forall i, j, p_{ij} = Max(p_{ij}, p_{ji})$$

The propagation matrix of our example is updated as follows:

$$
\begin{array}{c}
\begin{array}{cccccc}
m_1 & m_2 & m_3 & m_4 & m_5 & m_6
\end{array} \\
\begin{array}{c}
m_1 \\ m_2 \\ m_3 \\ m_4 \\ m_5 \\ m_6
\end{array}
\left(
\begin{array}{cccccc}
1 & 0.9 & 0.9 & 0.9 & 0.9 & 0.9 \\
0.9 & 1 & 1 & 1 & 1 & 1 \\
0.9 & 1 & 1 & 1 & 1 & 1 \\
0.9 & 1 & 1 & 1 & 1 & 1 \\
0.9 & 0 & 1 & 1 & 1 & 1 \\
0.9 & 0 & 1 & 1 & 1 & 1
\end{array}
\right)
\end{array}
$$

Based on this updated propagation matrix, the members are classified into free leak clusters using a clustering algorithm.

Algorithm 2. D-MAX DISCOVERING TEAMS ALGORITHM

---

    print $Clusters = (Clust_1, Clust_2, ..., Clust_{Cluster})$: Teams constitution
    {I}nitializations
2: $Distances, Centroid, MaxDistancesMax = 0$
    $ClustNB = -1$
4: $Threshold = \eta$
    for each $ClustersinCluster$ **do**
6:    $Clusters_i = member_{m_i}$
      $Distances_i = 0$
8:    $Centroid_i = 0$
    end for
    {I}terations
10: **for** each members m_i in G **do**
      for each members m_i in $Clusters$ **do**
12:      $Distances_i \leftarrow P_{member,member_i}$
        if $Distances_i \succ Centroid_i$ **then**
14:        $Centroid_i \leftarrow Distances_i$
        end if
16:    **end for**
      for each $Centroid_i$ **do**
18:      **if** $Centroid_i \succ MaxDistancesMax_i$ **then**
        $MaxDistancesMax_i \leftarrow Centroid_i$
20:        **if** $MaxDistancesMax_i \succ \eta$ AND $ClustNB \neq -1$ **then**
          $Clusters_i = FUSION(Clusters_i, Clusters_{NB})$
22:        **end if**
        $ClustNB = i$
24:        $Clusters_{ClustNB} \leftarrow member_i$
      end if
26:    **end for**
    end for
28: **return** $Clusters$

Our clustering algorithm is a k-means algorithm [5]. The principle of the algorithm is that for each cluster C_k and each member m_i, the member m_i is calssified to the cluster C_k if there is high data propagation between the member m_i and at least one of the members of C_k. The algorithm uses the following specific distance called D_{max}:

$$D_{max}(C_k, m_j) = \underset{m_i \in C_k}{Max} \; p_{ij}$$

where Max is the maximum function, C_k a cluster to be built, m_j is member that can be clustered into the cluster C_k, and m_i is a crowd member in C_k, p_{ij} is the propagation value between m_i and m_j given in te propagation matrix.

The clustering algorithm prossess as follows:

- Inputs: the data disclosure threshold η, number of clusters.
- Initialization: The initialization of the clusters is done by assigning arbitrarily a member to each cluster.
- Iterations: For each candidate member m_i and a cluster C_j, if $D_{max}(C_j, m_i) \geq \eta$ then m_i is added to C_k
- Stop: The algorithm is deemed to have converged when the assignments of members to clusters no longer change.

Moreover, in the case that a candidate member has a strong communication with more than one cluster, we will merge the clusters with whom the candidate member has a high data propagation and assign it to the new merged cluster. Because it may probably disclosure the data of the cluster to the other clusters. For instance if $d(C_1, m_k) = 0, 8$ and $d(C_2, m_k) = 0, 7$, then we will merge the cluster C_1 and C_2 and K will integrate the cluster C_{12} result of the C_1 and C_2 fusion. The algorithm is presented as Algorithm 2.

5   Conclusion and Future Work

In this paper, we proposed a Friendlysourcing framework as a clustering based approach for data leak aware discovering competitive teams during the crowd-sourcing process in social network. First, the Markov model is used to estimate the hidden relationships between crowd members in the social network. Given the results of the previous step, then, the clustering approach groups the crowd members into data leak aware competitive teams.

In the future works, we plan to evaluate efficiency of the proposed approach by means of data leakage and time consumption. Regarding the social network complexity, the current approach provides several classifications of competitive teams but not easy to choose the best one. Then, we will study how to take into consideration more constraints specifically user preferences.

References

1. Daren, C.: Brabham. Crowdsourcing as a model for problem solving: An introduction and cases. Convergence: The International Journal of Research into New Media Technologies 14(1), 75–90 (2008)

2. Deutch, D., Greenshpan, O., Kostenko, B., Milo, T.: Using markov chain monte carlo to play trivia. In: ICDE, pp. 1308–1311 (2011)
3. Fang, L., Kim, H., LeFevre, K., Tami, A.: A privacy recommendation wizard for users of social networking sites. In: ACM Conference on Computer and Communications Security, pp. 630–632 (2010)
4. Hggstrm, O.: Finite markov chains and algorithmic applications. London Mathematical Society Student Texts. Cambridge University Press (2000)
5. MacQueen, J., et al.: Some methods for classification and analysis of multivariate observations. In: Proceedings of the Fifth Berkeley Symposium on Mathematical Statistics and Probability, California, USA, vol. 1, p. 14 (1967)
6. Sellis, T.K., Miller, R.J., Kementsietsidis, A., Velegrakis, Y. (eds.): Proceedings of the ACM SIGMOD International Conference on Management of Data, SIGMOD 2011, Athens, Greece, June 12-16. ACM (2011)
7. Skopik, F., Schall, D., Psaier, H., Treiber, M., Dustdar, S.: Towards social crowd environments using service-oriented architectures. IT - Information Technology 53(3), 108–116 (2011)
8. Talukder, N., Ouzzani, M., Elmagarmid, A.K., Elmeleegy, H., Yakout, M.: Privometer: Privacy protection in social networks. In: ICDE Workshops, pp. 266–269 (2010)
9. Weise, T.: Global Optimization Algorithms - Theory and Application (June 2009)

Big Data and Cloud

Surya Nepal[1] and Athman Bouguettaya[2]

[1] CSIRO ICT Centre Marsfield, NSW, Australia
surya.nepal@csiro.au
[2] RMIT University Melbourne, NSW, Australia
athman.bouguettaya@rmit.edu.au

The rise of multimedia, social media, sensor networks and the Internet of Things will fan the flames of data explosion in many application domains. These emerging Big Data applications present enterprises with very high data-to-compute ratio. For example, in Square Kilometre Array (SKA), a telescope generates more data than the whole Internet with a single file often reaching multiple terabytes in size. As data volumes continue to increase exponentially, the challenge is not only how to store and manage data but also to effectively analyse the data to gain insight knowledge to make smarter decisions. One of the key challenges is transforming the raw data available to your business into business value and strategic advantage. The better management and analysis of Big Data will become the next frontier of innovation, competition and productivity. For example, according to a McKinsey Global Institute study, a retailer exploiting the full potential of big data could increase its operating margin by more than 60 percent; efficient and effective use of big data could save more than $300 billion dollars for US government in healthcare alone. Therefore, there is a need of effective and efficient management and analysis of big data.

Recently, Cloud computing has emerged as a promising technology towards handling Big Data. The driving forces behind cloud computing are: (a) significantly reduced Total Cost of Ownership (TCO) of the required IT infrastructure and software including (but not limited to) purchasing, operating, maintaining and updating costs; (b) high Quality of Service (QoS) provided by cloud service providers such as availability, reliability and Pay-As-You-Go (PAYG) based low prices; and (c) easy access to organizational information and services anytime anywhere. In a nutshell, cloud computing provides a new paradigm for delivering computing resources (e.g., infrastructure, platform, software, etc.) to customers such as utilities (e.g., water, electricity, gas, etc.) on demand. Despite its own shortcomings, cloud becomes an attractive technology platform for developing and deploying Big Data analytics.

The international workshop on Big Data and Cloud (BDC 2012) held in conjunction with WISE 2012 provided the scientific community a dedicated forum for discussing state-of-the-art research, development, and deployment efforts of Big Data in Cloud. We have selected two papers to be presented at the workshop. The first paper *"A Service Oriented Framework for Animating Big Spatiotemporal Datasets"* proposes a service oriented distributed system framework for animating big spatio-temporal vector datasets. The second paper *"Searching frequent itemsets by clustering data: towards a parallel approach using MapReduce"* proposes a new algorithm for searching frequent itemsets in large data bases.

A. Haller et al. (Eds.): WISE 2011 and 2012 Combined Workshops, LNCS 7652, p. 237, 2013.
© Springer-Verlag Berlin Heidelberg 2013

A Service Oriented Framework
for Animating Big Spatiotemporal Datasets

Ahmet Sayar

Department of Computer Engineering, Kocaeli University
Izmit/Kocaeli, 41380, Turkey
ahmet.sayar@kocaeli.edu.tr

Abstract. We propose a service oriented distributed system framework for animating big spatiotemporal vector datasets. The aim is exploiting the patterns in spatial data dynamically changing over time. Animation consists of successively played and temporally related still map images. Each map image in the animation is a satellite map enriched with (or plotted over) spatiotemporal datasets. The system components are designed as web services. We also extend open standards' GIS web services definitions with topic-based publish-subscribe paradigm, which best suits to the animation requirements.

Keywords: Distributed Systems, animation, GIS, animated map, spatiotemporal data.

1   Introduction

Vast amounts of data related to earth are time-series and spatial in nature. Spatial data are preferably represented and displayed as map layers. When you have the same layer in several different moments along the time, it is better to display them as part of a movie. This is called time-series animation, which is a visualization technique ideally suited for the display and analysis of spatiotemporal and geographic data sets. Map animations enable scientific analysis and results to be understood not only by the scientist but also the public and policymakers from different domains and education level. Animated maps can be interpreted more easily than their static representations by the users.

We propose a service oriented distributed system framework [1] for map animations. Maps are created from spatiotemporal datasets. Services in the system are defined with standard bodies (Open Geospatial Consortium (OGC) [2] and ISO-TC211). Standardization offers advantages for data sharing, for combining software components and for overlaying graphical outputs from different sources. However, standardization comes with its costs. Costs mostly come from the fact that web services are based on XML based SOAP over HTTP protocol and data to be processed are encoded in Geographic Markup language (GML) [3] which is an XML-based format. To overcome such problems in a distributed system framework requiring large scale XML-encoded geographic feature sets, we have investigated the possibilities of using topic-based publish-subscribe paradigms (which is mostly used in P2P systems) for

A. Haller et al. (Eds.): WISE 2011 and 2012 Combined Workshops, LNCS 7652, pp. 238–250, 2013.

exchanging data payload between web services. To do so, we have integrated NaradaBrokering [4] to the communication between OGC compatible standard GIS (Geographic Information Systems) Web Services. Web Services are well-known and widely used RPC (Remote Procedure Call) systems. In basic RPCs, services (client/server) tightly coupled in terms of space and time. This is an example of synchronous communication in which client and server must be active at the same time. The proposed system has a couple of advantages over using pure Web Services. It enables map creation from the partially retuned data and gets rid of SOAP (Simple Object Access Protocol) message creation overheads. Moreover, by using Narada-Brokering (event and topic-based publish/subscribe system) we utilize network level quality of services provided in it.

After developing an efficient data transfer protocol between standard GIS [5] web services, we propose an animation web service extended from Web Map Services (WMS) [6]. WMS have capabilities of animating temporally related map images one by one dated in a vertical frame as a film. Images are created from the spatiotemporal data provided by Web Feature Services (WFS) [7].

2   Related Work

In the last decade renewed interest in animation has emerged due to technological developments. Because of these developments it is nowadays relatively easy and inexpensive to construct animations. This has led to an increase in the number, variety and complexity of animations produced. One of the areas that animations have been successfully applied is map animations.

Map animations have been studied by various disciplines such as computer sciences, remote sensing and pattern recognition. Most of the applications [8-12] are central (i.e. desktop) in which data and services are physically located in the same machine and the analyses are carried out in the same place. These early works are on recognition of temporal changes in spatial datasets through animation, but our focus is creating animating web services enabling sharing and collaboration of animated spatial data among the virtual organizations through the distributed systems.

After the invent of the internet and advancements in distributed systems there has been great dissemination possibilities, and it has become easier to access data and processing services and coupling them for the application purposes. In this context, there some client-centric and server-centric solution approaches. MathWorks's mapping toolbox [13] and GeoServer's animator toolbox [14] can be given as examples of client-centric approaches. They connect to remote WMS (OGC compatible) and fetch the map images to create an animation. They build animations as a set of frames, and each frame is a separate WMS getMap call, similar to the others in the set, but with a different value in one of the parameters. Map images are stored into local file system at client side, and animations are created as animated GIF or movies in AVI format. Köbben [15], Becker [16] and Esri's ArcGIS [17] can be given as examples of server-centric map animation approaches, but they do not develop their services in accordance with the service oriented architectures. Furthermore, they do not consider the issues of large scale data transfer over the network for the real time animations.The animations they produce are mostly in the form of moving window (bounding box) or zooming in and out.

Compared to the related works, our solution approach is considered as server-centric. The proposed system not only supports zooming or moving animations but also animates spatiotemporal changes in large scale feature collections for a specific bounding box. To do that we take data rendering and overlaying issues into considerations, and propose a distributed system framework for collaborating and sharing the animated maps as streaming data through Real-time Transfer Protocol (RTP). This enables animations to be played in collaborative and Grid [18] environments.

For the efficiency and applicability, we adopted publish-subscribe system to the communication between OGC compatible standard GIS Web Services. This removes the deficiencies of synchronous RPC communications in GIS Web Services. We could also use some other alternative paradigms such as shared spaces approach and message queuing approach [1] for the same purposes. In shared space approach, producers insert messages asynchronously into a container, and consumers pulls (i.e. reads synchronously) from the container. In this approach services (client/server) tightly coupled in terms of space and time. In case of basic message queuing systems, producers insert messages asynchronously into FIFO and consumers reads synchronously (pull) from FIFO. This approach also provides time and space decoupling.

As you realize shared spaces and message queuing approaches are time and space decoupled. On the other hand, publish/subscribe is an event-based system and provides decoupling in terms of time, space and synchronization as well.

3   Architecture: Streaming Map Movies

This paper proposes a distributed service oriented architectural framework for binding map based geo-data animations to the distributed services by adopting and implementing OGC's WFS and WMS services in accordance with the publicly available standards. OGC also has a discussion paper [19] on animations as an extension to WMS. It discusses how WMS specifications can be extended to allow animations that move in space over time. In this case, the only parameter changing is the bounding box in successive map images. On the other hand, we focus on overlay layers which are rendered from feature data collections represented as GML and provided by WFS. We take both geometrical and non-geometrical attributes of spatiotemporal data sets into account and animate their changing values over time. We produce map animations in the form of streaming map videos similar to ones you see in the weather cast web sites.

The work presented here looks into the possibilities of extending the web based map services with time-series data as animated maps, and provide a framework enabling integration of animation services to the distributed systems. In the proposed framework, for the service level interoperability, in terms of request and response types, and definition of animated layer descriptions we use OGC defined standards [6,7], but for the data transfer we propose a novel approach based on integration of OGC specifications with web services [18] and topic-based publish-subscribe paradigms.

This section starts with explaining topic-based publish/subscribe system and then present the system components to create a map image from spatiotemporal datasets and illustrates the extension (Fig. 2) to OGC standards by using Naradabrokering, and then elaborates on WMS and its capabilities of being an animation service, and finally presents a service oriented framework for animating spatiotemporal datasets (Fig. 5).

3.1   Publish/Subscribe Paradigm

Publish/subscribe systems provide a useful paradigm for selective data dissemination and most of the complexity related to addressing and routing is encapsulated within the network infrastructure. Publish/subscribe systems utilize multicast networking facilities (also at data link level). It is actually a broker layer overlay network, which is based on transport level connections between nodes.

Subscribers register their interest in an event and are subsequently asynchronously notified of events generated by publishers. Publishers are generators of events and subscribers are consumers of events [20].

Publish/subscribe systems enable loosely coupled form of computer communication and interaction. Information generation and consumption is independent from each other. This feature is also called decoupling in terms of time, space and synchronization. Parties don't need to be active at the same time. Publisher can generate events when a subscriber is disconnected. Subscriber can be notified when publisher is disconnected (see Fig. 1).

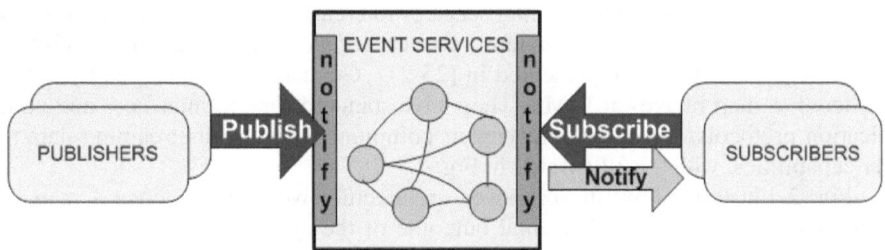

Fig. 1. Event-based Publish/Subscribe paradigm

There are three types of publish/subscribe paradigm, topic-based, content-based and type-based. In a topic-based system, messages are published to "topics" or named logical channels. Subscribers in a topic-based system will receive all messages published to the topics to which they subscribe, and all subscribers to a topic will receive the same messages. The publisher is responsible for defining the classes of messages to which subscribers can subscribe. Topics are usually expressed in a URL-like notation. It is similar to creating a group. When subscriber subscribe to a topic it becomes a member of an event group. Topic is a pre-defined criterion and does not say anything about the specific content of an event. Example topic is "course2012".

In a content-based system, messages are only delivered to a subscriber if the attributes or content of those messages match constraints defined by the subscriber. The subscriber is responsible for classifying the messages. Topics are formed by considering content of events, e.g. internal attributes of data structures. Application developers might use SQL, XPath and some other string API tools to handle complex topic strings. Example topic is "Course=DSMWare and Grade>10".

Type-based system is an extension to content-based system. It ensures type safety at compile time. It does not filter according to a string topic but type as in the notion of programming languages.

Selecting different variant of publish/subscribe system depends on expressiveness and performance criteria. Content-based approach has higher expressiveness than topic-based approach and topic-based approach has higher performance than content-based approach. Application developers need to consider tradeoffs in scalability, expressiveness and quality of services depending on used architecture and implementation/protocols. In our implementation domain, which is GIS, we use topic-based approach. WFS are publishers and WMS are subscribers. We use NaradaBrokering as publish/subscribe system. The proposed system is explained in the following chapters.

There are some examples of publish subscribe systems. Among these are SIFT (Stanford – Stanford Information Filtering Tool) [21], Microsoft 's Herald [22], SCRIBE [23] and IBM's Gryphon [24]. Scribe is example of topic-based publish/subscribe systems. Gryphon is a content-based publish/subscribe system. Many applications such as stock quotes, network management systems, RSS feed monitoring, already benefit from this paradigm.

3.2   Systems Components to Create a Map

WMS and WFS are the fundamental services to create a still map images from geodata sets according to the open standards. Our earlier works on developing Web Service based WMS and WFS are presented in [25-27]. Geodata sets are served by WFS and rendered as map images at WMS. They have standard service interfaces and communication protocols. Here, we extend their communications with streaming data transfer capabilities, which is illustrated in Fig. 2.

Fig. 2 illustrates how the proposed architecture works to produce a map image from geographic features. The final outcome of the system (Fig. 2) is a map image. WMS clients request that image through WMS's getMap service interface. Once WMS get this request, they create corresponding getFeature requests to WFS to get the feature data in GML format. After getting the data, WMS check and extract all the geometry elements such as points, line-strings, polygons etc., and converts them into appropriate image formats. WMS developers use any kind of graphics tools to create map images from those geometric features of data.

Developing WMS and WFS as web services enables them to be discoverable and used in third party distributed systems. However, efficient data transportation capability still remains as a challenge, because of the fact that web services are based on XML based SOAP over HTTP protocol. In order to overcome such problem in the proposed distributed system framework requiring large scale XML-encoded geographic feature sets, we have investigated the possibilities of using topic-based publish-subscribe paradigms (which is mostly used in P2P systems) for exchanging data payload between web services. WFS using NaradaBrokering are called streaming WFS. When the NaradaBrokering is used, WFS are still queried with standard SOAP messages (requests) (arrow 1 in Fig. 2). However, the responses are published (i.e. streamed) to an NB topic as they become available (arrow 3 in Fig. 2). Arrow 2 shows the subscription stage. WFS send the "IP", "topic" and "port number" to which the results will be streamed. The clients (WMS) subscribed to the same topic can receive the streams. WFS use MySQL in background and stream the results row by row (consider the relational tables) instead of waiting the calculation of the result set to be completed. The streaming enables I/O and CPU jobs to be overlapped and ends up with performance gains. The performance results are presented in our earlier work [26].

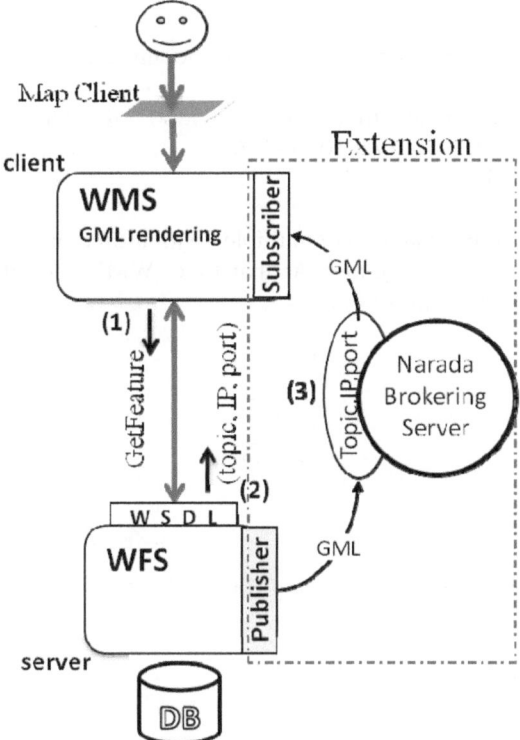

Fig. 2. Systems Components

3.3   WMS as an Animation Server

We think that WMS are presently the most suited candidate for building a map server providing with the animated maps in accordance with the domain standards and web service standards. Although the standardization for animations is not mature yet [19], WMS's specifications offer an appropriate standard for the sharing of data containing a temporal dimension. WMS provide same service interface called getMap for both animation and still map images, but with a little different parameter set for each.

WMS create static maps in pictorial formats from geo-data provided by WFS and store them in memory as an image array. Map animation is basically showing those images dated in a vertical frame one by one as a movie. If a WMS is capable of providing animation services for a layer (spatial data sets), then, some attributes are added under the layer definition in its capabilities file. Capabilities file is metadata whose standards are defined as XML schema by OGC. If a layer (i.e. a map) is provided in time intervals this layer might then have a tag named "<Dimension>". According to the standards, multi-dimensional data objects are described with <Dimension> tag. Time is one of the dimension name defined in WMS capabilities file. If a time dimension is defined for a specific layer, that means geodata used to create that layer are spatiotemporal data and available in some time intervals and collected with a pre-defined periodicity. Examples of periodicities are once every second, once every year etc.

<Dimension> tag has some properties (name, units etc.) and a format as given below.

    <Dimension name="time" units="ISO8601" default="2000-08-22">
        2001-04-10/2010-10-11/P1Y
    </Dimension>

One layer might be available in multiple disjoint time intervals and those intervals might have different periodicities. At that time, WMS add additional lines to time dimension element as displayed below.

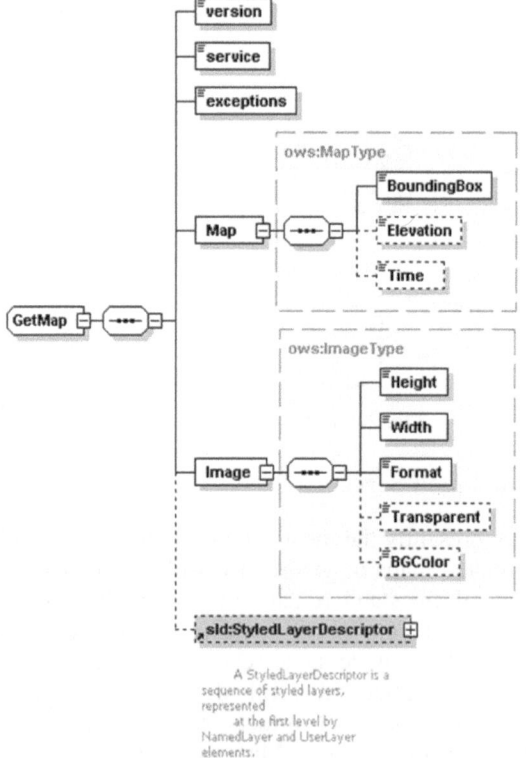

Fig. 3. GetMap Request schema to be able to create streaming map movies

```
<Dimension name="time" units="ISO8601" default="2000-08-22">
       2001-04-10/2010-10-11/P1D
       1990-01-01/1998-08-22/P1Y
</Dimension>
```

Since we developed WMS as web services, we created XML schema for getMap requests in accordance with the standard definitions (see Fig. 2). The requests created according to this schema are inserted into SOAP messages to invoke animation web services. Fig. 4 is an instance of the standard getMap request created over the schema.

Table 1 shows a sample parameter set and their values for getMap request created with the standard schema given in Fig. 2.

Table 1. Parameter Values for Sample Animation Request

Parameters and sample values
VERSION = 1.3.1
REQUEST = GetMap
LAYERS = earth
CRS = CRS:64
BBOX = -90,-45,90,45
WIDTH = 800
HEIGHT = 400
TIME=2008-6-01/2010-6-01/P1Y
FORMAT= video / mpeg

The remainder of this section explains what a WMS does when it gets an animation (or map movie) request as given in Fig. 4.

In the values of time parameter given as examples above, first date defines the starting date and second date defines the end date of the available data collection.

The last value (ex. "P1D") defines the periodicity of data collection. According to last value in parameter time, WMS cut the time into multiple values and for each time interval it makes a request to WFS to get feature data in GML. See the successive requests (req_i) in Fig. 5, gml_i represent corresponding responses from WFS for the requests req_i. img_i are images created from corresponding gml_i.

User interaction with the system is achieved through browser based WMS clients. MapClient predefines the animation format in a particular style, define in what date ranges and in what time slices animation is needed. The appropriate query is created (e.g. Fig) and sent to WMS thorough its getMap web service interface. Once WMS get this query, it creates successive queries (req_1, req_2, .., req_n) based on the time parameter in getMap request. Each of those queries is responded with gml (gml_1, gml_2,..., gml_n). For each gml a still map is created. Every still map corresponding to a time slice is stored in memory as a part of an image-array and displayed in a vertical sub-window simulating a camera film.

```
<?xml version="1.0" encoding="UTF-8"?>
<GetMap xmlns="http://www.opengis.net/ows">
        <version>1.1.1</version>
        <service>wms</service>
        <exceptions>application_vnd_ogc_se_xml</exceptions>
        <Map>
                <BoundingBox ts="">-124.85,32.26,-113.56,42.75</BoundingBox>
                <Elevation>5.0</Elevation>
                <Time>01-01-1987/12-31-1992/P1Y</Time>
        </Map>
        <Image>
                <Height>400</Height>
                <Width>400</Width>
                <Format>video/mpeg</Format>
                <Transparent>true</Transparent>
                <BGColor>0xFFFFFF</BGColor>
        </Image>
        <ns1:StyledLayerDescriptor version="1.0.20" xmlns:ns1="http://www.opet/sld">
                <ns1:NamedLayer>
                        <ns1:Name>Nasa:Satellite</ns1:Name>
                        <ns1:Description>
                                <ns1:Title>Nasa:Satellite</ns1:Title>
                                <ns1:Abstract>Nasa:Satellite</ns1:Abstract>
                        </ns1:Description>
                </ns1:NamedLayer>
                ....
                <ns1:NamedLayer>
                        <ns1:Name>World:Seismic</ns1:Name>
                        <ns1:Description>
                                <ns1:Title>World:Seismic</ns1:Title>
                                <ns1:Abstract>World:Seismic</ns1:Abstract>
                        </ns1:Description>
                </ns1:NamedLayer>
        </ns1:StyledLayerDescriptor>
</GetMap>
```

Fig. 4. Sample GetMap request for WMS to create streaming map movies

3.4 Publishing Animated Map Images

After having the temporally related still map images created, we need to publish them as movie streams. WMS are access points for the distributed systems or any client to use map animation services. The first step in an animation is the creation of series of temporally related successive map image, which is explained in the previous chapter. The second step is playing these still map images as an animation, which is explained in this chapter.

WMS need successive map images for a period of time, in order to be able to create an animation. The time period and the periodicity of the movie frames are defined by parameter called "time" in GetMap request (Fig. 4). The number of frames to be played depends on the time period and periodicity of the time intervals. To assemble the individual frames into an animation, there are different approaches according to application requirements. In the Internet world, widely used and well-known approach is using animated GIF89a files. The browser will be enough to open the animation, but the users do not have any control on the image such as stopping, pausing or play-backing the animation. The animated gif will play only in one direction. There is no way to make it play backward. PROC GMAP [28] is an example of gif animations. The Java Media Framework (JMF) [29] and Quicktime [30] provide with other approaches to assemble individual frames into an animation. JMF uses RTP sessions and need Java virtual machine installed on the client machines. That is used on Java based applications. OGC standards do not specifically and clearly define how to transfer and display animation. It only specifies the interfaces in terms of standard queries and output formats. In the proposed framework, we preferred to use IP multicast approach with JMF technologies.

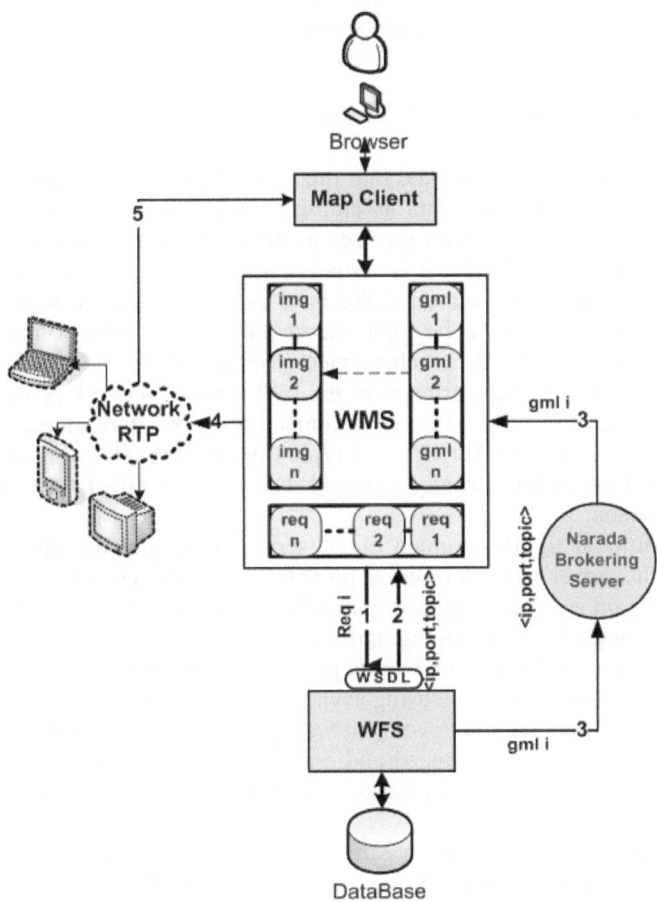

Fig. 5. Detailed Streaming Map Movies architecture

An animation is produced as video streams. Map images are converted into sequence of video streams and published to an RTP session [31]. RTP sessions are formalized as <IPAddress, PortNumber> pairs. There are various video stream formats. The framework uses H.263 and H.261, which are well-known and widely used formats. Those animated video streams can be played and displayed on video-conferencing systems and collaborative environments such as AccessGrid (http://www.accessgrid.org/), but they are supposed to support H.263 and H.261 formats. The produced streams are published to multicast or unicast RTP sessions. Video streams can be delivered to a variety of platforms such as RealPlayer, Polycom and Access Grid [32]. The published video streams can also be displayed by any client building his own custom system and services to display the map video streams. The easiest way to display the map movie stream is connecting to RTP sessions by using a JMF Client.

The quality of streams depends on some configurable parameters such as video format, frame rate and update rate. These parameters are set at the creation time, depending on the data and the application specific requirements.

4 Conclusion and Future Work

The work presented here has looked into the possibilities of extending the web based map services with time-series data as animated maps, and introduced a framework enabling integration of animation services to the distributed systems. At the core of the framework there is a WMS. It is actually an access point for the distributed systems to use map animation services. We have also extended the standard GIS web service communications with the topic based publish-subscribe communication approach. In the framework, GIS web services uses standard interfaces for the handshake, the actual data is transferred over the P2P overlay network provided by Naradabrokering. After the handshake, communicating peers start transferring the data through the agreed upon broker (IP) and topic (any string). This approach enables us to create map images for partially returned data and get rid of the SOAP message creation overheads.

In the traditional tightly coupled client–server paradigm, the client cannot post messages to the server while the server process is not running, nor can the server receive messages unless the client is running. In case of transferring and rendering big spatial data through RPC messaging and web services, blocking IO and session time outs degrades the usability of web based applications. The proposed system gets rid of such communication latencies by using asynchronous communication.

The proposed framework is developed with open standards and Java technologies. Therefore, it can easily be enhanced and extended for application specific purposes, deployed on any platforms and integrated to the third party distributed system applications.

In the proposed system, movie streams (for map animations) are created on demand from the archived data sets. In the future we plan to enhance the system with the capability of archiving map animations. In that case, each archived map animation needs to be annotated with some parameters enabling them to be searched.

These parameters might be "temporal data layers from which movie streams are created", "frame rates", "starting-ending time/dates of the animation" and "periodicity of the data frames".

References

1. Tanenbaum, A.S., Steen, M.V.: Distributed Systems: Principles and Paradigms, 2nd edn. Prentice Hall (2006)
2. OGC (1994), http://www.opengeospatial.org/ (accessed February 14, 2008)
3. Cox, S., Daisey, P., Lake, R., Portele, C., Whiteside, A.: OpenGIS® Geography Markup Language (GML) Encoding Specification, vol. 02-023r4. Open Geospatial Consortium (OGC) (2003)
4. Pallickara, S., Fox, G.: NaradaBrokering: A Distributed Middleware Framework and Architecture for Enabling Durable Peer-to-Peer Grids. In: Endler, M., Schmidt, D. (eds.) Middleware 2003. LNCS, vol. 2672, pp. 41–61. Springer, Heidelberg (2003)
5. Peng, Z.-R., Tsou, M.-H.: Internet GIS: Distributed Geographic Information Services for the Internet and Wireless Networks. John Wiley & Sons, New Jersey (2003)
6. Beaujardiere, J.D.L.: OGC Web Map Service Interface, vol. 06-042. Open GIS Consortium Inc. (OGC) (2006)
7. Vretanos, P.A.: Web Feature Service 2.0 Interface Standard, vol. 09-025r1 and ISO/DIS 19142. Open Geospatial Consortium Inc. (2010)
8. Blok, C., Kobben, B., Cheng, T., Kuterema, A.A.: Visualization of relationships between spatial patterns in time by cartographic animation. Cartography and Geographic Information Science 26(2), 139–151 (1999)
9. Peterson, M.P.: Interactive and animated cartography. Prentice Hall, Englewood Cliffs (1995)
10. Harrower, M.: Tips for designing effective animated maps. Cartographic Perspectives 44, 63–65 (2003)
11. Harrower, M.: Visualizing change: Using cartographic animation to explore remotely-sensed data. Cartographic Perspectives 39, 30–42 (2002)
12. Frihida, A., Marceau, D.J., Theriault, M.: Extracting and Visualizing Individual Space-Time Paths: An Integration of GIS and KDD in Transport Demand Modeling. Cartography and Geographic Information Science 31(1), 30–42 (2004)
13. MathWorks (2012), http://www.mathworks.com/products/mapping/ (accessed March 03, 2012)
14. GeoServer (2011), http://docs.geoserver.org/stable/en/user/tutorials/animreflector.html# (accessed February 2, 2012)
15. Köbben, B.: SVG and Geo Web Services for visualization of time series data of flood risk. In: SVG Open 2008, Nürnberg, Germany, August 26-28 (2008)
16. Becker, T.: Visualizing Time Series Data Using Web Map Service Time Dimension and SVG Interactive Animation. University of Twente (2009)
17. GoogleEarth: GoogleEarth (2011) (accessed August 21, 2011)
18. Atkinson, M., DeRoure, D., Dunlop, A., Fox, G., Henderson, P., Hey, T., Paton, N., Newhouse, S., Parastatidis, S., Trefethen, A., Watson, P., Webber, J.: Web Service Grids: An Evolutionary Approach Concurrency & Computation. Practice & Experience 17(2-4), 377–389 (2005)
19. LaMar, E.: Proposed Animation Service Extension, vol. 06-045r1, p. 23. Open Geospatial Consortium Inc. (OGC) (2005)

20. Eugster, P., Felber, P., Guerraoui, R., Kermarrec, A.-M.: The Many Faces of Pub-lish/Subscribe. ACM Computing Surveys 35(2), 114–131 (2003)
21. Molina, H.G., Yan, T.W.: Stanford Information Filtering Tool (SIFT). In: USENIX Technical Conference, New Orleans, Louisiana, USA, February 16, pp. 177–186 (1995)
22. Cabrera, L.F., Jones, M.B., Theimer, M.: Herald: Achieving a Global Event Noti-fication Service. In: Eighth Workshop on Hot Topics in Operating Systems (HotOS-VIII), Elmau/Oberbayern, Germany. IEEE Computer Society (May 2001)
23. Castro, M., Druschel, P., Kermarrec, A.-M., Rowstron, A.: SCRIBE: A large-scale and decentralized publish-subscribe infrastructure. IEEE Journal on Selected Areas in Communications 20(8), 21–29 (2002)
24. Strom, R., Banavar, G., Chandra, T., Kaplan, M., Miller, K., Mukherjee, B., Sturman, D., Ward, M.: Gryphon: An Information Flow Based Approach to Message Brokering. In: The Ninth International Symposium on Software Reliability Engineering Paderborn, Germany. IEEE (1998)
25. Pierce, M.E., Fox, G.C., Aktas, M.S., Aydin, G., Qi, Z., Sayar, A.: The QuakeSim Project: Web Services for Managing Geophysical Data and Applications. Pure and Applied Geophysics (PAGEOPH) 165(3-4), 635–651 (2008), doi:10.1007/s00024-008-0319-7
26. Aydin, G., Sayar, A., Gadgil, H., Aktas, M.S., Fox, G.C., Ko, S., Bulut, H., Pierce, M.E.: Building and Applying Geographical Information Systems Grids. Concurrency and Computation: Practice and Experience 20(14), 1653–1695 (2008)
27. Aktas, M., Aydin, G., Donnellan, A., Fox, G., Granat, R., Grant, L., Lyzenga, G., McLeod, D., Pallickara, S., Parker, J., Pierce, M., Rundle, J., Sayar, A., Tullis, T.: iSERVO: Implementing the International Solid Earth Research Virtual Observatory by Integrating Computational Grid and Geographical Information Web Services. Pure and Applied Geophysics (PAGEOPH) 163(11-12), 2281–2296 (2006), doi:10.1007/s00024-006-0137-8
28. Eberhart, M.: Make the Map You Want with PROC GMAP and the Annotate Facility. In: Philadelphia Department of Public Health (2008)
29. JMF (2011), http://www.oracle.com/technetwork/java/javase/tech/index-jsp-140239.html (accessed January 09, 2011) (2008)
30. Quicktime (2011), http://developer.apple.com/quicktime/
31. Schulzrinne, H., Casner, S.L., Jacopson, F.V.: RTP: A Transport Protocol for Real-Time Applications (2003)
32. Wu, W., Fox, G.C., Bulut, H., Uyar, A., Altay, H.: Design and Implementation of a Collaboration Web-Service System Neural. Parallel & Scientific Computations 12(3) (2004)

Searching Frequent Itemsets by Clustering Data: Towards a Parallel Approach Using Mapreduce

Maria Malek and Hubert Kadima

EISTI-LARIS Laboratory,
Ave du Parc, 95011 Cergy-Pontoise, France
{maria.malek,hubert.kadima}@eisti.fr

Abstract. We propose a new algorithm for searching frequent itemsets in large data bases. The idea is to start searching from a set of representative examples instead of testing the 1-itemset,the k-itemset and so on. A clustering algorithm is firstly applied in order to cluster the transactions into k clusters. The set of the k representative examples will be used as the starting point for searching frequent itemsets. Each cluster is represented by the most representative example. We show some preliminary results and we then propose a parallel version of this algorithm based on the MapReduce Framework.

Keywords: Data Mining, large transaction bases, frequent itemsets, clustering algorithm, MapReduce.

1 Introduction

Association rules mining is a very known data mining techniques that aims to find relationships between items in large data bases that contain transactions. The problem of frequent itemsets has been introduced by Agrawal in 1993 and the well known Apriori algorithm has been proposed in 1994 [1]. This algorithm is based on the downward closure propriety: if an itemset is not frequent, any superset of it will not be frequent. The Aprioiri algorithm performs a breadth-first search in the search space by generating candidates of length K+1 from frequent k-itemsets. This algorithm is based on storing database in the memory. Unfortunately, when the dataset size is huge, both the memory use and the computational cost still be expensive.

In 2000, The FP-growth algorithm has been proposed [3], the idea is to use a frequent pattern tree data structure to achieve a condensed representation of the data transactions. This algorithm is based on a divide-and-conquer algorithm approach in order to decompose the problem into a set of smaller problems. The performance study shows that the FP-growth method is efficient and scalable for mining both long and short frequent patterns and faster than the Apriori algorithm.

In [5], authors discuss a depth first implementation of Apriori, the algorithm builds a tree in memory that contains all frequent itemsets: all sets that contained in at least minsup transactions from the original database, by adding one item

A. Haller et al. (Eds.): WISE 2011 and 2012 Combined Workshops, LNCS 7652, pp. 251–258, 2013.

at a time. Every path corresponds to a unique frequent itemset. Authors study the theoretical complexity of this algorithm and show encouraging experiment results.

In [8] Authors propose an novel algorithm for mining complete frequent itemsets. In this algorithm transactions "ids" are mapped and compressed to continuous transaction intervals in a difference space and the counting of itemsets is performed by intersecting these interval lists in a depth-order along the lexicographic tree. Authors shows interesting experiment results in comparison with FP-Growth algorithm.

In [6] a massively parallel FP-Growth algorithm is presented, the proposed algorithm allows to eliminate virtually communication among computers. the algorithm is expressed with the MapReduce framework. Authors demonstrate that through empirical study on a large dataset of web pages and tags the algorithm can achieve virtually linear speedup.

In this paper, we propose a new algorithm for searching frequent itemsets. The idea is to start searching from a set of representative examples instead of testing the 1-itemset,the k-itemset and so on. A clustering algorithm is firstly applied in order to cluster the transactions into k clusters. Each cluster is represented by the most representative example. We show some preliminary results and we then propose a parallel version of this algorithm based on the MapReduce Framework.

2   A New Algorithm for Frequent Itemsets Searching: Sequential Version

We propose a novel algorithm for searching frequent itemsets. The idea is to start searching from a set of representative examples instead of testing the 1-itemset,the k-itemset and so on. A clustering algorithm is firstly applied in order to cluster the transactions into k clusters. Each cluster is represented by the most representative example. We currently use the k-medoids algorithm in order to cluster the transactions [4].

The set of the k representative examples will be used as the starting point for searching frequent itemsets. We present in the next two versions of our algorithm: the sequential version and the parallel one implemented using the MapReduce framework. The idea of the algorithm is to test firstly a representative example e of length l, if this example is found frequent then all examples included in e whose length values are lower than l will be frequent; otherwise all examples that include e will be no-frequent.

2.1   Algorithm Parameters, Structures and Definition

The proposed algorithm takes as an input a set of transactions called D, two lists of itemsets are computed:

A list named *accepted* that contains the retained frequent itemsets.
A list named *excluded* that contains the retained no-frequent itemsets.

The algorithm parameters are the same ones as the k-medoids and the Apriori algorithms:

K is the initial number of clusters.
minSupp is the threshold used for computing frequent itemsets.

We need an intermediate list that we call *candidates* containing the itemsets to test. These itemsets will be sorted by their decreasing lengths.

We have to differentiate between a *local frequent itemset* and a *global one*. Here are the two definitions:

Global Frequent Itemsets. Let D be a set of transactions. Let L_i be an i-itemset of length i, L_i is **a global frequent itemset** iff it is frequent in D.

Local Frequent Itemsets. Let D is a set of transaction segmented on k disjoints clusters. Let L_i be an i-itemset of length i, L_i is **a local frequent itemset** iff it is frequent in the cluster to whose it belongs.

The algorithm tests firstly if a given example e is a local frequent itemset, if yes the list called *accepted* is updated, otherwise the algorithm tests if e is a global frequent itemset, if yes the list *accepted* is updated, otherwise the list *excluded* is updated.

2.2   Algorithm Description: Sequential Version

- Apply the k-medoids on D (the transactions base) and stock the k representative examples as well as the K clusters.
- Let $C_1, C_2, .., C_k$ be the k representative examples sorted by their decreasing lengths, If two examples have the same length, then they will be ordered by their occurrence values (their supports) in D. The list *candidates* is initialized to $C_1, C_2, .., C_k$.
- While the list *candidates* $\neq \Phi$ do
  - Let C_i be the first element of *candidates*:
  - If $C_i \not\subseteq accepted$ et $C_i \not\subseteq excluded$ then
    1. If C_i is a **local frequent itemset** then *update-accepted(C_i)*, exit.
    2. If C_i is a **global frequent itemset** then *update-accepeted(C_i)*, exit.
    3. else, *update-excluded (C_i)*, and add frequent itemsets included in C_i to the *candidates* list.

Update-Accepted(C_i) : Add to the list *accepted*, the itemset C_i and all the itemsets included in it.
Update-Excluded(C_i) : Add to the list *excluded*, the itemset C_i and all the itemsets that include C_i.

We note that this algorithm as described above, does not guarantee the obtention of *all the frequent itemsets*. The algorithm needs to be completed. We will discuss this idea later.

3   Preliminary Results

The data set is composed of a set of navigation logs extracted from the Microsoft site. The site is composed of 17 pages with some links between each others that we present by the set of the characters : $\{A, B, C, .., P, Q\}$. The initial data logs file contained navigations paths of 388445 users. By keeping only users who have sufficient paths length the users number is reduced to 36014. Next, we call this set of transactions D.

Table 1 shows the application of k-medoids with different value of K. We notice that from k=5, we obtain a small cluster (whose cardinality is 5% of the cardinality of D). We aim now to compute the support of the representative examples in order to construct the two lists *accepted* and *excluded*.

Table 2 shows the frequent itemset of length higher than 4 that are obtained by applying only the Apriori algorithm. Notice that 8 frequent itemsets are found, among them 4 have been found as the representative examples by the k-medoids when k=5.

As we mentioned above this part of algorithm is not complete. It does not guarantee the obtention of all frequent itemsets but only those whose are related to the results of the k-medoids by the relations of inclusion.

The figure on the left in table3 shows the cardinality of each cluster when k=4 and the frequency of the representative example in the cluster which varies between 7% and 39%. The figure on the right in table 3 shows the global and local support of the k representative examples, notice that the local support varies from 51% to 74%.

This explains the fact that representative example are frequent and shows that the local support computation can be enough to determine if this example is frequent or not in this case.

Table 1. This table shows the results of applying k-medoid on the transactions base D, k is the given number of cluster, we report for each value of k the representative example Ei of each cluster Ci as well as the cardinality of the cluster.

K	C_1		C_2		C_3		C_4		C_5	
	E_1	C_1	E_2	C_2	E_3	C_3	E_4	C_4	E_5	C_5
k=2	ABFG	55%	ABDLK	45%						
k=3	ABFFG	46%	ABDKL	23%	ABCF	31%				
k=4	AFG	35%	ABDKL	34%	ABCFJ	17%	ABDFG	14%		
k=5	AFGJ	11%	ABDKL	42%	ABCFJ	19%	ABDFG	23%	BDFGN	5%

Figure 1 shows the number of frequent itemsets found locally in function of k, we note that when k=4 most of the found frequent itemsets are locally found. The above experiment helps us to determine the better value of k that allows the obtention of the maximal number of locally frequent itemsets. We show next in the parallel version of our algorithm, that members of the same cluster are stocked at the same worker. this is why we aim to maximise the number of local computations in clusters in order to minimise also communication costs.

Table 2. This tables shows all frequent itemsets whose supports are higher than 1576 and whose length is 4 or 5. Four of the eight itemsets have been found by the novel algorithm.

Itemset	support	found by the novel algorithm
AFGJ	2406	yes
ABCJL	2406	no
ABCFJ	1576	yes
ABCKL	1922	no
ABDFG	2628	yes
ABDGL	1813	no
ABDKJ	1735	yes
ABFGJ	1834	no

Table 3. This figure shows the number of transactions in each cluster when k=4, and for each k the frequency rate of the found representative example. This figures shows the local and global supports for representative examples when k=4.

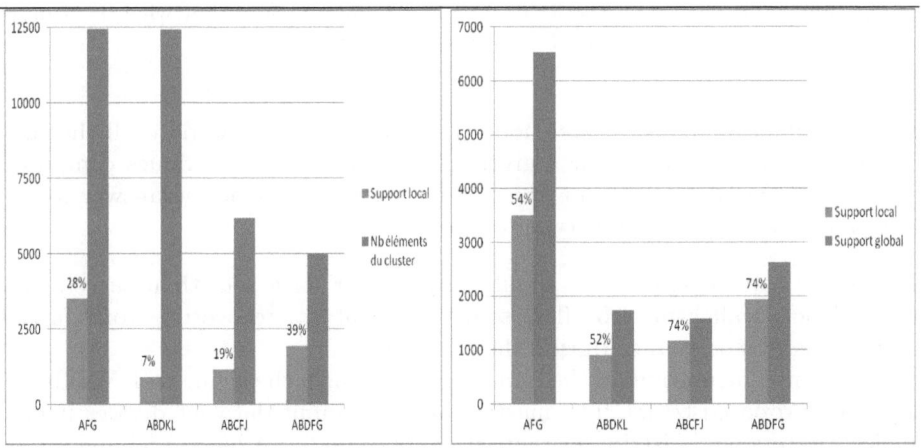

Finally, figure 2 shows the execution time evolution(micro seconds) in function of K.

4 Towards a Parallel Version Using MapReduce

MapReduce is a framework for parallel and distributed computing that has been introduced by Google [2] in order to handle huge data sets using a large number of computers (nodes), collectively referred to as a cluster.

MapReduce is based on two stages:

1. The Map step: the master node takes the input, divides it into smaller sub-problems, and distributes them to worker nodes. The worker node processes the smaller problem, and passes the answer back to its master node in the form of list of key-values couples.

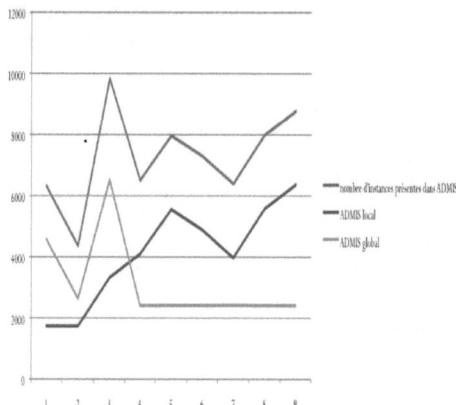

Fig. 1. The red curve shows the number of frequent itemsets found locally in function of the k value, the green one shows the number of frequent itemsets found globally, and the orange one shows all the found frequent itemsets. Note that when k=4 most of the found frequent itemsets are locally found.

2. The Reduce step: the master node then collects the answers to all the sub-problems and combines for a given key the intermediates values computed by the different mappers in order to form the output or the answer to the problem it was originally trying to solve.

MapReduce can be applied to significantly larger data sets than servers can handle. The parallelism also offers some possibility of recovering from partial failure of servers or storage during the operation.

Several implementations of the k-means algorithm in the framework MarReduce have been proposed [7] [9]. The master takes the input data set divides it into smaller sub-sets, and distributes them to mapper nodes. A list containing k representative exemples which are randomly chosen is sent to all mappers. The master launches then a MapReduce job for each iteration until the algorithm convergence (until stabilization of representative examples). A MapReduce job consists of the following two steps:

1. The Mapper function computes for each example the closer representative example, the example is then assigned to the associated cluster.
2. The reducer function: for each representative example, the reducer collects the partial sum of the computed distances from all mappers, and then re-compute the new representative examples list.

We propose a parallel implementation of our new algorithm as follows:

- Apply the above parallel version of the k-means algorithm (or the k-medoid one) in order to obtain the the data set segmented into k clusters.
- Initialize the list of candidates to the k representative exemples.

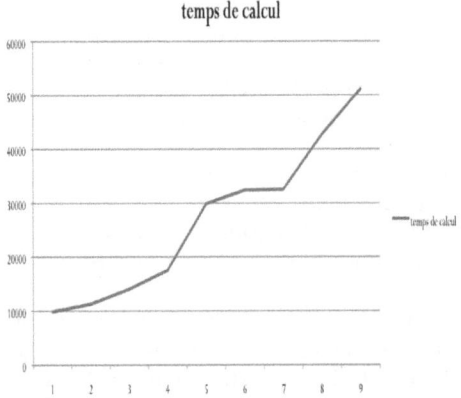

Fig. 2. This figure shows the execution time evolution(micro seconds) in function of K

- Re-distribute the k obtained clusters on k mappers,
- **repeat**
  1. Send the list of candidates to the k mappers.
  2. Each mapper computes the local support of each candidate.
  3. The reducer collects all local supports for each candidate and compute for some candidates the global supports if necessary.
  4. The master updates accepted, excluded and candidates lists.
- **until** the list of candidates is empty.

We are now implementing this version and comparing performances with other parallel implementation like the massively parallel FP-Growth algorithm presented in [6].

5   Conclusion

In this paper, we propose a new algorithm for searching frequent itemsets in large data bases. The idea is to start searching from a set of representative examples instead of testing the 1-itemset, and so on. The k-medoid clustering algorithm is firstly applied in order to cluster the transactions into k clusters. Each cluster is represented by the most representative example. We have shown according to experimental results that beginning the search of frequent itemsets from these representative examples leads to find a significant number of them in a short time. We have then presented a parallel version of this algorithm based on the MapReduce Framework in order to optimise performance. We are now implementing this version and comparing performances with other parallel implementations.

References

1. Agrawal, R., Srikant, R.: Fast algorithms for mining association rules in large databases. In: VLDB, pp. 487–499 (1994)
2. Dean, J., Ghemawat, S.: Mapreduce: Simplified data processing on large clusters. In: OSDI, pp. 137–150 (2004)
3. Han, J., Pei, J., Yin, Y.: Mining frequent patterns without candidate generation. SIGMOD Rec. 29(2), 1–12 (2000)
4. Kaufman, L., Rousseeuw, P.J.: Clustering by means of medoids. In: Dodge, Y. (ed.) Statistical Data Analysis Based on the Norm and Related Methods, pp. 405–416. North-Holland (1987)
5. Kosters, W.A., Pijls, W.: Apriori, a depth first implementation. In: Proc. of the Workshop on Frequent Itemset Mining Implementations (2003)
6. Li, H., Wang, Y., Zhang, D., Zhang, M., Chang, E.Y.: Pfp: parallel fp-growth for query recommendation. In: Proceedings of the 2008 ACM Conference on Recommender Systems, RecSys 2008, pp. 107–114. ACM, New York (2008)
7. Owen, S., Anil, R., Dunning, T., Friedman, E.: Mahout in Action, 1st edn. Manning Publications (January 2011)
8. Song, M., Rajasekaran, S.: A transaction mapping algorithm for frequent itemsets mining. IEEE Transactions on Knowledge and Data Engineering 18, 472–481 (2006)
9. Zhao, W., Ma, H., He, Q.: Parallel k-means clustering based on mapreduce. In: Jaatun, M.G., Zhao, G., Rong, C. (eds.) Cloud Computing. LNCS, vol. 5931, pp. 674–679. Springer, Heidelberg (2009)

Introduction to the Proceedings of the 5th International Workshop on Personalization in Cloud and Service Computing (PCS) 2011

Jian Yu[1], Hong-Linh Truong[2], and Yanbo Han[3]

[1] Faculty of ICT, Swinburne University of Technology, Melbourne, Australia
[2] Distributed Systems Group, Vienna University of Technology, Vienna, Austria
[3] Research Center for Cloud Computing, North China University of Technology, Beijing, China
jianyu@swin.edu.au, truong@infosys.tuwien.ac.at, hanyanbo@ncut.edu.cn

Introduction

Following the success of the International Workshop on Personalization in Grid and Service Computing (PGSC. Inaugural: Urumchi, Xinjiang, China; 2nd: Shenzhen, China; 3rd: Lanzhou, Gansu, China; 4th: Nanjing, Jiangsu, China), the 5th International Workshop on Personalization in Cloud and Service Computing (PCS) in 2011 was held in conjunction with the WISE 2011 conference in Sydney, Australia. This year, we engineered a brand new name for the workshop and expanded the initial research areas in service and grid computing to embrace the rapidly growing cloud computing research community.

Personalization and context-awareness researches in the area of service, grid and cloud computing are emerging as an important topic with the transformation of Internet and Web from traditional linking and sharing of computers and documents (i.e., "Web of Data") to current connecting of people and things (i.e., "Web of People", "Web of Things", and "Web of People and Thing"). Adopting Service-Oriented Computing as a basic paradigm, the Grid and Cloud Computing approach focuses on the virtualization and transparent provisioning of software and people resources. Under such paradigm, information and service supply faces a number of new challenges. For example, how to smartly deal with large amounts of services based on user's personalized needs, how to handle personalized service composition in a dynamic environment, and how to customize business processes according to stakeholder's preferences. Personalization for processes and services in such a dynamic environment is one of the most exciting trends in Grid/Cloud Computing today that holds the potential to enrich user experiences and make our daily life more productive, convenient, and enjoyable.

We hope that this workshop will promote the discussion around the personalization and context-awareness issues in the service, grid, and cloud computing fields, ranging from theoretical foundations, supporting infrastructures, engineering approaches, to applications and case studies, and lead to new solutions to smarter services for the future ubiquitous world.

The three full papers chosen for PCS 2011 represent a range of relevant topics. The papers were selected after a rigorous peer-review by the workshop Program Committee members and external reviewers.

The paper "Towards A Taxonomy Framework of Evolution for SOA Solution: From a Practical Point of View" proposes a taxonomy framework for evolutionary SOA solution changes. A case study on a SOA-based configurable logistics management system by using this taxonomy has been conducted with a conclusion that the configurable urban logistics delivery management system is positioned as a SOA solution that supports for design-time, process schema layer, and automatic evolution.

The paper "An Auxiliary Storage Subsystem of Storage Space Hidden for User Data to Distributed Computing Systems" proposes a new model of remote storage hidden to a cloud-based system following the iSCSI standard and evaluates the improved user data transmission rate.

The paper "A Hierarchical Representation for Indexing Data Condensed Semantically from Physically Massive Data out of Sensor Networks on the Rove" presents a new data abbreviation approach to reducing the amount of data to be transmitted in sensor data aggregation and integration.

We sincerely thank the PCS 2011 Program Committee members and external reviewers for their time in providing valuable reviews for the workshop.

Jian Yu,
Hong-Linh Truong,
Yanbo Han

PCS 2011 Workshop Chairs

Towards a Taxonomy Framework of Evolution for SOA Solution: From a Practical Point of View

Zaiwen Feng[1,2,3], Patrick C.K. Hung[4], Keqing He[1,2], Yutao Ma[1,2],
Matthias Farwick[5], Bing Li[1,2] and Rong Peng[1,2]

[1] State Key Lab of Software Engineering, Wuhan University, China
[2] School of Computer, Wuhan University, China
[3] Hubei Provincial Key Laboratory of Intelligent Robot,
Wuhan Institute of Techonology, China
[4] Faculty of Business and IT, University of Ontario Institute of Technology, Canada
[5] Institute of Computer Science University of Innsbruck, Austria
{Zwfeng,hekeqing}@whu.edu.cn, patrick.hung@uoit.ca

Abstract. Presently, the research on evolution of SOA solution is becoming more and more important in industry. It is necessary to extract a taxonomy framework for evolution of SOA solution because by using it, evolution of SOA-based systems can be analyzed and compared comprehensively and objectively. In this paper, a taxonomy framework for evolution of SOA solution is proposed that is illustrated from four perspectives: (a) motivations for evolutionary changes (*why*), (b) locations where evolutionary changes happen (*where*), (c) times when evolutionary changes happen (*when*), and (d) support mechanisms in the process of evolutionary changes (*how*). Furthermore, the taxonomy framework is applied on analyzing a SOA-based configurable system for urban logistics delivery management, as an application of taxonomy framework.

Keywords: Taxonomy, Service Oriented Architecture (SOA), Change Management.

1 Introduction

Services are subject to constant change and variation. Change, after all, focuses businesses on realizing the benefits of Service-Oriented Architecture (SOA), because the primary reason for embarking on a SOA project is to improve business agility while reducing IT costs in the face of today's brittle, inflexible, and tightly coupled system [19].

SOA is an Information Technology (IT) architectural approach that supports the creation of business processes from functional units defined as services. It is noted that the business process itself is a service that is named *process service*, which is regarded as the process description encapsulated within a final Web service provided for the service consumer [12]. Service-Oriented Modeling and Architecture Modeling Environment (SOME-ME) is the first framework for the model-driven design of SOA solution that is proposed by IBM Corporation [17].

A. Haller et al. (Eds.): WISE 2011 and 2012 Combined Workshops, LNCS 7652, pp. 261–274, 2013.
© Springer-Verlag Berlin Heidelberg 2013

Around ten years ago, Chapin *et al.* [15] propose the classification for software evolution and maintenance from perspective of change motivation. Afterwards, Buckley *et al.* [4] complement taxonomy from other perspectives, i.e. how, when, what and where, of software change. Comparing to Chapin *et al.*'s taxonomy, Buckley' taxonomy focuses on characteristics of software change mechanisms and the factors that influence these mechanisms. As one of application for software change taxonomy, Simmonds *et al.* apply taxonomy on analyzing different types of versioning tools or refactoring tools [20]. But, Chapin and Buckley's taxonomy is towards the context of traditional software paradigm, like object-oriented software.

Existing software engineering techniques do not support development of SOA solution entirely due to lack of comprehensive understanding service activities in networked computing system. The significant differences between conventional software solution and SOA solution has been summarized by Yau *et al.* [8], like uncontrollable service components provides by the third-party, across organizations, and multiple candidate service resource. Therefore, it is necessary to give taxonomy for changes of SOA solution.

In this paper, we will propose a taxonomy framework for SOA solution changes, which is illustrated from four perspectives: (a) motivations of SOA solution changes (*why*), (b) locations of SOA solution changes (*where*), (c) times of SOA solution changes (*when*) (d) support mechanisms in the process of SOA solution changes (*how*)? Furthermore, as an application the taxonomy, we will apply the framework into a *configurable SOA-based urban logistics delivery management system* to evaluate system the degree of support for change.

The remaining of paper is described as follow. The framework of proposed taxonomy is introduced in Section 2. The concrete descriptions of taxonomy are presented in Section 3. A case study is described in Section 4. The related works is analyzed in Section 5. Finally, the conclusions and future work are drawn in Section 6.

2   Framework of Taxonomy

Four angles are considered for SOA solution changes, which includes *why, where, when and how* changes of service. The first angle for SOA solution change is the motivation of changes (*why*). We summarize four motivations that is perfective motivation, corrective motivation, performance motivation and cost motivation.

The second angle for SOA solution change is the location that changes happen (*where*). We can further consider location of changes from three properties that are artifact, transparency, and abstraction degree.

The third angle for SOA solution change is the time that changes happen (*when*). We can further consider time of changes from two properties that are happening time of change and frequency.

The last angle for SOA solution change is the support mechanism for changes (*how*). We can further consider the support mechanism for changes from two properties that are degree of automation and degree of change impact. Figure 1 is the taxonomy framework for SOA solution evolves.

3   Taxonomy of Evolution for SOA Solution

In this section, detailed descriptions for the taxonomy framework of SOA solution change are proposed from perspective of *why, where, when and how*.

3.1   Motivations of SOA Solution Evolution

Motivations of evolution answer the question of *why* SOA solution changes. We think that service change is mainly caused by four aspects. I.e. perfective motivation, corrective motivation, performance motivation and cost motivation. They are described below.

1)   Perfective Motivation
In SOA, service is designed to meet business demands. Change of business demands is regarded as the perfective motivation of service, which include the following scenarios:

- Enhance capability of SOA solution: In this type, a new service will be inserted into the original service compositions before or after an existing service.
- Reduce capability of SOA solution: In this type, an existing service will be deleted from the original service workflow before or after another service.

Another perfective motivation is the compliance to regulations change. Business rules often need to change to cope with change of regulations. Business rules are usually expressed either as constraints or in the form of "if condition then action". Different types of constraints of business rules include static, dynamic and hybrid [33]. Change of business rules will bring change of judgment condition in the form of condition then action.

2)   Corrective Motivation
Sometimes service is composed of finer-granularity services that are orchestrated via business processes. These finer-granularity services is called *service component* in this paper. Service components may be open to public and their versioning cannot be controlled by service designer. Coping with it, we have to redesign services, search for another substitutable service component, or fix the interface mismatches via adaptors [21]. Those motivations that refer to service component versioning are regarded as *corrective motivation* of service change. Corrective motivation consists of several scenarios:

- Service components interface versioning: It refers to the WSDL document of service components changes that is not backward compatible. These versioning mainly consists of *operation name change, inline types*, and *change input/output parameter types*.
- Service components policy versioning: A policy is a set of rules that apply to any number of services. Policy typically governs under what conditions consumers are entitled to access service functionality [14]. These scenarios of policy versioning consist of security/reliability message policy changes or value domain mapping to input/output message changes.

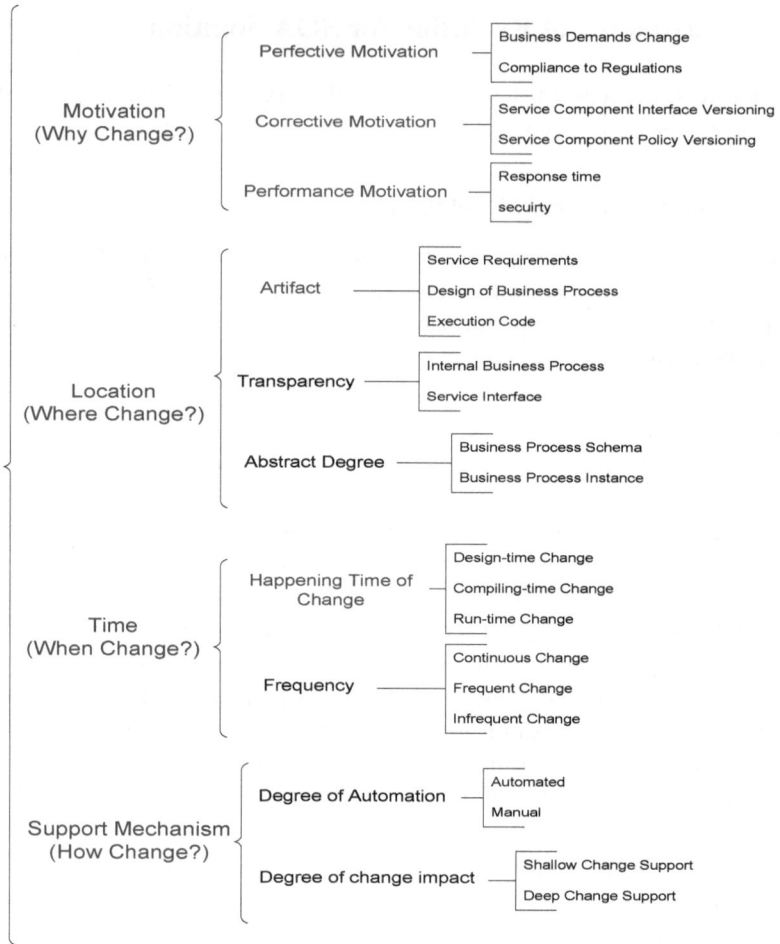

Fig. 1.   Taxonomy Framework for SOA solution changes

3) Performance Motivation

As we know service is often composed of loosely coupled and Internet based service components, thereby affecting adaptability and robustness. Moreover, service consumers always hope service better quality. To enhance performance, we can replace service components with better performance like response time or security such as the work at [6].

4) Cost Motivation

Cost of service is due to cost of service components. We prefer to search for lower-price service with the same functionality to replace old service components.

3.2   Location of SOA Solution Evolution

This dimension in our taxonomy addresses the *where* question. We will discuss the location where evolution of SOA solution happens.

1) Artifact
Similar to the evolution of software of conventional paradigm, many kinds of software artifacts are inclined to change in the process of SOA solution evolution, which includes *design of abstract business processes, service requirements documents, executable codes or client code* [5] and so forth. Changes of one of these artifacts maybe affect some others.

2) Transparency
A service, which represents a unit of logical computing, is published in a service registry. Logical computing can be represented as business process. Business Process Execution Language (BPEL) is often used to code this business process. Service can be not only regarded as a function box that has interface consumed by the consumer, but also can be viewed as a business process that orchestrates services into a composite one. Service, which is represented as either *service interface* or *internal business process*, depends on transparency we see it.

When we see service change from perspective of *service interface*, those change scenarios are often equal to modifications of WSDL document. Fokaefs *et al.* summarize some change scenarios for changes of WSDL documents via observing 18 historical versions of the Amazon Elastic Compute Cloud (Amazon EC2) Web service [31].

When we see service change from perspective of *internal business process*, the actions of change often refer to change patterns of business process. Process change is defined that a change operation modifies the initial process model by altering the set of activities and their order relations [23]. Weber *et al.* summarize a set of 18 change patterns for process-aware information system [35].

3) Abstraction Degree
Change of SOA solution can be divided into change of *business process schema* and change of *business process instance* in terms of *abstraction degree*. In SOA, the information system is seen as a set of connected services. A Process-Aware Information System (PAIS) can be realized using such architecture. The integration of SOA and PAIS is illustrated by emerging standards such as BEPL and Business Process Modeling Notation (BPMN) [25].

Substantial work has been done in the related areas in supporting change of *business process schema*. One distinguished way is so called *business process configuration*. Configurable process models enable the sharing of common processes among different organizations in a controlled manner [37]. Given a configurable process model, analysts are able to define a configuration of this model by assigning values to its variation points based on a set of requirements. Once a configuration is defined, the model can be individualized [18].

Summed up and said, there are two means for *business process instance* change. One is *process instance migration*. Process schema allows for the creation of multiple process instances during run-time. One of the most challenging issues for business process evolution is the handling of running instances when their schema is modified.

Process instance migration [32, 26] is regarded as the mainstream technique that deals with the dynamic business process evolution. By using the technique of process instance migration, active process instances are migrated to the modified process model, so that they can benefit from the optimized process model.

Another means of process instance change is that process instance sometimes needs to deviate from the standard way of working which is defined by process schema. Xu *et al.* focus a framework for handling runtime ad hoc execution modifications based on an artifact-centric business process model. With the support of rules and declarative constructs such as *retract*, *skip*, *add* and *replace* ad-hoc changes can be applied to execution at anytime depending on runtime data and the instance status gathered through the use of artifacts. Run-time execution variations can be specified as execution modification rules, which lead to deviations to the normal business process model [36].

3.3   Time of SOA Solution Evolution

This dimension answers the *when* questions when the evolution of SOA solution happens.

1)   Happening time of change
Change of SOA solution can be divided into *design-time change, run-time change and compiling-time change* in terms of happening time of change.

Business designer will change process schema generally by using some SOA development tools when business demands or rules changes. This change or restructuring is aiming to static process schema, which we name *design-time* change.

The features of loose-coupling and late-binding of SOA solution make it possible for activities of business process to dynamically (re)bind real Web service in the run-time of process execution, which we name *run-time change*. Run-time change is often related to change of process instance. System supporting for run-time change is shown in [7, 8, 9].

Evolution of SOA solution can also happen at *compiling-time*. For example, configurable process fragments are weaved into the BPEL-based process at compiling-time via weaving technique of Aspect-Oriented Programming (AOP) [13].

2)   Change Frequency
We divide service change into *infrequent change, frequent change* and *continuous change* in terms of change frequency. *Service change frequency* refers to the ratio of numbers of service versioning to a span of time.

•   Infrequent change
Some enterprise Web services address itself to a stable service for the public so that those services *changes infrequently*. Amazon Elastic Compute Cloud[1] (Amazon EC2) is a Web service that provides resizable compute capacity in the cloud. Fokaefs *et al.* observe that the total number of versioning from 6/26/2006 to 8/31/2010 is only about 18 [31].

---

[1] http://aws.amazon.com/ec2/

- Frequent change

Flexibility is an important and desirable property for service because of dynamic business environments. In order to realize this agile solution people often use business process language to describe the logic order within the service and a group of finer-granularity services are orchestrated in terms of this business process. Some open-sourcing or enterprise modeling tools like activiebpel Vos, Eclipse bpel designer, Netbean SOA designer etc. can help us to design and redesign the business process. This change is often on-demand and frequent. Therefore we insist that this type of change is *frequent change* of service.

- Continuous change

Rapid change of dynamic business environments sometimes need that service shall react to change of business quickly. Some variation points in services can be enabled through the use of a BPEL constructs such as flows with a transition condition based on a configuration parameter. Koning *et al.* propose VxBPEL that supports dynamic variability for Web Services in BPEL [22]. Comparing to BPEL, the advantage of VxBPEL is that the choices for the variation point are dynamic and extensible so that new variants can be introduced at run-time.

3.4   Support Mechanisms of SOA Solution Evolution

The dimension answers *how* service change happens. Change supports for service are analyzed in this section.

1) Degree of automation

Degree of automation refers to the degree that machine can support for change of service. We express this degree with *automated* and *manual*.

Automatic service change refers that machine can completely support service of change without any manual intervention. That is to say, process of service change is transparent to the user of service. Automatic service change techniques can be used to support Quality of Service (QoS) assurance by use of late-binding. The principle of automated late-binding framework is to search and recommend appropriate service component automatically and then automatically substitute old one to realize local or optimal performance optimization of SOA solution [8, 9]. Automated business process reconfiguration often happens in the context of exception handler of service-based application, such as [11]. It supports the detection of various web service faults and their automatic recovery based on well-prepared self-healing policies.

Manual change support refers to human interference during changes, such as modifying the invoking code in BPEL document or redesigns business process manually.

2) Degree of Change impact

There are two types of service changes *shallow* versus *deep service changes* proposed by Papazoglou in terms of degree of change impact [3]. With *shallow changes* the change effects are localized to a service or are strictly restricted to the clients of that service. *Deep changes* cause cascading types of changes which extend beyond the clients of a service possibly to entire value-chain.

A change support mechanism can either be for *shallow change* or *deep change* of services. *Shallow changes* characterize both singular services and business processes.

Therefore we need a robust versioning strategy to support service compatibility, compliance, conformance and substitutability. Andrikopoulos *et al.* propose a formal framework, type safety criteria and algorithms which control and delimit the evolution of services to assist service developers in controlling and managing service changes in a uniform and consistent manner [34]. Ryn *et al.* identify properties that can be used as requirements in determining which conversations can be migrated to a new protocol when an old one has been changed [16].

Deep changes of services rely on the assistance of a change-oriented service life cycle methodology to respond appropriately to changes [3]. The initial phase focuses on identifying the need for change and scoping its extent. The second phase focuses on the actual analysis, redesign or improvement of the existing services. Some works have been done, for example, Wang *et al.* developed the graph-based service dependency matrix and inter service relation matrix to calculate the service cohesions and impact effects [24]. The third phase is to assess the impact of changes. Deep changes are a challenging and open research problem in the context of service engineering.

4   Implementation and Experiment

In this section we illustrate our prototype *configurable urban logistics delivery management system*. This management system is SOA-based and process-awareness. We describe the architecture of our configurable logistics management system. Given a usage scenario of the system, we will apply our taxonomy framework on the system. We expect that the taxonomy will help the developer of system to find flaw or deficiency of the *configurable urban logistics delivery management system*.

4.1   Prototype Implementation

In order to demonstrate configurable service evolution, we performed a proof-of-concept implementation. Using this implementation, we can configure service composition to realize scalable business requirements. The architecture of configurable service evolution platform is depicted in Figure 2.

System Management Interface. The *system management interface* offers registration and log in interface for system user.

Business Process Model Editor. *Business process model editor* provides user a panel on which the business process model stored in the logistics domain knowledge repository can be loaded and shown.

Business Process Evolution Editor. *Business process evolution editor* offers a configuration interface for adding, deleting, and modifying the original business process model.

Business Process Evolution Manager. *Business process evolution manager* can modify the original business process model in terms of the configurable requestor.

Business Process Model in Logistics Domain. By using BPEL, we set up two business process model for logistics delivery and warehousing management. The basic process model for logistics delivery consists of 11 Web services. The basic process model for warehousing management consists of 9 Web services.

Business Process Execution Engine. The *business process execution engine* is responsible for executing business process. We select ActiveBPEL engine[2] as our business process execution engine.

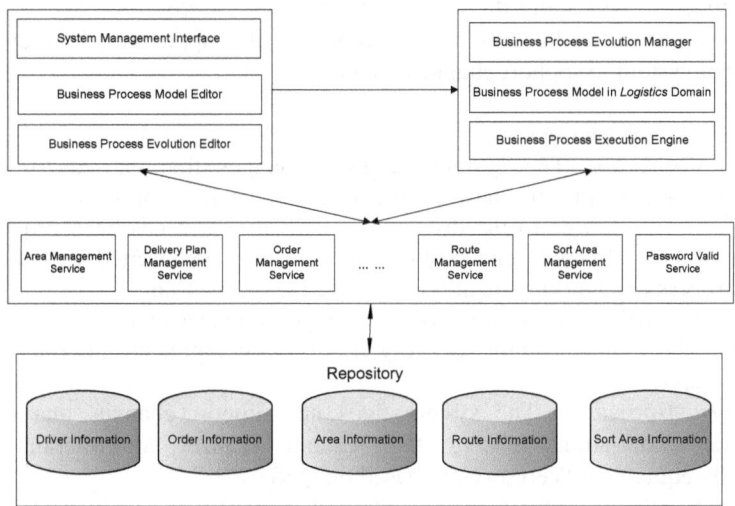

Fig. 2. Configurable SOA Evolution Platform

4.2   A Usage Scenario of the System

Let us apply *urban logistics delivery management* as our usage scenario of the system. With the problem of urban traffic jam, it becomes important on how to deliver cargo on time. Our SOA-based management system offers logistics delivery management service.

The main process for urban logistics delivery management consists of 11 activities. These 11 tasks are realized by Web services. As the inputs of *urban logistics delivery* service, cargo_name, delivery_location and delivery_time are firstly provided by the users. The system will integrate the orders in terms of delivery location, search for all the routes between warehouse and the delivery location, and inquiry drivers who have spare time. After that, the delivery routes are arranged and orders are re-sorted for car loading plan. Next, the system will select the vacant sorting area and give car loading plan. Finally, sorting form, car loading form and delivery form are produced, which is the outputs of *urban logistics delivery* service.

Sometimes we need to evolve the functionalities of *urban logistics delivery* service. For example, we need to have password authentication before integrating orders. Also, we need to send a mobile message to the driver after the delivery form has been produced. We can configure the *urban logistics delivery* process model to evolve the functionalities of the *urban logistics delivery* service. For it, we add a new

---

[2] http:// www.activevos.com/

activity named *password authentication* to the original business process model. After that, when user invokes the urban logistics delivery service, he/she has to input password to complete authentication.

4.3   Application of Taxonomy

In this section we will apply this taxonomy on *configurable urban logistics delivery management system* for illustration. The purpose of that application is to evaluate the efficiency of system to support change of service.

4.3.1 Motivation of Change
Perfective Motivation. The system supports changes that caused by perfective motivation. For example, the new business demand will be proposed, for example, that *send mobile message* to the driver after delivery form has been produced. The new activity that takes on send mobile message will be inserted into the original business process to support realization of perfective motivation.
Corrective Motivation. The system does not support changes that caused by corrective motivation. The system cannot recovery if service components interface or policy versioning happen.
Performance Motivation. The system does not support changes that caused by performance motivation. The system cannot realize hot-swapping between semantically equivalent Web services based on QoS.
Cost Motivation. The system does not support changes that caused by cost motivation. The system cannot realize hot-swapping between semantically equivalent Web services based on cost of service.

4.3.2 Location of Change
Artifact. The artifacts for changes of system are the execution codes of business process and the client code.
Transparency. It can be observed that the system interface changes in the process of system evolution from perspective of close world. Moreover, the internal business process changes can also be observed when we see the system from perspective of open world.
Abstraction. Up to now the system change can only happen in the process schema layer. Process instance changes are not realized since *instance migration* and run-time modification techniques are not available in the system now.

4.3.3 Time of Change
Happening time of change. The system now supports *design-time change* as developer can configure the business process via Graphic User Interface (GUI). The system does not support run-time change so that it needs to re-deploy the modified process schema before executing the new process.
Frequent change. The system support *frequent change* since business process of system can be configured to meet dynamic business environments. But the variation analysis of business process have not realized yet.

4.3.4 Support Mechanism of Change

Degree of automation. The system supports automated change since the new activities can be automatically added to the proper location of the original business process with help of the domain rules.

Degree of change impact. Up to now the system only supports for shallow change. The system has not yet considered the new added/deleted/modified activity has impact on other activities when process is changed.

4.3.5 Discussion

Using the taxonomy, the *configurable urban logistics delivery management system* is positioned as a SOA solution that supports for design-time, process schema layer, and automatic evolution. The positive side of the system is that it can provide automatic configuration in terms of configurable demands. From the negative side, the system is short of capability for process instance evolution. It is difficult for the system to response to individualized change demands for each instance. Missing the capability of late-binding makes it difficult to deal with exception when service component versioning and policy versioning happens.

5   Related Works

Related works are discusses from two angles: (i) taxonomy of SOA solution evolution, (ii) Change support mechanisms and change types for evolution of SOA solution.

To our best knowledge, we have not found the systematic and elaborate typology for evolution of SOA solution. The works shown in [4, 15] are classic typologies for software evolution. Both do not cover the SOA-based software paradigm. The works shown in [15] proposed a classification for software changes based on change motivation. As a complementary of [15], the works shown in [4] focus more on technical aspects on software evolution such as taxonomy of characteristics of software change mechanisms and the factors that influence these mechanisms. The works shown in [10] give us taxonomy for flexible business process. But SOA solution changes are not limited to change of business process. Different from the above, the taxonomy of this paper is aiming to change of SOA solution.

Some literatures discussed how to adapt service component to make sure the whole quality of SOA solution such as [8, 9, 21], which can be viewed as support mechanisms at service level. Some works shown in [7, 11, 13] realize business process change with the help of AOP technique. From the viewpoint of enterprise, the works shown in [29, 30] give some solutions coping with service or business process versioning. The works shown in [2, 22] are the variation analysis for business process. The works shown in [1, 27, 28] describe some types of SOA versioning. These works above are the theoretical base for describing our taxonomy framework for SOA solution evolution.

6   Conclusions and Future Work

In this paper, we have proposed a taxonomy framework for evolutionary SOA solution changes. Then, as a case study, we do analysis on a SOA-based configurable logistics management system by using this taxonomy. The experiment result shows that the *configurable urban logistics delivery management system* is positioned as a SOA solution that supports for design-time, process schema layer, and automatic evolution. Its capability for changing is limited on not supporting late-binding and process instance evolution.

The future work will be carried from the following points. First, we will apply the taxonomy framework to evaluate more SOA solutions, SOA solution development tools and versioning management tools to evaluate their capability of supporting changes. Moreover, we will improve continuously our taxonomy framework in terms of the feedbacks of evaluation.

Second, we will do a research topic on the logic rules for these four perspectives of the framework. For example, this human-based evolution is caused by this reason (why) so that the person (who) have to conduct this change at this place (where) at what time (when) by which method (how). We think that these logic rules will be transformed to change analysis template, and can be used for analyzing kinds of change scenarios of SOA solutions.

Acknowledgement. This work is supported by National Natural Science Foundation of China undergrant 61100017, 61100018, 61170026, 60970017, 61272115, Natural Science Foundation of Hubei Province (No. 2010CDB08503) and the Fundamental Research Funds for the Central Universities (3101007), Hubei Provincial Key Laboratory of Intelligent Robot (HBIR 201002), Specialized Research Fund for the Doctoral Program of Higher Education (No.20090141120020).

References

1. Brown, K., Ellis, M.: Best practices for Web services versioning, Keep your Web services current with WSDL and UDDI (June 2004),
   http://www.ibm.com/developerworks/webservices/library/ws-version/
2. Zhang, L.-J., Arsanjani, A., Allam, A., Lu, D., Chee, Y.-M.: Variation-Oriented Analysis for SOA Solution Design. In: IEEE International Conference on Service Computing (SCC 2007), pp. 560–568 (2007)
3. Papazoglou, M.P.: The Challenges of Service Evolution. In: Bellahsène, Z., Léonard, M. (eds.) CAiSE 2008. LNCS, vol. 5074, pp. 1–15. Springer, Heidelberg (2008)
4. Buckley, J., Mens, T., Zenger, M., et al.: Towards a taxonomy of software change. Journal of Software Maintenance and Evolution: Research and Practice 17, 309–332 (2005)
5. Ouederni, M., Salaun, G., Pimentel, E.: Client Update: A Solution for Service Evolution. In: IEEE International Conference on Service Computing (SCC 2011), pp. 394–401 (2011)

6. Cibran, M.A., Verheecke, B., Vanderperren, W., et al.: Aspect-oriented Programming for Dynamic Web Service Selection. In: Integration and Management, World Wide Web, vol. 10, pp. 211–242 (2007)
7. Charfi, A., Dinkelaker, T., Mezini, M.: A plug-in Architecture for Self-Adaptive Web Service Compositions. In: Proceeding of 2009 IEEE International Conference on Web Services (ICWS 2009) (2009)
8. Yau, S.S., Ye, N., Sarjoughian, H.S., et al.: Toward Development of Adaptive Service-Based Software Systems. IEEE Transaction on Service Computing 2(3) (2009)
9. Shen, Q.Z., Benatallah, B., Maamar, Z., et al.: Configurable Composition and Adaptive Provisioning of Web Services. IEEE Transaction on Service Computing 2(1) (2009)
10. Regev, G., Soffer, P., Schmidt, R.: Taxonomy of Flexibility in Business Process, http://ftp.informatik.rwth-aachen.de/Publications/CEUR-WS/Vol-236/paper2.pdf
11. Subramanian, S., Thiran, P., Narendra, N.C., et al.: On the Enhancement of BPEL Engines for Self-Healing Composite Web Services. In: Proceeding of the International Symposium on Applications and the Internet (SAINT), pp. 33–39 (2008)
12. Erl, T.: Service-Oriented Business Processes with BPEL (2012), http://www.whatissoa.com/soaspecs/bpel4ws.php
13. Braem, M., Verlaenen, K., Joncheere, N., Vanderperren, W., Van Der Straeten, R., Truyen, E., Joosen, W., Jonckers, V.: Isolating Process-Level Concerns Using Padus. In: Dustdar, S., Fiadeiro, J.L., Sheth, A.P. (eds.) BPM 2006. LNCS, vol. 4102, pp. 113–128. Springer, Heidelberg (2006)
14. Web Service Policy 1.2 – Framework (WS-Policy), W3C Member Submission (2006), http://www.w3.org/Submission/WS-Policy/
15. Chapin, N., Hale, J.E., Khan, K.M., et al.: Types of software evolution and software maintenance. Journal of Software Maintenance and Evolution: Research and Practice 13, 3–30 (2001)
16. Ryu, S.H., et al.: Supporting the dynamic evolution of web service protocols in service-oriented architectures. ACM Transactions on the Web 1(1), 1–39 (2007)
17. Zhang, L.J., Zhou, N., Chee, Y.M., et al.: SOMA-ME: A platform for model-driven design of SOA solutions. IBM System Journal 47(3) (2008)
18. Rosa, M.L., Dumas, M., ter Hofstede, A.H.M.: Jan Mendling, Configurable Multi-Perspective Business Process Models. Information Systems 36, 313–340 (2011)
19. Grappling with SOA Change and Version Management, http://www.zapthink.com/2006/05/19/grappling-with-soa-change-and-version-management/
20. Simmonds, J., de Nantes, E.D.M., Mens, T.: A Comparison of Software Refactoring Tools, http://citeseerx.ist.psu.edu/viewdoc/download?doi=10.1.1.1.9162&rep=rep1&type=pdf
21. Kongdenfha, W., Nezhad, H.M., Benatallah, B., et al.: Mismatch patterns and adaptation aspect: A foundation for rapid development of web service adapters. IEEE Transaction on Service Computing 2(2), 94–107 (2009)
22. Koning, M., Sun, C.-A., Sinnema, M., Avgeriou, P.: VxBPEL: Supporting variability for Web services in BPEL. Journal of Information and Software Technology 51, 258–269 (2009)
23. Fahland, D., Mendling, J., Reijers, H.A., Weber, B., Weidlich, M., Zugal, S.: Declarative Versus Imperative Process Modeling Languages: The Issue of Maintainability. In: Rinderle-Ma, S., Sadiq, S., Leymann, F. (eds.) BPM 2009. LNBIP, vol. 43, pp. 477–488. Springer, Heidelberg (2010)

24. Wang, S., Capretz, M.A.M.: A Dependency Impact Analysis Model for Web Service Evolution. In: Proceeding of 2009 IEEE International Conference on Web Services (ICWS 2009), pp. 369–377 (2009)
25. van der Aalst, W.M.P.: Process-Aware Information Systems: Lessons to be Learned from Process Mining. In: Jensen, K., van der Aalst, W.M.P. (eds.) ToPNoC II. LNCS, vol. 5460, pp. 1–26. Springer, Heidelberg (2009)
26. Wei, S., Xiao-Xing, M., Hao, H., Jian, L.: Dynamic Evolution of Processes in Process-Aware Information System. Chinese Journal of Software 22(3), 417–438 (2011)
27. Parachuri, D., Mallick, S.: Service Versioning in SOA, Part I: Issue in and approaches to Versioning. Technical Report, http://www.infosys.com/offerings/IT-services/soa-services/white-papers/Documents/service-versioning-SOA-1.pdf
28. Parachuri, D., Mallick, S.: Service Versioning in SOA, Part II: Handling the impact of Versioning. Technical Report, http://www.infosys.com/offerings/IT-services/soa-services/white-papers/Documents/service-versioning-SOA-2.pdf
29. Leitner, P., Michlmayr, A., Rosenberg, F., et al.: End to End Versioning Support for Web Services. Technical Report (2008),
http://www.zdnetasia.com/whitepaper/end-to-end-versioning-support-for-web-services_wp-2220647.htm
30. Versioning Business Process. Technical Report (2008),
http://download.oracle.com/docs/cd/E13214_01/wli/docs92/bpguide/bpguideVersion.html
31. Fokaefs, M., Mikhaiel, R., Tsantalis, N., Stroulia, E., Lau, A.: An Empirical Study on Web Service Evolution. In: 2011 IEEE International Conference on Web Services (ICWS 2011), pp. 49–56 (2011)
32. Casati, F., Ceri, S., Pernici, B., Pozzi, G.: Workflow evolution. Journal of Data & Knowledge Engineering 24, 211–238 (1998)
33. Cibran, M.A., Verheecke, B.: Dynamic Business Rules for Web Service Composition. In: Proceedings of the Second Dynamic Aspects Workshop (DAW 2005), pp. 13–18 (2005)
34. Andrikopoulos, V., Benbernou, S., Papazoglou, M.P.: On The Evolution of Services. IEEE Transaction on Software Engineering (March 2011)
35. Weber, B., Reichert, M., Rinderle-Ma, S.: Change patterns and change support features – Enhancing flexibility in process-aware information systems. Data & Knowledge Engineering 66, 438–466 (2008)
36. Xu, W., Su, J., Yan, Z., Yang, J., Zhang, L.: An Artifact-Centric Approach to Dynamic Modification of Workflow Execution. In: Meersman, R., Dillon, T., Herrero, P., Kumar, A., Reichert, M., Qing, L., Ooi, B.-C., Damiani, E., Schmidt, D.C., White, J., Hauswirth, M., Hitzler, P., Mohania, M. (eds.) OTM 2011, Part I. LNCS, vol. 7044, pp. 256–273. Springer, Heidelberg (2011)
37. van der Aalst, W.M.P.: Business Process Configuration in The Cloud: How to Support and Analyze Multi-Tenant Processes? In: 2011 Ninth IEEE European Conference on Web Services (ECWS 2011), pp. 3–10 (2011)

A Hierarchical Representation for Indexing Data Condensed Semantically from Physically Massive Data Out of Sensor Networks on the Rove

MinHwan Ok

Korea Railroad Research Institute,
Woulam, Uiwang, Gyeonggi, Korea
panflute@informatics.krri.re.kr

Abstract. Many of conveniences are nowadays on the way to be smart; mobile phones, cars and power stations. Among them, mobile conveniences generate much of sensor data in company with the persons. Some of sensor data are required for processing at distant places in which the sensor data are aggregated, for capabilities of smartness. In the case of vehicles the sensor data are transmitted for malfunction detection and health monitoring of the vehicle in near future. The sensor data are substantially large in the amount of one vehicle's data by multiple kinds of sensors, and the amount of a number of vehicles' data gathered is huge to be received concurrently at some server. Further when the gathered data are to be aggregated in one system the management of the enormous data could determine the functionality of the system. In this work, a data abbreviation diminishes the amount to be transmitted, and the data negating a valid extent consist the majority of data to be aggregated exploiting the semantics of the sensor data gathered. This method is far different from the conventional compressions. The aggregated data are managed and displayed when necessary in one system tracing faulty cars in a region. Although the aggregated data are in a condensed form, complete ones are retrievable from the original server.

Keywords: Internet of Things, Sensor Data Integration, Cyber-physical system.

1   Introduction

The smart devices are evolving to help human life much comfortable. The smart phone seems to become a smart assistant with its computation capability. The vehicle is being equipped with the computation capability to be a smart car in near future. For the smart car, the safety in driving is one importance in the capability of smartness. As cars share the road in driving, a faulty car is liable to cause traffic accident. This car should be early found and need be assisted, apart from the road. It is not appropriate solutions either notifying a problem by the driver or detecting most of all sort of the faultiness within the car. Notification by the driver might be too late and detecting most of the faultiness should be too heavy and expensive functionality in one car. In this work the sensor data a vehicle generated are transmitted to a system of

A. Haller et al. (Eds.): WISE 2011 and 2012 Combined Workshops, LNCS 7652, pp. 275–280, 2013.
© Springer-Verlag Berlin Heidelberg 2013

distributed databases for malfunction detection and health monitoring of the vehicle. As the data are aggregated in the system, reducing the amount of the enormous data is processed in two methods aggregating the data. The topmost server becomes the head index to the sensor data of a faulty car.

A platform named *BeTelGeuse* is proposed for gathering and processing situational data[1]. BeTelGeuse is an extensible data collection platform for mobile devices that automatically infers higher-level context from sensor data. The system is able to provide context data of mobile phones. This work is similar in system organization for *Internet of Things* to, but different by the perspective toward *Cyber-physical System* from, BeTelGeuse. For reduction of amount of data in the system, GAMPS[2] was proposed. It is a general framework that addresses reduction of data so that the data is efficiently reconstructed within a given maximum error. While GAMPS aims for the system suchlike BeTelGeuse and the proposed system of this work, the complete data required are delivered between databases in the proposed system, and polynomial time approximation schemes of GAMPS is not adequate for timely processing of the sensor data.

2   Distributed Databases of Sensor Data from Vehicles

The vehicle traverses a number of areas to arrive at its destination. Monitoring and detecting an abnormal state of the vehicle could be designed by tracking the vehicles however the privacy would be easily violated. Thus the servers covering their areas gathers sensor data from individual vehicles identifying each one by only IDs registered to the system in this work. The traversal is not traced unless someone collects the recorded data from all the area servers. The system organization is shown in Fig.1.

Once the vehicle at its time slot transmits a dataset of a set of sensors in a reduced form by a *data abbreviation*[3], the area server receives the abbreviated data and restores the complete data and records the dataset. The area server calculates with the dataset to find out the data out of the boundary value meaning a sensor detected the abnormal state in the vehicle. The complete data are *condensed semantically* according to a predefined classification[4] and then the classificatory data in a reduced amount is transmitted, with the data indicating the abnormal state if found, to a server covering the region constituted with areas. The region server aggregates the condensed data of the vehicle from the area servers. The procedure with the abnormal state found is described in the next section in detail.

Data generated at a vehicle are transmitted to the server of the area, and data abbreviation reduces their amount to be transmitted. RPM of the engine, for instance, is digitized into RPM data. The RPM is checked several times within one collection period in the vehicle, and thus the values of RPM data changes discretely. Before the next data collection, the values are split into two groups, one group of the displacements above the average and the other group of displacements below the average. Merely differences from previous displacements are recorded in either group spilt by the average of RPM, together with the average of RPM in that period.

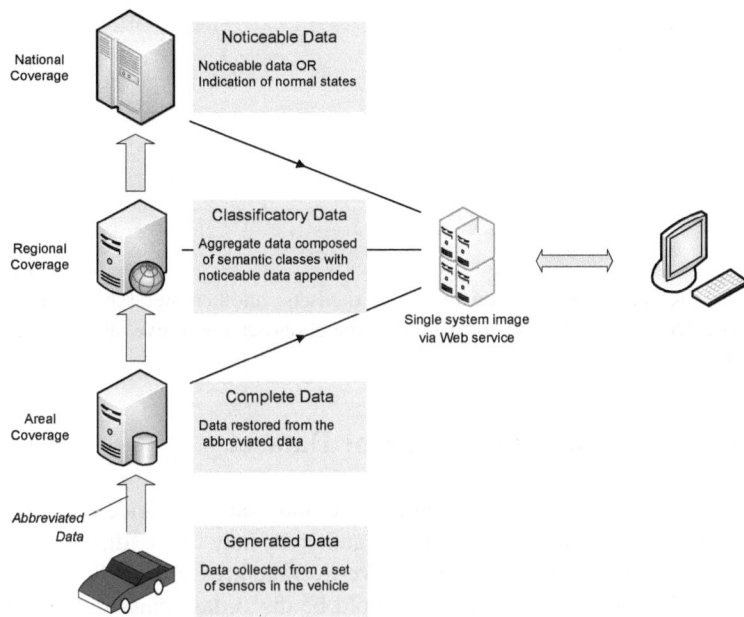

Fig. 1. The system organization with the process and flow of the sensor data originated from the vehicle. The rightmost console can access the data at either server of its coverage.

This technique reduces the data in bits and called *Scale* among abbreviation techniques[3]. The condensed data are recovered into the complete data, the digitized RPM data, at an area server.

From the area servers to the nation server, the complete data are condensed to be aggregated in the server at a higher level. Fig. 2 shows the hierarchy of servers managing (condensed) datasets. The essence of semantic condensing is to change data with wider bits into data with narrower bits, exploiting the semantic meaning of the value. Suppose the original RPM data ranges from 0 to 110 (8-bits) in an area server of the congregational level. To be aggregated in a region server of the regional level, the data are changed into one of {**Under the Extent**, Near the Low-boundary, Low, Below the Average, Above the Average, High, Near the High-boundary, **Over the Extent**} (3-bits). If the vehicle of this data is found to be noticeable, the condensed data are further condensed into one of {Normal, **Abnormal**} (1-bit). The condensed dataset are aggregated in the nation server of the central level, together with an appended complete datum(8-bit) relayed from the area server to record the abnormal state with the original data. A client could trace a vehicle noticeable from the central level as servers of the central level and the regional level maintains indexes to the complete data at the congregational level.

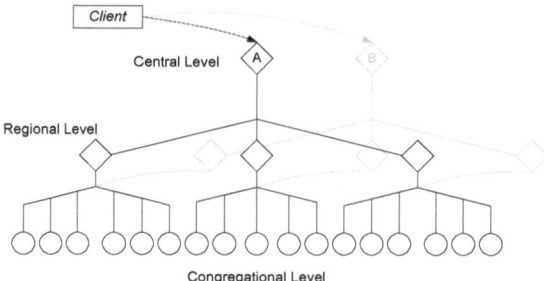

Fig. 2. A hierarchical representation of sensor datasets; another hierarchy could be built for malfunction detection with the sensor dataset at the congregational level at which the complete data are stored.

3   Discordant Pairs in the Sensor Dataset

The abnormal state of a vehicle could be monitored and detected in the dataset received from the vehicle. The dataset gathered may have discordant pairs; in deceleration of a vehicle, for example, above-average pressure on the brake pedal but not-corresponding decrease of velocity should be discordant pairs in their values. For another case, above-average pressure on the brake pedal together with high RPM of the engine should mean a strange event with accelerating and decelerating simultaneously. These data are discordant in their values and even contradictory. Comparing pairs of relevant data would find out those pairs discordant in the values and they are discordant in semantics.

Data series of a single sensor could indicate an abnormal state in the dataset; the temperature of a tyre as an example, a drastic escalation of the internal temperature intimates imminent explosion of the tyre. In this case, the drastic escalation is discordant in semantics. Detecting the abnormal state with series of a single property (sensor) is relatively simple. The internal temperature should have a steady escalation and a limit predefined, far ahead its explosion in the example. The gradient is simply calculated with data previously recorded.

For detecting the abnormal state with pairs of properties (sensors), a boundary is applied to one property according to the other property directly related. In the acceleration of a vehicle, there are direct relations in the pairs of properties: degree of the acceleration pedal to RPM of the engine, for example. RPM is raised up to a reasonable boundary according to the degree of the acceleration pedal. The RPM over or under the boundary with the current degree indicates an abnormal state. This range is called a *valid extent*, from the low boundary to the high boundary. The boundary values are equal among the vehicles of the identical model for the pairs of properties the vehicle internal.

The data negating its valid extent is appended to the condensed data transmitted from the area servers. This appended data is named the *noticeable data*. The region server received the noticeable data marks the vehicle *noticeable* and starts collecting the data of those properties. As the route tracking is not conducted for the privacy reason, the region server requests the recorded data of the vehicle to all the area servers under the region server as depicted in Fig. 3. The collected recorded data is reported to the monitor person and the person responds to this report; i.e. informs the cars on the same road of the vehicle noticeable. The region server transmits further condensed dataset with the noticeable data to the nation server.

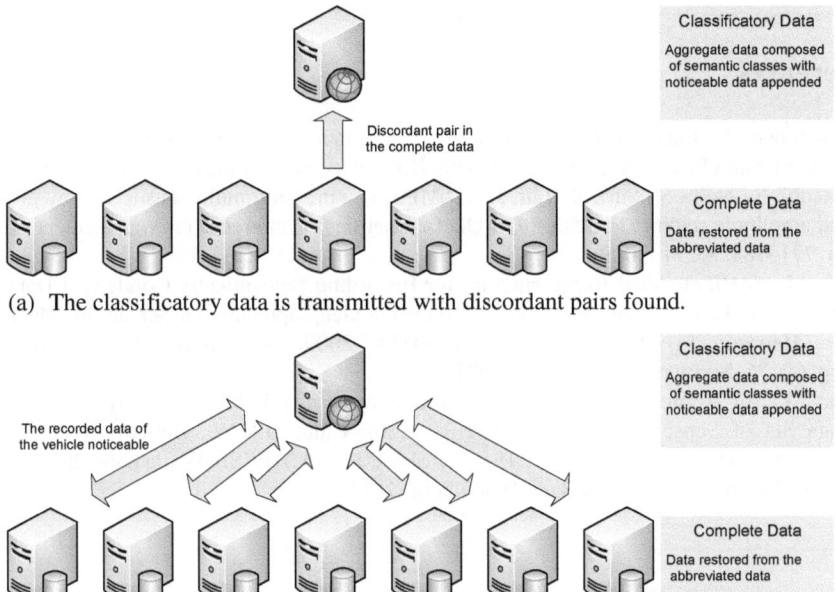

(a) The classificatory data is transmitted with discordant pairs found.

(b) The recorded data are collected from the area servers the vehicle passed by.

Fig. 3. The region server collects data to decide whether the vehicle noticeable is a faulty car

4   Summary with Future Work

As a faulty car is liable to become the cause of any traffic accident possible, the objective of the system is distinct; keeping cars with normal states safe by taking the faulty car apart from the road, and then let the faulty car be served with a professional support. Although not explained in this work, the system could trace the car *probably* noticeable. For RPM in the example, if the classificatory value of Near the High-boundary is continuous for some duration, the vehicle could be sorted into 'probably noticeable'. Yet much of work should be conducted for this.

There are pairs of properties that complicate to clarify direct (or indirect) relations. For the relation of RPM of the engine to the velocity of the vehicle, the velocity increases according to RPM in proportion to the level of the automatic transmission. The level of the automatic transmission is changed according to the velocity. The level of the automatic transmission intervenes however, the circumstance conditions such as the slope of the road should also intervene between RPM and the velocity. For this reason, the area server compares the pair of the properties to the data recorded previously with other vehicle of the identical vehicle model, in the case the circumstance condition is also related, such like RPM and the velocity; the same pair of other vehicle passed by this location on that road recently, with a similar RPM. Since the comparison is to find out pairs discordant in the values, historically averaged values could be used for a selection from the candidates for a valid extent. This would be our future work.

References

1. Kukkonen, J., Lagerspetz, E., Nurmi, P., Andersson, M.: BeTelGeuse: A Platform for Gathering and Processing Situational Data. IEEE Pervasive Computing 8(2), 49–56 (2009)
2. Gandhi, S., Nath, S., Suri, S., Liu, J.: GAMPS: compressing multi sensor data by grouping and amplitude scaling. In: 35th SIGMOD International Conference on Management of Data, pp. 771–784. ACM (2009)
3. Ok, M.: A Hierarchical Representation for Recording Semantically Condensed Data from Physically Massive Data Out of Sensor Networks Geographically Dispersed. In: Meersman, R., Herrero, P., Dillon, T. (eds.) OTM 2009 Workshops. LNCS, vol. 5872, pp. 69–76. Springer, Heidelberg (2009)
4. Ok, M.: An Abbreviate Representation for Semantically Indexing of Physically Massive Data out of Sensor Networks on the Rove. In: Chiu, D.K.W., Bellatreche, L., Sasaki, H., Leung, H.-f., Cheung, S.-C., Hu, H., Shao, J. (eds.) WISE Workshops 2010. LNCS, vol. 6724, pp. 343–350. Springer, Heidelberg (2011)

An Auxiliary Storage Subsystem of Storage Space Hidden for User Data to Distributed Computing Systems

MinHwan Ok

Korea Railroad Research Institute,
Woulam, Uiwang, Gyeonggi, Korea
panflute@informatics.krri.re.kr

Abstract. The handheld computing device demonstrates its mobility nowadays, being furnished by distributed computing resources in a form similar to the thin client. The distributed computing systems such as Cloud supports user data with their storages implying that the users let their data be under control of the Cloud system. However it is not an acceptable premise to many users. For these users, an auxiliary storage subsystem is proposed in a new model of remote storage to mobilize the user data of external storages in this work. The iSCSI devices could be connected to the distributed computing system through the storage subsystem as needed. iSCSI is a recent, standard, and widely deployed protocol for storage networking, and the subsystem is able to accommodate many types of storages outside of the distributed computing system. The proposed subsystem could supply the user data at places near the distributed computing systems and showed an improved transmission of user data in the experimentation.

Keywords: Distributed Computing System, Storage Subsystem, Shadow Storage, iSCSI, External Storage.

1 Introduction

Hardware for computing are developing in their capability and capacity. Processing speed is getting faster and storage space is getting larger. In the case of mobile devices, the tablet computer is the emerging de facto standard in the handheld computing. The flash memory is already de facto standard of personal handheld storage. While these mobile devices play a handheld-assistant role, most of heavy tasks are requested to distributed computing systems in large scale. The heavy tasks accompanied with a large amount of data require enough storage space for the data and computing power corresponding to the storage space.

Those large amounts of data do not have mobility by various reasons. Among the reasons the security concerns are difficult to deal with, and this reason of security restrains the mobilization of the large amount of data although the data are necessary for processing at distant places. For a small amount of data, it would be a viable alternative that encrypting the data before transmission to a distant place, decrypting them when initializing the processes at the distant place, following encrypting when finalizing the processing at the distant place and decrypting after transmission to its

A. Haller et al. (Eds.): WISE 2011 and 2012 Combined Workshops, LNCS 7652, pp. 281–291, 2013.

primary place. However, this coupled encrypting and decrypting should harass the system and consumes essential computing resources including processor and main memory, and the additional resource consumption would persist for the duration in proportion to the amount of data. In this work, a storage subsystem for the large amount of data is proposed to make the data mobile without additional consumption of the essential resources. The storage could become functional only with storage information provide by a new model of storage with mobility.

The user data of large amount is necessarily moved to some location near the application is running at. The concept of Cloud or Grid assumes that the distributed computing system possesses user data for processing but it is not an acceptable premise to many users. The architecture of an auxiliary storage subsystem is proposed for the problem in the previous work [1]. However the user's approval was required for storing the user data in the architecture and a new model of remote storage would replace the user's approval.

2   Auxiliary Storage Subsystem to Distributed Computing Systems

Almost all organizations have data to be secured in their organizational storage. The organizational storage contains massive data collected since the establishment of the organization. Scientific researches often have massive data generated during their experiments. Grid computing provides computing power to process those data and the storage subsystems containing the experiment data are consolidated into the Grid in their specific manner, for an example in [2]. Cloud computing would require the user data transmitted into the Cloud for processing. However it is not the case the data owner and the owner of the processing facility are special groups such as scientific research teams and professional computing teams. Further, all the owners do not like to let their data under control of the Cloud system. For this reason an auxiliary storage is required to the distributed computing systems including Cloud and Grid for general purposes, analogous to the removable devices of personal computers.

Since the storages to be attached or detached are outside of the distributed computing system, only a registered user should be able to attach an external storage to the distributed computing system. An auxiliary storage subsystem would take the trustworthiness part on the side of the distributed computing system. The storage subsystem should also play a manager role on the external storages. The storage subsystem is depicted in Fig. 1.

The storage subsystem is an additional trustworthiness part and the owner should register an external storage in which the user data located, as a registrant of the distributed computing system. It has also a manager role on the external storage including connection, disconnection, and loading data, which is an additional one to the basic storage management of the distributed computing system. The procedure to access user data is shown in Fig. 2.

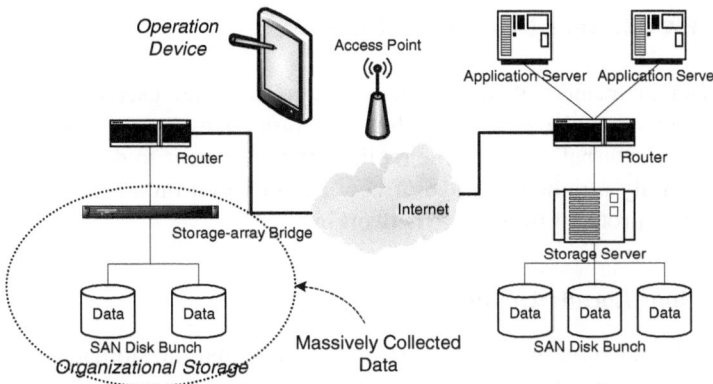

Fig. 1. Massive data is to move to other computing site for processing

(1) Application Client	Login to the distributed computing system
(2) Application Client	Login to the auxiliary storage subsystem
(3) Replicating Server	Connect to the external storage
(4) Application Server	Load data from the external storage or Preloaded data from the replicating server
(5) Replicating Server	Disconnect from the external storage
(6) Replicating Server	Logout from the auxiliary storage subsystem
(7) Application Client	Logout from the distributed computing system

Fig. 2. An application client log-ins the auxiliary storage subsystem for its external storage

The user data is transmitted as files to the replicating server for processing in the application server. The application server opens files required in processing from the replicating server. Note that the user data is stored in the replicating server only. Necessary blocks are transferred into the main memory or virtual memory of the application server and those blocks does not constitute a regular file in the application server. When there are updated blocks in the files open, the blocks are first stored in the replicating server and relayed to the external storage of the owner later in Fig. 3.

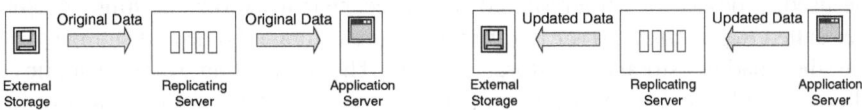

Fig. 3. The replicating server supplies the original data to the application, and relays the updated data to the external storage

3   A New Model of Remote Storage

The user data is located at an external storage, and the user carries a handheld computing device to place an order for data processing in the application server distant from the external storage. The replicating server near to the application server is placed to supply a remote storage of new model based on a recent, standard, and widely deployed protocol for storage networking.

3.1   Partitions Locally Unknown

iSCSI[3] has founded on the idea of a remote storage that the storage is located away and the access to that storage is conducted remotely. The remote storage becomes accessible through a login process with ID and password admitted to the storage. This primitive lockup method is for the security of user data. Without proper login to the storage, the storage is locally a partition (or file) of *unknown* type or an iSCSI type, and the data of the files in the remote storage is not accessible.

Volume management would be exploited to reinforce the security, as the pairs of ID and password have not proved it the ultimate security method. The storage space could be divided into partitions and/or integrated into one partition (or volume) by the volume management. The storage space consists of divided partitions is addressable after integration of those partitions.

The basic operation of iSCSI storage. Once an authenticated user login to the remote storage, a list of files stored in the storage is provided. When a file is opened and loaded into an application of the authenticated user, blocks are transmitted to the application from the file of the remote storage. In this transmission, those blocks are loaded into the main memory (or virtual memory) in which the application resides but not stored as a regular file. Modified blocks are transmitted to remote storage and the file is updated.

The accessibility to the storage and the address-ability to the space of the storage constitute a new model of hidden storage space, named *shadow storage* in this work. As shadow shows only its silhouette, the shadow storage doesn't show its data to the unauthenticated. The Fig. 4 illustrates the process of the shadow storage working. The location the user's original data is stored is called *Storage Home*. The user data is required for any processing in an application at some location, to say, an application server on which the application is running. The location the user data should visit for any processing is called *Storage Visiting*. Since the shadow storage would be mounted to be used, a device or server should be placed near to storage visiting. A storage server called *Replicating Server* intervenes between storage home and storage visiting to be the shadow storage of storage home. The replicating server supplies the application with the user data far from storage visiting, at a location near to storage visiting. The user data is *preloaded* into the shadow storage of the replicating server.

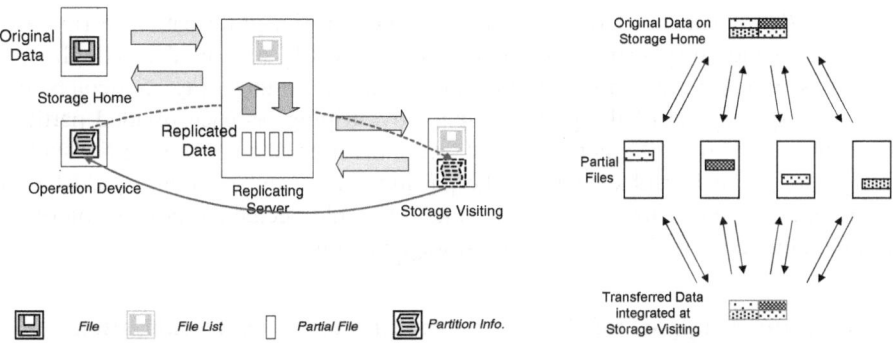

Fig. 4. Shadow storage model   **Fig. 5.** Displacement of partial files

Preloading procedure of the shadow storage is described in Fig. 6. A partition should be prepared for user data to be stored, and this partition is called partition of shadow. This partition of shadow is constructed by commands from storage visiting and the *partition information* is placed in the main memory of storage visiting. Once constructed, the partition information is transmitted to and written in the *operation device*, application client. Note that the partition information is a specific one in this work, different from the generic information for an iSCSI partition, and not written at storage visiting. All the partition spaces are addressed via a specific storage management when data is stored, and the storage space is not addressable by a mere log-in to a single iSCSI partition without the partition information. The file opened at storage visiting is transmitted from storage home, stored in the partition of shadow and forwarded to storage visiting. When the connection to the replicating server is re-established, the partition information could be read from the operation device. Without the partition information, nobody could address the storage space.

For the security of user data being stored at some location, the partition space could be managed by dividing and integrating in the form of *horizontal displacement*

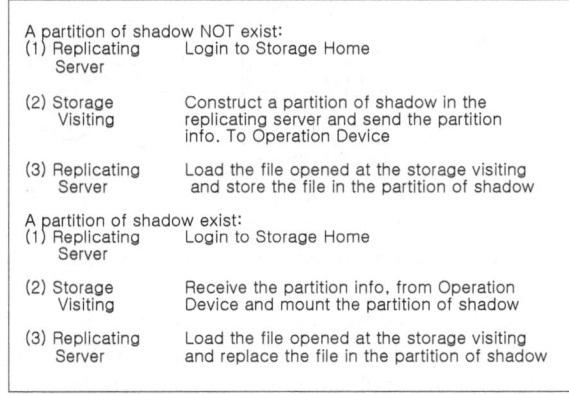

Fig. 6. Procedure of constructing a new partition of shadow or mounting a partition of shadow

or *vertical displacement*, which is recorded in the partition information. Striping of the partition yields displacement of partial files. The file of user data is split into partial files and separated to multiples of divided partitions. Fig. 5 depicts this horizontal displacement of the partition space by striping, as each divided partition contains the partial files of other files. In the case the user data is one long file such as a binary file, vertical displacement of the partition space by spanning would be the alternative. For a higher level of security, the user could decide the destruction of the partition on the replicating server leaving storage visiting.

3.2   Structural Support for Shadow Storage in the System Organization

The iSCSI protocol can make clients access the SCSI I/O devices over IP network, and a client can use the remote storage transparently[4]. The owner's personal storage is recognized as a part of the storage server's local storage by a particular network device driver[5]. Once the software requests a file to the network device driver, *iSCSI Initiator*, it relays the request to the other network device driver, *iSCSI Target*, of the storage or the iSCSI device itself. The target storage starts to transfer the file to the storage server via the network device drivers.

When the target storage transfers data to the storage server, data blocks are delivered via iSCSI Target/Initiator or iSCSI device/iSCSI Initiator. For the performance reason, block I/O was adopted[6] that provides necessary blocks of an opened file to the storage server and updates corresponding blocks when modification occurs. Block I/O outperforms file I/O and does not adhere to a certain file system.

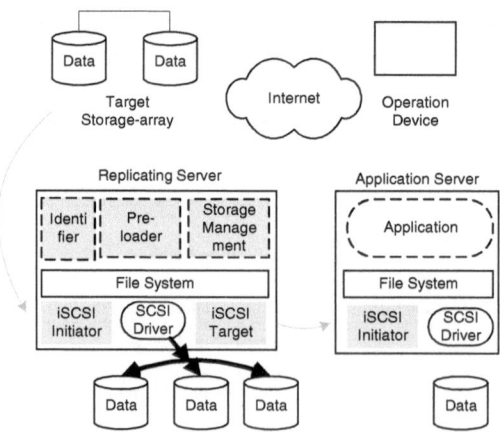

Fig. 7. The replicating server connects to an external storage by iSCSI protocol

The system modules for shadow storage are shown in Fig. 7. The storage server plays the role of replicating server. *Identifier* module monitors connections with the external storage and with the application server, and maintains each connection, i.e. connection re-establishment and transfer recovery. It also manages user login/logout from/to the auxiliary storage subsystem and from/to the distributed computing system.

The authenticated user would have the partition information in the operation device and could mount the partition of shadow for the application server. *Storage Management* module integrates its disk spaces and makes up volumes for preloading. The disk spaces may be integrated into one volume, in effect of striping. The user among the registrant of the distributed computing system is allowed to transmit data by the auxiliary storage subsystem. *Preloader* module caches the user data from the external storage for the application server. On arrival a new job invokes its application but waits for data to process, while other jobs are processing their data. Pre-loader loads data from an external storage and stores it in the shadow storage of replicating server once the job has started and becomes asleep waiting for the data to be loaded. Further on the Preloader is detailed in our previous work[7].

4   Performance Evaluation with the External Storage

A pair of iSCSI initiator and target comprises the storage networking between a storage client and a storage server. The replicating server, a storage server for shadow storage, is a storage client against an external storage, and also a storage server against an application server. The file transmission over IP network is performed using the implemented replicating server. Since there are two network connections, transmission rates of both connections are evaluated. The former connection is between the external storage and the replicating server. The latter connection is between the replicating server and the application server. For the equivalent level of hardware performance, two workstations are in places of the external storage and the application server. Hardware specifications are summarized in Table 1. HDDs listed are ones engaged only in the experimentation and OSs are installed on distinct HDDs not listed.

Table 1. Two workstations are connected to the replicating server in the experimentation

	Replicating Server	'Acting' External Storage	'Acting' Application Server
OS	RedHat-Linux kernel 2.6	Windows Server2003 R2	Windows 7
CPU	Xeon 3.0GHz 1 core/ Dual Processors	Pentium4 3.4GHz 1 core/ Single Processor	Xeon 2.8GHz 1 core/ Single Processor
RAM	2GB	2 GB	1 GB
HDD	Ultra320 SCSI 10,000RPM 2 drives	Ultra320 SCSI 10,000RPM 2 drives	SATA 7,200RPM 1 drive
NIC	Gigabit Ethernet	Gigabit Ethernet	Gigabit Ethernet

Either connection was established on the network with other traffic. The former connection is geographically about 8.5*km* away outside the site of the replicating server. The latter connection is geographically about 0.5*km* apart from the replicating server in the site. A file of 100MB is received through the former connection and cached in the replicating server, and then the cached file is transmitted through the latter connection. The two transmissions are performed separately for comparison shown in Fig. 8. A netbook takes the part of an operation device in the experimentation.

The experimentation scenario is as follows: the user logs in the acting application server first and the replicating server second with the operation device. The user partitions are mounted in the acting external storage by the commands of an initiator at the replicating server. The user constructs partitions of shadow connected to the external storage in the replicating server by the commands of an initiator at the application server, and this partition information is saved as a *local* file of the operation device. The user makes an application running and opens a file of the external storage with the application, and the file is preloaded into the partition of shadow in the replicating server. For the comparison purpose in the experimentation, the preloaded file is transmitted to the application server but this transmission of a whole file is unnecessary in practice. Requested blocks are transmitted into the main memory for the use of an application running.

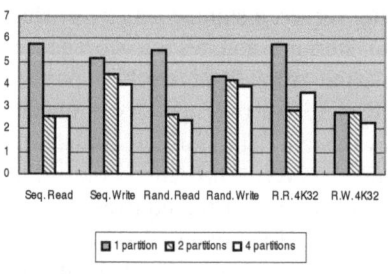 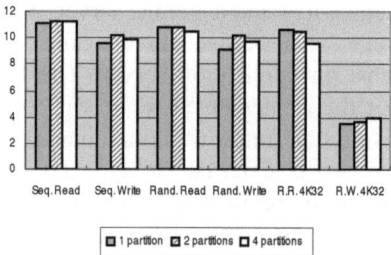

(a) From an external storage outside the site (b) To an application server inside the site

Fig. 8. Transmission rates of the two connections in MB/s

There are 3 transmission types according to the number of partitions containing the file to be transmitted. One partition could constitute the partition of shadow without any displacement of the partition space. Two or four partitions could constitute the partition of shadow to be integrated at the application server. Each partition generates its transmission of a partial file. Serialized transmissions got lower transmission rates in receiving the striped partitions of shadow in effect. In the transmissions of long distance, the serialization affected much and resulted in much lower transmission rates, as shown in Fig. 8(a). In the transmissions of short distance, the random-write was less effective in a block size of 4kB with the queue-depth of 32 in Fig. 8(b). The horizontal displacement of storage space is unnecessary in the external storage, in reality. One partition of shadow should be sufficient in the external storage and would be efficient for the transmission of long distance. In the experimentation, preloading into the partition of shadow improves the transmission of requested blocks hence the essential computing resources including processor and main memory could be efficiently used in the distributed computing system, with the shadow storage hidden to other users.

5   Related Works

Despite many decades of research in distributed file systems, none are well suited for deployment on a computational Grid. Even those file systems designed to be "global" are not appropriate for use in Grid computing systems, because they cannot be deployed without intervention by the administrator at both client and server, and do not provide consistency semantics or security models needed by Grid applications. Chirp[8] distributed file system allows an ordinary user to easily deploy, configure, and harness distributed storage without requiring any kernel changes, special privileges, or attention from the system administrator at either client or server. The desired properties of file or storage system are well listed up in [8] for Grid computing; rapid unprivileged deployment, support for unmodified applications, support for both large and small I/O, flexible security policies, and tunable performance tradeoffs. Chirp could be used as a personal file bridge, the shared Grid file server, or a cluster file system. The functionality of a personal file bridge is the most similarity to this work, excepting Chirp only supports the file systems based on Unix file system. Further the capacity and performance of a Chirp server is constrained by the properties of the underlying kernel-level file system. The storage subsystem of this work is founded on iSCSI technology. The storage space prepared by iSCSI is not adherent to a specific file system, but the owner of the storage space determines which file system at the time the storage space is prepared.

The availability and the integrity of user data are supported by data replication including fault tolerance in the Cloud storage. Data replication is more complicated and expensive than that of this work as the magnitude of data or the multitude of users scales that should be managed by the system. Since the user data are stored in the storages in Cloud, an approach such as HAIL (High-Availability and Integrity Layer)[9] is necessitated for remote file-integrity assurance in a system. It is a distributed cryptographic system that allows a set of servers to prove to a client that a stored file is intact and retrievable. The cost for the assurance of remote file-integrity is the redundancy of user data, and this cost reveals an intrinsic defect of the Cloud storage from the situation that independent users' data are coerced to be stored in a shared place. While the approach claimed the larger storage than the storage required in storing the exact user data, SKUTE (Scattered Key-valUe sTorE)[10] is an approach designed to provide low response time on read and write operations, to ensure replicas' geographical dispersion in a cost-efficient way and to offer differentiated availability guarantees per data item to multiple applications, while minimizing bandwidth and storage consumption. The approach dynamically finds the optimal resource allocation that balances the query processing overhead and satisfies the availability objectives in a cost-efficient way for different query rates and storage requirements. The approach omits maintaining data consistency among replicas and thus not so cost-effective since it is a cost requisite to data replication in a shared place, the Cloud storage.

6   Conclusion

All the users do not like to let their data under control of the distributed computing system such as Cloud. The concept of Cloud or Grid would not appeal to these users since possessing their data is not an acceptable premise. This is a big blockage in the developments of applications for the distributed computing system. A new model of remote storage hidden to the system is proposed and evaluated for the improved transmission of user data stored at a distant place. Without the operation device in which the partition information is written, the partition of shadow does not be mounted by the unauthenticated, and he cannot even login to the shadow storage without ID and password in the iSCSI protocol.

The part of disk drives allotted to the partition of shadow should be under storage space management of the replicating server. The partition of shadow could be constructed as a file instead of a partition in iSCSI protocol, and this special file would be manageable in the server management, for efficient space utilization, for instance. The special files are also not accessible without ID/Password nor addressable without the partition information of the operation device. The new model of remote storage, shadow storage, could be an alternative to the unacceptable premise stated.

As the amount of user data grows the total capacity of storages should expand in the distributed computing system, even though all the users accept letting their data be under control of the distributed computing system such as Cloud. This expansion should be troublesome in the system administration. Further all the data stored are not continuously used in the distributed computing system and in this sense the Grid or the Cloud is not efficient on storage. The shadow storage is also the alternative for efficient storage utilization. The operation device might be a tablet computer in the future.

References

1. Ok, M.: An Auxiliary Storage Subsystem to Distributed Computing Systems for External Storage Service. In: Hsu, C.-H., Yang, L.T., Park, J.H., Yeo, S.-S. (eds.) ICA3PP 2010, Part I. LNCS, vol. 6081, pp. 246–253. Springer, Heidelberg (2010)
2. Beberg, A.L., Pande, V.S.: Storage@home: Petascale Distributed Storage. In: Int. Symposium on Parallel and Distributed Processing, pp. 1–6. IEEE (2007)
3. Hufferd, J.L.: iSCSI: the universal storage connection. Addison-Wesley Professional (2002)
4. Lu, Y., Du, D.H.C.: Performance study of iSCSI-based storage subsystems. Communication Magazine 41, 76–82 (2003)
5. Ok, M., Kim, D., Park, M.-S.: UbiqStor: Server and Proxy for Remote Storage of Mobile Devices. In: Zhou, X., Sokolsky, O., Yan, L., Jung, E.-S., Shao, Z., Mu, Y., Lee, D.C., Kim, D.Y., Jeong, Y.-S., Xu, C.-Z. (eds.) EUC Workshops 2006. LNCS, vol. 4097, pp. 22–31. Springer, Heidelberg (2006)

6. Block Device Driver Architecture,
   `http://msdn.microsoft.com/library/en-us/wceddk40/`
   `html/_wceddk_system_architecture_for_block_devices.asp`
7. Ok, M.: A Sharable Storage Service for Distributed Computing Systems in Combination of Remote and Local Storage. In: Hua, A., Chang, S.-L. (eds.) ICA3PP 2009. LNCS, vol. 5574, pp. 545–556. Springer, Heidelberg (2009)
8. Thain, D., Moretti, C., Hemmes, J.: Chirp: A Practical Global File System for Cluster and Grid Computing. J. Grid Comp. 7(1), 51–71 (2009)
9. Bowers, K.D., Juels, A., Oprea, A.: HAIL: A High-Availability and Integrity Layer for Cloud Storage. In: 16th ACM Conference on Computer and Communications Security, pp. 187–198. ACM, New York (2009)
10. Bonvin, N., Papaioannou, T.G., Aberer, K.: A Self-Organized, Fault-Tolerant and Scalable Replication Scheme for Cloud Storage. In: 1st ACM Symposium on Cloud Computing, pp. 205–216. ACM, New York (2010)

Introduction to the Proceedings of the Workshop on User-Focused Service Engineering, Consumption and Aggregation (USECA) 2011

Hye-Young Paik[1], Ingo Weber[2], and Marek Kowalkiewicz[3]

[1] CSE, UNSW, Sydney, Australia
hpaik@cse.unsw.edu.au
[2] NICTA, Sydney, Australia
ingo.weber@nicta.com.au
[3] SAP Research, Singapore
marek@kowalkiewicz.net

Introduction

The first User-Focused Service Engineering, Consumption and Aggregation workshop (USECA) in 2011 was held in conjunction with the WISE 2011 conference in Sydney, Australia. Web services and related technology are a widely accepted standard architectural paradigm for application development. The idea of reusing existing software components to build new applications has been well documented and supported for the world of enterprise computing and professional developers. However, this powerful idea has not been transferred to end-users who have limited or no computing knowledge. The current methodologies, models, languages and tools developed for Web service composition are suited to IT professionals and people with years of training in computing technologies. It is still hard to imagine any of these technologies being used by business professionals, as opposed to computing professionals.

There are three areas of focus of the workshop: service engineering, service consumption, and service aggregation. In the space of service engineering, the issues are in creating and deploying services from the point of view of non-technical end users. When service consumption is considered, aspects such as user experience, multimodal interaction, multitude of possible consumption devices, operating systems, and user interaction styles need to be addressed. Service aggregation can be understood in multiple ways. Here, we want to address not only the issue of getting various services together for the purpose of building more advanced applications, but also for the purpose of personalising, customising and sharing these applications and services.

We hope that the workshop will contribute towards promoting discussions for the problems mentioned above, leading to solutions that are more consumer-centric, and ultimately successful on the market.

The four full papers chosen for USECA 2011 represent a range of relevant topics. The papers were selected after a thorough peer-review by the workshop Program Committee members and external reviewers.

The paper titled "Data Visualisation Techniques for Exploratory Analysis Processes" by Riley Perry and Fethi Rabhi (UNSW, Australia). is concerned with the design of user-centered environments for supporting the exploration of large datasets through the concept of visualisation instances (VI). A visualisation instance offers an interactive way for the data to be dynamically generated while exploring the datasets.

The paper "Form Annotation Framework for Form-based Process Automation" by Sungwook Kim (UNSW, Australia) proposes simple annotations, such as tagging, to help end users organise and manage business forms/documents. Furthermore, the annotations are used to automate form-based business processes and the supporting framework is designed with little or no technical background knowledge workers in mind.

The paper "A Web Services Variability Description Language (WSVL) for Business USers Oriented Service Customization" by Tuan Nguyen, Alan Colman and Jun Han (Swinburne Univ. of Technology, Australia) presents an extension of WSDL to facilitate flexible representations of Web service customisation options. The language does not require people who perform customisation to have knowledge of Web service technologies.

The paper "Towards a Service Framework for Remote Sales Support via Augmented Reality" by Ross Brown and Alistair Barros (QUT, Australia) offers a solution to remote online sales services via Augmented Reality services in which a customer is allowed to project 3D images of products onto his own physical space at home (e.g., lounge room) to perform in-situ analysis of purchases.

We sincerely thank the USECA 2011 workshop Program Committee members for their time and support throughout the reviewing period.

Hye-young (Helen),
Ingo,
Marek

USECA 2011 Workshop Chairs

1   Workshop Program Committee Members

Boualem Benatallah, University of New South Wales, Australia
Fabio Casati, University of Trento, Italy
Jinjun Chen, University of Technology Sydney, Australia
Alan Colman, Swinburne University of Technology, Australia
Florian Daniel, University of Trento, Italy
Keith Duddy, Queensland University of Technology, Australia
Alexander Dreiling, SAP Research, Australia
Karl M. Goeschka, Vienna University of Technology, Austria
Hakim Hacid, Alcatel-Lucent Bell Labs, France
Christian Janiesch, Karlsruhe Institute of Technology, Germany
Tomasz Kaczmarek, Poznan University of Economics, Poland
Dimka Karastoyanova, University of Stuttgart, Germany

Ryszard Kowalczyk, Swinburne University of Technology, Australia
Hamid Motahari, HP Labs, USA
Surya Nepal, CSIRO, Australia
Anne Ngu, Texas State University, San Marcos, USA
Fethi Rabhi, University of New South Wales, Australia
Chris Smith, University of New Castle, UK
Jian Yu, Swinburne University of Technology, Australia
Julien Vayssiere, SAP, Germany

Data Visualisation Techniques
for Exploratory Analysis Processes

Riley Perry and Fethi Rabhi

School of Computer Science and Engineering
The University of New South Wales, Sydney 2052, Australia
riley.perry@gmail.com, f.rabhi@unsw.edu.au

Abstract. The paper is concerned with the design of user-centered environments for the exploration of large datasets. The specific focus is on high frequency financial news and market data. In previous work, the ADAGE SOA based framework which allows users to model and execute analysis processes has been proposed. One disadvantage is that although much of a business process may be automated, there still remain many steps that are user driven, particularly at the visualization stage. This paper examines this problem in more detail and proposes the concept of a visualization instance (VI) that captures the four essential elements of most analysis processes: data sources, business processes, UI controls and display types. VIs offer an interactive way for the data to be dynamically generated while exploring the datasets. By providing the ability to compose several VIs, complex visualizations can be generated. The paper describes an implementation that allows the processing of Thomson Reuters news data based on an existing ADAGE platform.

Keywords: Visualisation, ADAGE, SOA, User Interface.

1   Introduction

eResearch is being adopted worldwide across all research disciplines, harnessing high-capacity and collaborative information and communications technology to improve and enable research that cannot be conducted otherwise. eResearch is mainly driven by non-IT specialists analyzing vast amounts of ad-hoc data from different sources such as Web logs, network traffic messages, periodic sampling of data from sensors and financial reports. Whilst there are many tools to support particular aspects of the analysis process (e.g. accessing data, data mining, text processing), there are few approaches that support all analysis activities in an integrated way. The main difficulty in providing integrated tools is dealing with different types of datasets and varying requirements between different communities of users. One promising approach has been to leverage concepts of a Service-Oriented Architecture (SOA) that acts as an intermediate layer between data and computing resources (at the infrastructure layer) and analysis processes (at the business process layer). An SOA has many advantages: it enables different technologies (e.g. Web services) to provide facilities such as run time messaging, service composition and choreography, process modeling, discovery, semantic support and run time management. In addition,

A. Haller et al. (Eds.): WISE 2011 and 2012 Combined Workshops, LNCS 7652, pp. 295–306, 2013.

the functionality provided by existing software programs (e.g. legacy systems or application packages) need not be redeveloped; the software program's functionality can be exposed as a service through some wrapper code.

Despite such advantages, the SOA approach still presents many challenges when dealing with eResearch analysis processes. Unlike enterprise processes which are predictable and repetitive, such processes often represent the exploration of a complex data space. A traditional BPM-style of development would not be adequate as the user would need to frequently modify such processes as a result of each exploration step. This paper investigates this problem further and describes a novel approach aimed at assisting the user in making effective use of an underlying SOA-based analysis platform via some visualization abstractions.

The rest of the paper is structured as follows. Section 2 describes some related work in this area. Section 3 presents our approach followed by a case study and a description of our current implementation in Section 4. Section 5 concludes this paper.

2   Related Work

From a computer science perspective, conducting complex data analysis is in the field of knowledge discovery which has its roots in databases and data mining. Knowledge discovery describes the overall process of finding patterns in data or making sense of data [1]. It encompasses tasks such as understanding data, data acquisition, data cleaning, data integration, model development, verification and evaluation. The knowledge discovery process not only involves data manipulation tasks, but also focuses on the understanding of business objectives and the deployment of any knowledge gained (e.g. see the process model of CRISP-DM: http://www.crisp-dm.org/). Our work is mostly concerned with the analysis of ad-hoc data, hence this paper centers on the analysis process, which is that part of the knowledge discovery process where data is involved.

eResearch brings about its own challenges in addition to those we find in knowledge discovery. eResearch encompasses collaboration between distributed research teams, managing and sharing of huge quantities of distributed data, access to shared computation resources, middleware that provide infrastructure support and networking, and dissemination of research via the Internet [2]. eResearch is primarily driven by non-IT experts from different application domains (e.g. biology, physics, health) who mostly use libraries and packages from a variety of sources and manage the analysis process by themselves. Due to the wide choice of libraries available, navigating this large design space has become its own challenge. For example, finance analysts use software packages like S-PLUS, RATS, SAS, and MATLAB. For complex analysis tasks, they have to use "macros" which involves learning scripting or programming languages. It has been long argued that many do not have the programming skills nor do they wish to acquire the programming skills just to use these packages effectively. Most researchers agree that the level of problem solving abstraction should be raised so that users can concentrate on specifying problems using terms and constructs from their domain of expertise, thereby enabling them to solve problems and perform analysis with higher efficiency [3]. The problem is then how to achieve this goal given the vast scope of application domains and types of users. Instead of adding more features to existing tools (which brings about more complexity), we believe that more attention should be devoted to assisting users in bringing together distributed data, integrating the tools they already use and allowing user communities to easily share storage and computational resources. In other words, more work should be done in assisting users in the process of data analysis for complex application domains.

"Problem-Solving Environments" (PSEs) [3] in computational science are seen as a mechanism to integrate different software construction and management tools, and application specific libraries, within a particular problem domain. One can therefore have a PSE for financial markets [3], for gas turbine engines [4], etc. Focus on implementing PSEs is based on the observation that scientists who used computational methods have to write and manage all of their own computer programs. However, PSEs suffer from the lack of flexibility and adaptability to problems beyond their original purpose.

There have been efforts at adapting workflow technology to control the interaction of computational components and provide a means to represent and reproduce these interactions in scientific workflow systems for grid computing such as Triana [5] and Taverna [6, 10]. Weber Reichert and Rinderle-Ma [7] have surveyed many scientific workflow systems and note they lack standardization in workflow specification syntax and semantics. On the other hand, due to the iterative and interactive nature of the analysis process, any workflow-like system needs to be flexible and support process change at run time. There has been much research on flexible Process-Aware Information Systems (PAIS). The term PAIS covers systems that allow for separating process logic and application code such as Workflow Management Systems and Business Process Management Systems (BPMS) [10]. Despite providing users with the flexibility to develop their own processes, users still have to manage and handle different data formats as well as find ways of integrating existing software packages and tools.

Earlier work has produced the Ad-hoc Data Grid Environment (ADAGE) framework which a general framework that can be adapted to many different types of analysis [9]. Its central idea is using a Service Oriented Architecture (SOA) as the underlying platform upon which user-defined analysis processes can be constructed. ADAGE consists of three components: the system architecture, a reference data model for managing ad-hoc data, and a set of services that performs the analysis. However, ADAGE does not specify how service composition should be performed by application users in the context of analyzing and exploring large datasets. In addition, modifying an existing service composition (for example to change a visualization output) is no trivial task.

3   Proposed Approach

In ADAGE, a typical analysis process is incrementally constructed until some dataset (called Event Set) of interest is created and visualized. The proposed work is motivated by the fact many analysts modify their business processes interactively based on what is presented to them visually in order to create a different event set.

Some steps in the process are tied to a data source (usually exposed by a Web Service). We now introduce some concepts aimed at capturing the essential characteristics of this type of analysis processes.

3.1 Visualization Instance

A Visualization Instance (VI) is a set of items which captures a visualisation case. This includes a Business Process, Data Sources, UI Controls and a Display Type defined as follows:

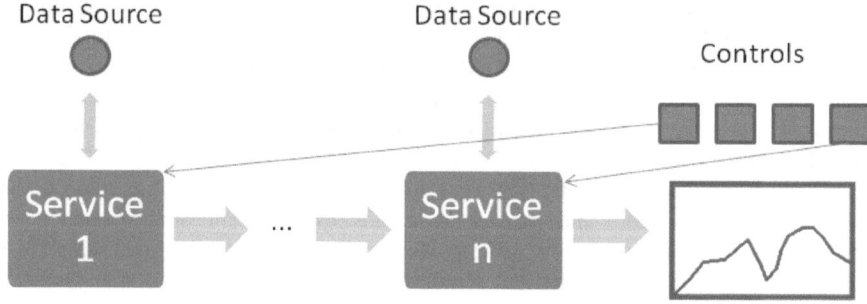

Fig. 1. Controls are bound to services and can change the event set

- **Business Process:** A series of services that either utilize data sources or process data. A business process eventually produces an event set.
- **Data Sources:** Provide data to business process steps and controls. These will usually harness a Web Service based SOA.
- **UI Controls:** Web interface controls like dials, radio buttons and sliders. These can be tied to the output of business process steps.
- **Display Type:** A particular method of displaying event sets. Examples are a line graph or a pie chart.

Currently, producing the right display is not an easy task for the user because a static business process must be defined beforehand with the right parameters to specify the desired output. It is proposed that the user can now control the visualization output in two ways:

- **Changing Display Type:** Controls can be used to change, filter and generally manipulate the way the event set data produced by the last step in the business process is visualized.
- **Changing Event Set:** Controls can be wired to the output of any step n of the business process causing service n+1 and all successive services to be executed with different or additional data. This causes the business process to re-execute from service n+1. A new event set is produced from the final service in the business process. This set of characteristics is called a Visualization Instance (VI).

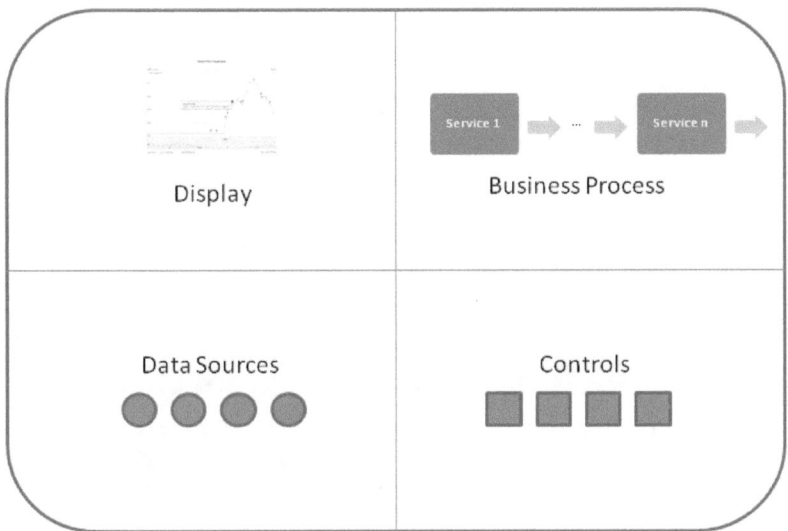

Fig. 2. A visualisation instance

3.2   Combining Visualization Instances

An Exploratory Analysis Process (EAP) is an open ended, usually iterative analysis of data exploration. EAPs often involve steps that cannot be predicted before the analysis starts. The user may for example find a pattern or data in a visualization step and then wish to produce another visualization based on conclusions made from that step. One solution to this problem is to introduce a system that involves several visualization steps and allows the user to arbitrarily display information. For this an overlaying system is introduced. The system comprises:

- **Active Overlays (AOs):** Traditional overlays are just graphical. They might for example, be a graph set to a certain opacity to augment another graph. An active overlay has a bound set of controls that work on the data for that particular VI. For instance, filtering out less relevant data or changing the input parameters to a service in a business process.
- **Interaction Panel:** A storage area for active overlays.

AOs can be categorized in two ways:

- **Instance Active Overlays (IAOs):** The AO is an instance of a VI. All controls are bound to data sources within that VI. Users can change to display subsets of controls for a particular AO. Overlays could also be reorderable , resizable, draggable and have different levels of opacity.
- **Composite Active Overlays (CAOs):** A CAO is a saved series of overlays. The user could for example combine several IAOs, adjust the control set and then save the entire set back into the interaction panel. IAOs (and CAOs) can be combined to produce a CAO. CAO data sources can be combined in several ways. For example data sources can be executed in the order they were originally overlaid. This paper only considers IAOs.

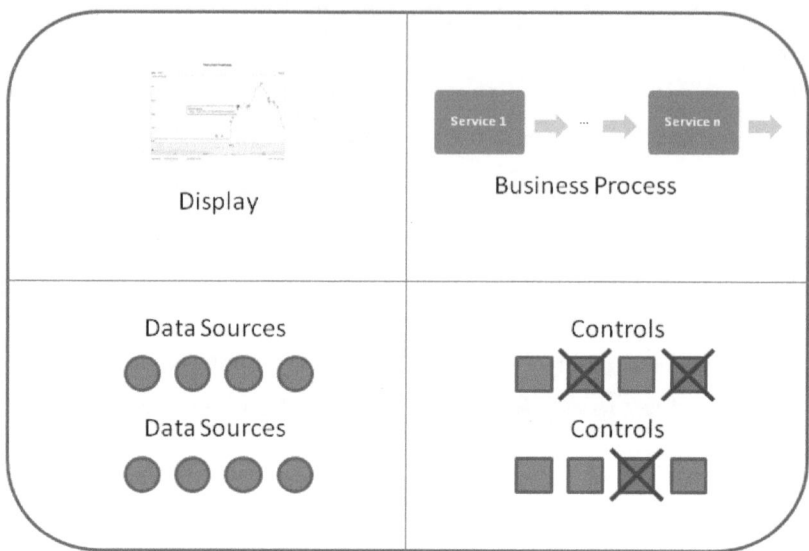

Fig. 3. Combined visualization instances

4   Case Study and Implementation

We now describe a case study in the area of processing large news datasets from Thomson Reuters, using the ADAGE-based News Processing Portal (N.PP)

4.1   The News Processing Portal (N.PP)

The News Processing Portal adopts the ADAGE approach [9] and takes a "service-oriented" view on news processing [11]. Any complex news processing application can be tackled by composing a set of judiciously designed services. In other words, users are presented with a set of services (i.e. a Service Oriented Architecture or SOA), each of which can be used as a building block in their own analysis business processes (this is in line with established SOA practice).

In the ADAGE approach, services are classified into these broad groups:

- Import Services: allow data to be read from a primary news data source. This could be done in real-time, periodically, or as a one-off exercise. Import services produce *news event datasets* that are consumed by the processing services.
- Processing Services perform a range of functions on news datasets e.g.,
  o Text processing services: perform analysis on news content to extract meaningful information;
  o News aggregation services: reduce the size of datasets by performing some kind of aggregation on the news;
- Export Services: convert datasets into forms (e.g., spreadsheets and graphs) that can be viewed or exported to other software tools.

The main source of data used in N.PP is from Thomson Reuters which was formed by the merger between Thomson Corporation and Reuters Group PLC in 2008. Thomson Reuters commanded 33.4% of the global market share for market data providers [12]. The news archive concerned in this project consists of the Reuters NewsScope Archive (RNA) which includes every iteration of stories transmitted by Reuters journalists over Reuters Integrated Data Network (IDN) since 2003, with timestamps accurate to the millisecond when they were transmitted [11].

4.2   Visual Instance Technologies for News

An example of how VI techniques can be harnessed for the N.PP platform is presented.

Definitions
The Thomson Reuters news archive contains a number of major categories of news. These are:

— Macroeconomic announcements: regular government financial press releases.
— Regulatory announcements: regular listed company announcements.
— Press announcements: non-regulatory listed company announcements.
— Analyst recommendations: ad hoc buy sell listed company recommendations by analysts, usually hired by investment banks.
— Journalistic commentary: various press releases by journalist related to listed companies.

Some examples of kinds of analysis that can be performed on news data include:

— Keyword search: raw search for articles based on a combination of keywords.
— Topic search: a search for articles based on a combination of keywords and a semantic analysis of the meaning of those keywords.
— Sentiment analysis: A semantic analysis of news stories aimed at gauging if the news is good or bad for profitability of a specific company.
— Impact analysis: A semantic analysis of news stories aimed at determining the impact on profitability (or relevance) of a news story to a specific company.
— Hypothesis testing: Testing of a particular market related hypothesis may incorporate Thomson Reuters news data is some way.

These can all be considered when deciding which display technique or controls to use.

Existing Implementation
End to end, the News Processing Portal implements a single VI. An example of this is using a merge processing service in the N.PP to combine the results of import services for news and price data (also available from Thomson Reuters) into an event set. The event set is then converted via an export service into a graph. The visualization instance concept also extends the ADAGE export service by introducing the concept of UI controls bound to data sources to dynamically change the display.

Problem: User Intervention Needed at Different Levels
There are two different UI issues:

— There is no consistent , user friendly way to use visual information across VIs (e.g. overlaying a previous graph on to a new one) and
— Current visualization techniques are limited. There are many ways to increase flexibility of each step.

The proposed solutions are presented in the following subsections.

New Display Techniques
Currently the visualization models just consider basic well known dimensions like price. A number of new dimensions could be used for news analysis, either directly or through combining/processing data. Impact for example, can be measured by an external service and could be represented as an opacity setting with the highest impact stories being the most solid. There could be two types of scale, one based on the highest current value and another on a preset value. A slider based filter control could also be used to filter out stories based on the position of the slider and the relative impact of the stories. Another dimension is sentiment. Tools for determining sentiment are the Harvard enquirer, SentiordNet, RapidSentilyzer, or a service provided by Thomson Reuters. These services could be integrated as data sources. Questions like "What are the ways to map a sentiment to changes in price?" could lead to new display techniques and control sets. A simple example is choosing the highest impact story of the day and then coloring the story green or red depending on whether the story is positive or negative.

Display Techniques for Grouping of News Items
News stories are related to each other in a number of ways. A number of extensions to a basic line style graph have been suggested. These include:

— Topic codes. E.g. the number of topics for a day or the topics codes displayed as a column at the top of a graph.
— Related news stories could all be colored the same
— A slider could range from "Keep all duplicates" to "Remove all duplicates" to control the number of duplicate stories shown.
— News story start and end time could be linked by a line
— Related RICS could be colored similarly
— News types could be grouped together
— News frequency. E.g. the number of stories on a given day could be shown as a number
— The amount of news (news frequency) could be mapped to price
— Keyword searches on news items mapped against a graph (i.e. remove all data points that don't contain that keyword)

Display techniques involve using a base chart or graph and then augment it with decorations around a data point. An extensive library of charts and graphs would also be available to the user. These could include Area charts, Dot graphs, Bar charts, Heat maps, Chronoscopes and more.

Functions

Applying functions where applicable is another possible extension. For example, using the price dimension we could apply a functional filter to a subset of the event set data and highlight "Abnormal returns" (prices that have moved abnormally) and "Volatility" (prices that are highly volatile).

Instance financial overlays might include functions like resistance, trend lines, moving averages, pivot points, moving averages, etc. Another idea is to use overlays to show known patterns like a logarithmic line, sine wave and so forth.

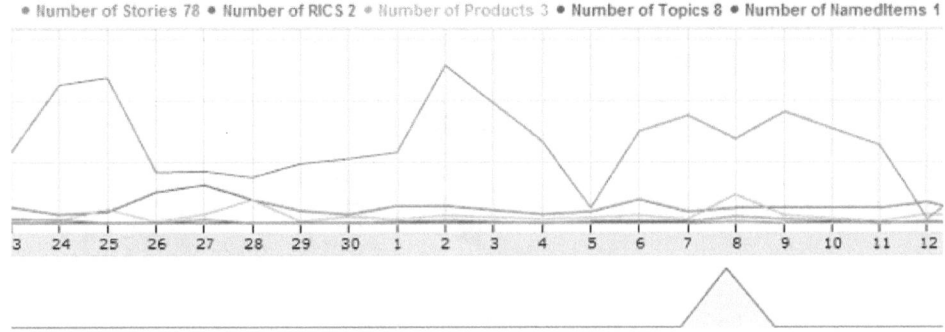

Fig. 4. Some new metrics

Controls for News Data

Examples of sets of controls for news stories that can integrate with existing data are:

— User defined filters. E.g. a series of radio buttons to filter out types of news
— UI sliders and dials. E.g. for the importance of a story
— Different sized circles on the y axis. E.g. showing relevance of story
— Search boxes that only display stories that match a certain string

Some controls will be tightly coupled with certain metrics or display techniques (due to the tool they rely on, e.g. sentiment analysis) but some (like a chart opacity slider for example) would be independent of the data.

Combining Visual Instances for News

IAOs from previous instances could be used to add information to the current instance diagram. There are obvious issues that arise. Examples are scale (for viewing complex data analysis), change of dimension between different analysis tools or types of data and diagram clutter. There are definite advantages; an example of an analysis process which can benefit from IAOs follows:

1) The user selects price vs number of related stories (for say BHP containing the name of a particular mine) via a merge, a graph is produced for iteration 1
2) The user drags the number of related stories data to the overlay canvas (by clicking on the red line or description). The scale is wrong to see price fluctuations so the user decides to generate another graph in iteration 2 with price vs time, which is normalised to the price (iteration 2).

3) The user then drags the overlay from iteration 1 which clearly shows price fluctuations against number of stories. It looks like there might be a correlation between the first two peaks but the third (and beginning of 4th) peak seems to rule this out.

4) The "slider" at the bottom of the graph is for sentiment and is instance based (acts on the overlay). When the user slides the slider to show just very positive stories the third and fourth peaks disappear, suggesting a correlation.

Note the sliders in the canvas section could be for opacity, size, etc. also - but their true power is when they work on a "related metric" like sentiment for stories.

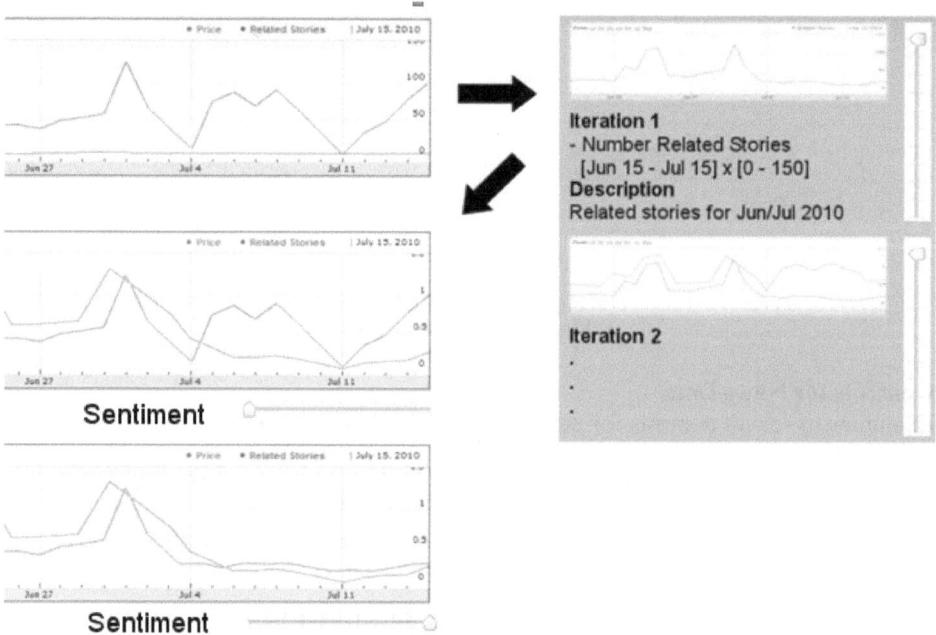

Fig. 5. The interaction panel with an example

4.3   Visual Instance Composer

To facilitate the building of VIs and IAOs a graphical tool is being developed. This tool allows the user to:

1. Build a business process (like a News/Pricing merge)
2. Associate controls with business process parameters
3. Allow the user to specify or automatically detect display and control types from event set data types
4. Move VIs on to the canvas (creating IAOs) and allowing them to be overlaid on to each other

Event sets would need mapping rules for each display type and controls types would be mapped to web service result data types (for example a Boolean could be mapped to an on/off switch).

4.4   Architecture and Implementation

The overall architecture is summarized in Figure 6. At the core of the proposed system is the Display adaptor for Exploratory Analysis Processes (DEAP) component which is used to create, modify and compose Vis. DEAP also allows the overlaying of IAOs and handles binding of controls to data sources. All DEAP (Interaction panel settings, VI state, etc) data is stored in a database.

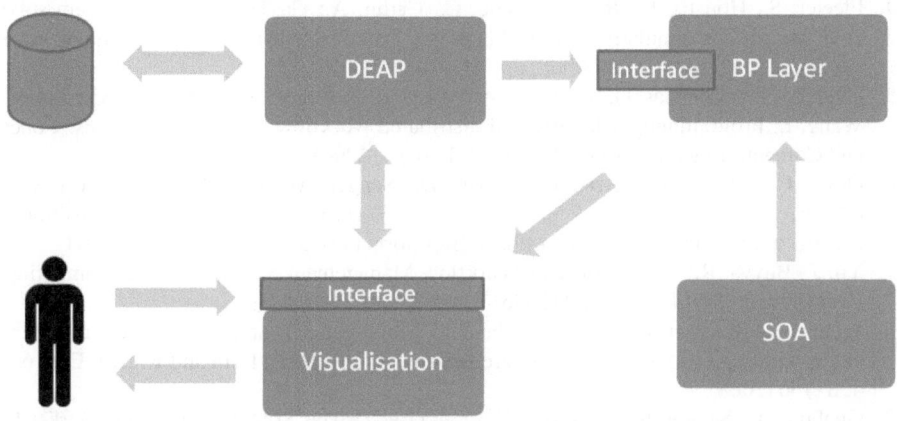

Fig. 6. DEAP Architecture

Two new interfaces link the DEAP framework to the existing Business Process layer and visualization system (N.PP). These interfaces are being built with HTML 5.0 and JQuery which utilize the Google visualization API.

5   Conclusion and Future Work

Given the trend towards the use of services in the area of eResearch analysis, this paper proposes a number of abstractions to facilitate the definition and execution of Exploratory Analysis Processes (EAPs). The VI concept, IAOs, CAOs and the DEAP architecture are introduced and demonstrated on a case study in the area of news and financial data analysis. These concepts can also be applied to domains other than finance as long as there is an underlying SOA available. For example overlaying would be useful in any situation that benefits from a line graph with two axes. Visual instance display filters such as dials, slider, and radio buttons could be used on many types of event sets, as can grouping techniques such as node coloring.

Acknowledgements. We wish to thank Sirca (www.sirca.org.au) for making financial news and market data available for this project. We also wish to thank Cat Kutay for helping on the visualization aspects of this paper.

References

1. Brachman, R.J., Khabaza, T., Kloesgen, W., Piatetsky-Shapiro, G., Simoudis, E.: Mining business databases. Commun. of the ACM 39(11), 42–48 (1996)
2. Appelbe, B., Bannon, D.: eResearch — paradigm shift or propaganda? J. Res. and Pract. in Inf. Technol. 39(2), 83–90 (2007)
3. Houstis, E., Gallopoulos, E., Bramley, R., Rice, J.: Problem-solving environments for computational science. IEEE Comput. Sci. and Eng. 4(3), 18–21 (1997)
4. Fleeter, S., Houstis, E., Rice, J., Zhou, C., Catlin, A.: GasTurbnLab: a problem solving environment for simulating gas turbines. In: Proc. 16th IMACS World Congr. on Sci. Comput., Appl. Math. and Simul., pp. 104–105 (2000)
5. Churches, D., Gombas, G., Harrison, A., Maassen, J., Robinson, C., Shields, M., Taylor, I., Wang, I.: Programming scientific and distributed workflow with triana services. Concurr. and Comput.: Pract. and Exp. 18(10), 1021–1037 (2006)
6. Oinn, T., Addis, M., Ferris, J., Marvin, D., Senger, M., Greenwood, M., Carver, T., Glover, K., Pocock, M.R., Wipat, A., Li, P.: Taverna: a tool for the composition and enactment of bioinformatics workflows. Bioinformatics 20(17), 3045–3054 (2004)
7. Yu, J., Buyya, R.: A taxonomy of Workflow Management Systems for grid computing. J. Grid Comput. 3(3-4), 171–200 (2006)
8. Weber, B., Reichert, M., Rinderle-Ma, S.: Change patterns and change support features - enhancing flexibility in Process-Aware Information Systems. Data and Knowl. Eng. 66(3), 438–466 (2008)
9. Guabtni, A., Kundisch, D., Rabhi, F.A.: A User-Driven SOA for Financial Market Data Analysis. Enterprise Modelling and Information Systems Architectures 5(2), 4–20 (2010)
10. Tan, W., Missier, P., Foster, I., Madduri, R., De Roure, D., Goble, C.: A comparison of using Taverna and BPEL in building scientific workflows: the case of caGrid. Concurrency and Computation: Practice and Experience 22, 1098–1117 (2010)
11. Robertson, C.S., Rabhi, F.A., Peat, M.: A Service-Oriented Approach towards Real Time Financial News Analysis. To appear in Consumer Information Systems (2011)
12. Burton-Taylor: Le secteur de l'information financière en pleine mutation (2010), http://www.burton-taylor.com/consulting/PR-LE-2-25-10.html

Form Annotation Framework for Form-Based Process Automation*

Sung Wook Kim

School of Computer Science and Engineering,
University of New South Wales
Sydney NSW 2052 Australia
skim@cse.unsw.edu.au

Abstract. Simple form of annotations, such as tagging, are proven to be helpful to end users in organising and managing large amount of resources (e.g., photos, documents). In this paper, we take a first step in applying annotation to forms to explore potential benefits of helping people with little or no technical background to automate the form-based processes. An analysis of real-world forms was conducted to design algorithms for tag recommendations and our initial evaluation suggests that useful tag recommendation can be generated based on the contents and the metadata of the forms. We also briefly present *EzyForms*, a framework for supporting form-based processes. The architecture supports an end-to-end lifecycle of forms, starting from its creation, annotation, and ultimately to its execution in a process.

1   Introduction

Adding metadata such as keywords or tags, is an effective way of categorising a large pool of resource for both personal and public purposes [1, 2]. The act of "tagging", due to its informality and flexibility in use, has been widely accepted by users for future browsing or searching in web-based systems like Flickr, Delicious, and Youtube . The motivations users have for sharing contents also attributes to the popularity of annotation. These include organisation for general public, communication with friends and family, and personal satisfaction [1]. A large body of research exists in various aspects of annotation such as tag usage [3, 4], tagging motivation [1, 4, 5] as well as tag recommendation strategies [6] and tooling supports for annotation [7–10] of online contents such as photos and social bookmarks. We leverage some of the lessons learned from the previous research work to design and implement a form annotation framework called *EzyForms* which collects domain knowledge from the form owners into the system's Knowledge Base (KB) to help modelling and execution of the form-based processes which are prevalent in every part of our daily life such as applying for a drivers licence, opening a new bank account, or applying for a grant or travel within an organisation. We envision that form annotations can ease many aspects of form-based process modelling (e.g., form selection, discovery, input field

---

* This work is supported by SDA Project, Smart Services CRC, Australia.

A. Haller et al. (Eds.): WISE 2011 and 2012 Combined Workshops, LNCS 7652, pp. 307–320, 2013.

mapping, and etc) and execution. However, we recognise the fact that manual tagging can be very time-consuming and tedious, and hence we present a list of tag recommendations as a part of the tooling support to ease the burden on the form owners.

EzyForms is an extension to our previous work called FormSys [11], in which PDF forms[1] are simply uploaded (like any other file-upload in a browser) to FormSys repository to automatically generate matching Web Services. These Web Services are then used by BPM designers to design and implement form-based processes. When a form is uploaded by the user, FormSys can dynamically generate two services:

- **soap2pdf**: receives data from an application, fills a form with it, and returns the form via email or an URL where the filled form is available.
- **pdf2soap**: extracts the data from a filled form, assembles and sends a SOAP message to an application.

The creation of Web Service is based on WSDL/SOAP standards and the procedure is completely automated and hidden to the end users. The idea in FormSys is that the end users can easily create Web Services out of forms used in their daily tasks - and some of them can be utilised in automation with the help of BPM professionals. Our current work aims to mature the tool further towards the end users by supporting a form annotation so that the form owners can get benefits of annotation in managing forms as well as process modelling.

Our contributions focus on supporting the tagging activities for forms. Specifically, we:

- Perform analysis on the forms being used in real-world and use it to generate tag recommendations.
- Design and implementation of tag recommendation strategy; specifically, for input field and form description annotation.
- Design and implementation of form annotation framework.
- Design and implementation of form-based process execution environment.
- Evaluation of input field tag recommendation strategy.

The remainder of the paper is structured as follows. More in-depth discussion on the form-based processes and why we are interested in automating the form-based processes are presented with an illustrating scenario in Section 2, followed by the analysis of the real-world forms in Section 3. In Section 4 we present tag recommendation strategies for input field and form description annotation which are based on the heuristics derived from the analysis results in Section 3. The prototype implementation and the evaluation is described in Section 5 followed by related work in Section 6. Finally, in Section 7 we come to the conclusions and discuss possible future works.

---

[1] To be precise, we use AcroForm, a sub-standard of PDF which contains editable and interactive PDF form elements.

2   Form-Based Processes

Forms enable cost effective and simple way of running and managing business processes. However, the use of the forms requires the end users like employees or students to do more manual work such as downloading the forms, typing in same information repeatedly, and getting approvals from several people from different organisation units. Moreover, the end users are often not familiar with the process and this costs them a lot of time and energy just to figure out which forms to fill in. The following real-world scenario outlines some of the drawbacks of using a form-based process.

Illustrating Scenarios: Academics and research students in the School of Computer Science and Engineering (CSE) at UNSW must complete a travel request process for work-related travels (e.g., attending a conference). The procedures and policies are described in a web page[2]. There are up to five forms involved in the process. Depending on the employment status or position held in the school, people need to choose a different set of forms (e.g., students are not required to fill in Teaching Arrangement Form) which are available from and managed by different business organisations in CSE/UNSW.

Despite these drawbacks, forms are still prevalent due to the fact that it requires heavy investment in IT and people in order to automate these processes with BPMS solutions or by implementing customised solutions, and many organisations do not see significant ROI in doing this [12].

Our current work aims to resolve this issue by supporting a **Form Annotation**. - We introduce an annotation step where the form owners annotate the forms with meaningful tags, and use this information to help the form owners to easily model and deploy new form-based process that the end users can select and execute without a hassle. To help the form owners with annotating the forms, we use various heuristics to generate tag recommendations which will be presented in more detail in Section 4.

Form Annotation: The form services are represented by WSDL, and currently there is no other mechanism to supplement how each form is described. From our own experience with the earlier version of the system, we have learned that there is a great need to enrich the form service description. For example, the text field names obtained from the PDF processing library[3] are not clean or complete to generate meaningful input field names in the corresponding SOAP messages. The rich business knowledge of the form owner about particularises about the usage of the form (e.g., in which process the form is used, who should use it) are not reflected in the automatically generated services.

To accommodate this, we designed a form annotation schema and the form owners are asked to perform various tasks to support the annotation process. Form owners are engaged in two separate levels of annotation activities for a form: input fields and form directives.

---

[2] CSE Travel Page, http://www.cse.unsw.edu.au/people/fipras/travel/

[3] iText, www.itextpdf.com

- **Input Fields**: the form owner chooses descriptive, human-readable names for the form fields using tags. This information is used to provide possible data mapping between different forms for input data sharing.
- **Form Directives**: the form owner chooses descriptive/representative tags for the form, she also annotates the form with conditions and rules that may apply to the usage of the form (e.g., approver, other forms that are used together). This information is used in form discovery/selection, and process execution.

In this paper, we focus on input field and form description annotation which are used by *EzyForms* for input field mapping recommendations and form discoveries respectively.

3 Analysis of Forms and Annotation Knowledge Base

For generating a tag recommendation list, we use the actual text appearing in the forms, as they are the most likely terms the form owner is going to find relevant for annotations. To understand the text extraction issue better, we studied 30 publicly available forms from 6 different organisations: UNSW FIPRAS[4], New South Wales Government Licence, Ohio State University Human Resources, Government of Ontario, Bank of New Zealand, and Australian Research Council.

Input Field Annotation: We first observe that the input fields in forms can be categorised as follows:

- *Personal Details*: Input fields that require personal information such as a name, address, or phone number of the form user.
- *Process Specific*: Input fields that are relevant to the business process the form is used for. For example, in a CSE travel form, this category includes fields such as a location, name, or a date of the conference.
- *Approval Related*: Input fields reserved for approval steps of the process. This includes the name of the approver and the date of approval.

For the input field annotation, we are interested in the first two categories.

Personal Details vs. *Process Specific* – The analysis showed (Table 1) that 56 out of 98 distinct input fields in the *Personal Details* category appear in other forms 184 times, which means that each input field were reused 1.88 times on average, compared to 0.11 times for the the input fields in the *Process Specific* category. This tells us that personal detail fields are much more likely to be reused when there are more than one form involved in a single process. Therefore the heuristics for generating a tag recommendation list target input fields in the *Personal Details* category in order to maximise the benefit.

---

[4] The Finance, Procurement and Assets Unit within School of Computer Science and Engineering.

Table 1. Different Categories of Input Fields: *Personal Details* vs. *Process Specific*

	Distinct	∼Reused	Reused	#Reused	Average
Personal Details	98	42	56	184	1.88
Process Specific	909	870	39	101	0.11

∼: Not, #: Number of times

Table 2. Types of Reused Personal Details Input Fields

	Forms	Reused	Textfield	Checkbox
UNSW FIPRAS	6	8	8	0
NSW Licence	3	11	9	2
OSU HR	5	6	6	0
Ontario	3	14	14	0
BNZ	7	9	9	0
ARC	7	8	8	0

Different Input Types – There are various types of input fields in forms including a textfield, checkbox, and radio button. Table 2 shows a number of times a textfield and a checkbox were reused for personal details input fields. It shows that majority of the input field being reused are of type textfield. This indicates that we can narrow down the extraction candidates to textfields related to personal details.

Positioning Text Labels – Next, we investigated where the relevant texts can be found within the forms for the input fields we are interested in. For each form, we analysed the positioning of the text we want to extract relative to each textfield used to for personal details.

The result (Fig. 1) shows that some organisations tend to be consistent in the label position they use. For example, reused personal input fields in UNSW FIPRAS forms always had text labels on the left(Fig. 1(a)), and for the OSU HR, majority of the input fields had a text label at the bottom (Fig. 1(b)). Of course, it is not always true for every organisations. For example, the text labels for NSW Licence department forms(Fig. 1(c)) were evenly placed on top, bottom, and left of each input field.

Form Directives Annotation: The form directive annotation we focus on for this paper is form description tag. There are two sources of information used to generate the list of tag form description tag recommendation which are: i) the words extracted from the form and the number of times the word appears in the document, and ii) the meta-data associated with the forms which contains information about the title and subject of the form.

Annotation Knowledge Base: Based on the observations, we first extract vocabularies that will form our initial "tag library" for input field annotation which are the labels related to personal details fields. This forms a part of KB that *EzyForms* uses to select a list of input field tag recommendation. When an input field is annotated with a new tag that did not exist previously in

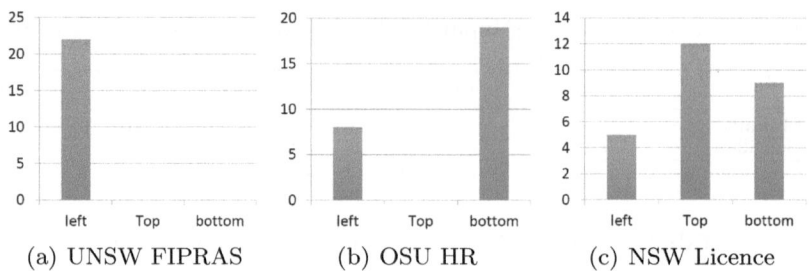

(a) UNSW FIPRAS   (b) OSU HR   (c) NSW Licence

Fig. 1. The position of texts relative to the text fields of personal details category

the tag library, it is added to the library so that they are available for tag recommendation for future forms.

4   Recommending Annotation Tags

4.1   Input Field Annotation

Once we have the text extracted for each input fields, we need to select a list of tags from the tag library to generate a tag recommendation list. The function *findTagRecommendation* listed in Table 3 returns a tag recommendation list for a given input field. The list is ranked using MS (Match Score).

Table 3. Overview of function findTagRecommendation

FUNCTION	findTagRecommendation
INPUTS	$t_i \in T$, $wo_i \in W$
	where, T is the set of all tags in the tag library, $T = \{t_1, t_2, t_3, \ldots, t_n\}$ and W is the set of words extracted from the form by the text extractor denoted by $W = \{wo_1, wo_2, wo_3, \ldots, wo_m\}$
OUTPUT	$R_i = \{(t_i, MS_i), (t_j, MS_j), (t_k, MS_k), \ldots, (t_l, MS_l)\}$
	where, R_i is the list of tag recommendation for input field i and MS_i is the Match Score calculated for the tag t_i ($MS_i \in [0,1]$)

The MS is calculated based on three different measures: Label Match (*Label-Match*), Position Match(*PosMatch*), and Id Match(*IdMatch*). *LabelMatch* provides linguistic similarity between the extracted texts and the tags in the tag library, *PosMatch* takes into account the position of the extracted texts relative to the input field, and *IdMatch* provides linguistic similarity between the PDF Acroform ID of the input field and the tags in the tag library. The MS is calculated as the weighted average of these three measures as shown in Equation 1.

$$MS_i = \frac{w1 * LabelMatch_i + w2 * PosMatch_i + w3 * IdMatch_i}{w1 + w2 + w3}$$

where, $(0 \leq w1 \leq 1)$ $(0 \leq w2 \leq 1)$ $(0 \leq w3 \leq 1)$ $(PosMatch \in [0,1])$ (1)

Weight w1, w2, w3 indicates the contribution of *LabelMatch, PosMatch*, and *IdMatch* respectively. The field id does not always have a meaningful name, but if the creator of the PDF form has assigned a meaningful id to the fields (e.g. words describing the field instead of randomly generated alphanumerics), it is a very useful source of information we can use. Therefore w3 is assigned twice as much weight than w1 and w2 by default.

Table 4. Weight values for calculating MS

Condition	w1	w2	w3
Default	0.25	0.25	0.5
Label Match == 1, PosMatch == 1	1	0	0
Label Match == 1, IdMatch == 1	1	0	0
Label Match == 1, PosMatch == 0, 0 < IdMatch < 1	0.5	0	0.5
Label Match == 0, PosMatch == 0, 0 < IdMatch	0	0	1
Label Match == 0, IdMatch == 0	0	0	0

***LabelMatch*:** The extracted texts are first tokenised based on the new line character. Then it removes unnecessary words from the list of string tokens, using a stop-word list. After the extracted texts are tailored, *EzyForms* enumerates all tags in the tag library and matches string tokens with the tags using *Ngram* algorithm. The *Ngram* algorithm calculates the similarity by dividing number of common N-grams by the total number of N-grams.

***PosMatch*:** We take an advantage of the fact that some business units tends to put the text label for personal information input fields in a consistent position for the forms they manage (as shown in Fig. 1). This predominant position each business units use to position the text label of the input field is referred to as a *regular* position for a given organisation. For example, the UNSW FIPRAS tends to put the text label on the left side of the textfield(Fig. 1(a)), therefore its *regular* position would be *left*. The term *irregular* position is used to denote the text labels that are not positioned in the *regular* position. For example, input fields that have a text label on the left of the textfield in case of OSU HR(Fig. 1(b)) are referred to as positioned in a *irregular* position. We assume that KB has information about where the *regular* position is for each business units so that we determine the *regular* position of each form by checking its form owner's business unit.

$$PosMatch_i = \begin{cases} 1 \text{ if } (pos \in regular \wedge 0 < labelmatch_i) \\ 0 \text{ if } pos \notin regular \end{cases}$$

$$where, pos = position \text{ of text extracted}$$

$$regular \in [left, top, bottom] \tag{2}$$

The Equation 2 shows that for a given input field, *PosMatch* function returns a score of 1 if the tag matches a text extracted from a *regular* position, and returns 0 otherwise.

IdMatch: The input field Id generated during the PDF form creation time is used to provide extra information for tag recommendation since we expect that many form developers will use reasonable Id that is proximate to the text label. *IdMatch* function also takes same approach as *LabelMatch* function. The mapper enumerates each textfield in the form and applies these rules to select tag recommendation from the tag library.

4.2 Form Directives Annotation

The function *findFormDescriptionTags* listed in Table 5 returns the tag recommendation list for form descriptions for a given form. It extracts words from the form, and then excludes unnecessary words using a stop-word list. After the extracted texts are tailored, *EzyForms* enumerates each word in the list and calculates the MS (Match Score) for each word.

Table 5. Overview of function findFormDescriptionTags

FUNCTION	findFormDescriptionTags
INPUTS	$wo_i \in W$ where, W is the set of words extracted from the form by the text extractor denoted by $W = \{wo_0, wo_1, wo_2, \ldots, wo_m \}$
OUTPUT	$R = \{(wo_i, MS_i), (wo_j, MS_j), (wo_k, MS_k), \ldots, (wo_l, MS_l)\}$ where, R is the tag recommendation list for form descriptions and MS_i is the Match Score calculated for the word wo_i

The MS for form description tag is calculated based on 3 different measures which are Subject Match(*SubMatch*), Title Match(*TitleMatch*), and Count Score(*CountScore*). The *SubMatch* and *TitleMatch* matches the word set W with the subject and title metadata of the form. A *SubMatch* score of 1 is given to the word that is found in the subject metadata, and a *TitleMatch* score of 1 is given to the word that is found in the title metadata. The *CountScore* is calculated based on the number of times a word is found in the form. The MS is calculated as the weighted average of these 3 measures as shown in Equation 3.

$$MS_i = \frac{w1 * SubMatch_i + w2 * TitleMatch_i + w3 * \frac{CountScore_i}{MaxCountScore}}{w1 + w2 + w3}$$
$$where, \quad w1 = 0.4, w2 = 0.4, w3 = 0.2$$
$$(SubMatch, TitleMatch \in [0,1]) \quad (3)$$

Weight w1, w2, w3 indicates the contribution of *SubMatch*, *TitleMatch*, and *CountScore* respectively. The subject and title metadata is not always present, but if the form developer has assigned a meaningful metadata to the fields, it is most significant source of information we can use. Therefore w1 and w2 is assigned a bigger weight than w3 by default. The *CountScore* for each word is divided by the maximum count score(*MaxCountScore*) obtained out of all available words in order to normalise a score between 0 to 1.

5   Implementation and Evaluation

The architecture we adopted is shown in Fig. 2. The forms uploaded to the Form Repository(1) are subject to annotations before they are deployed. FormSys Core Component(2) is responsible for the retrieval of forms and forms' metadata such as the position of the input fields for the Text Extractor(3) to extract the basis information. The tags that are available in Tag Library(6) are used by the Matcher(4) to generate the Tag Recommendation list(5). At form uploading and process modelling phase, the FormSys Web Front-End(7) provides an interface between the form owners and the system so that the forms are annotated and deployed as a service. At process execution phase, the FormSys Web Front-End interact with the Execution Engine(8) which relies on the process model to determine the flow of execution, and fills in the forms with end-users input data using soap2pdf(9) APIs. Upon the completion of the process execution, the filled-in forms are either emailed to a designated person for an approval or the URL is presented to the form user for downloading.

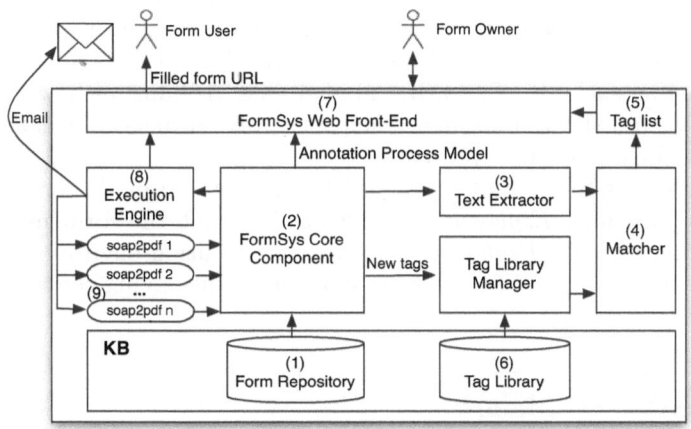

Fig. 2. *EzyForms-* Architecture

5.1   Case Study Implementation: CSE Travel Request Process

To demonstrate the applicability of our approaches to a real-world case, we took an example of form-based process from our own school which was described in Sect. 2. It can be demonstrated in two different perspectives. One from the perspective of the form owners who upload the forms and deploy processes, and from the perspective of the form user, in this case, to request a travel.

Figure 3(a) shows a tag recommendation list that is automatically generated by *EzyForms* for a input field *name* in the Teaching Arrangement form. In this case, the *EzyForms* came up with three tag recommendations which are *name*,

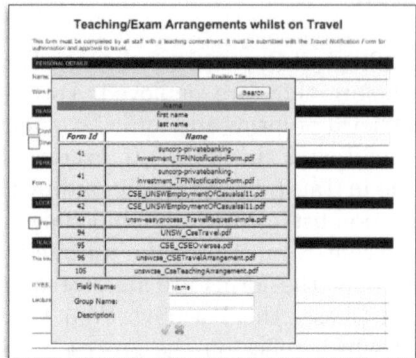

(a) Tag recommendation for an input field   (b) Same tag in another form

Fig. 3. Input field annotation performed by the form owner

first name, and *last name*. When a *name* tag is selected, it also show a list of forms that contain a *name* tag. When the name of these forms is clicked, *Ezy-Forms* opens the form image and shows the actual input field that is annotated with this tag as shown in Fig. 3(b). This gives a visibility of how the tags are being used within the system, hence helps the form owners to decide which tag name to use for a particular process.

Figure 4 shows the process executed by the form user. When a process is deployed successfully, *EzyForms* provides a link that the form user can click to initiate the process. *EzyForms* provides an execution environment as if the end-user is filling in the actual form document by displaying the form image in

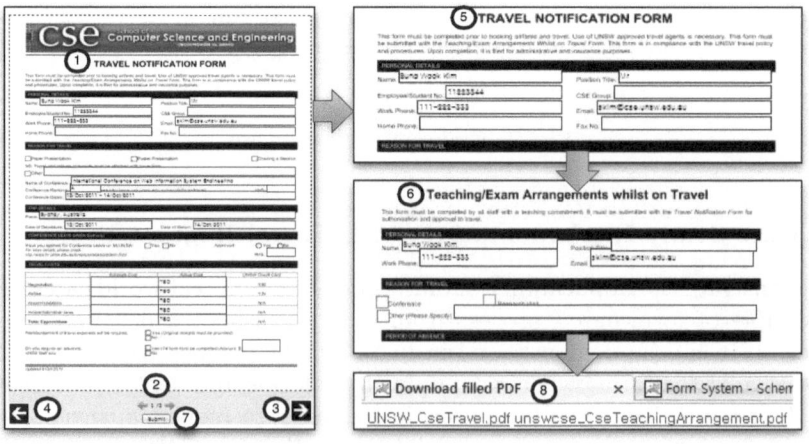

Fig. 4. Process Executed by the End-users

the background and overlaying textfields in appropriate positions for the form user to input data. *EzyForms* displays a single page at a time and provides a left and right button(2) to flip though the pages. When the first form of the process is completed, the form user can move on to the next form document by clicking next form button(3) or go back to the previous form by clicking the previous form button(4). When the form user moves to the next form, the input data entered in the previous form(5) is passed on to the next form and is used to fill in the input field with same tag name(6). Once all forms are completed (submit button(7)), the execution engine invokes each form service to fill in the form. Once the process is completed successfully, a link(8) to the filled in form documents is returned to the form user.

5.2   Evaluation of Text Extraction from Forms

For the evaluation of our text extraction approach, we focus on evaluating how well our text extraction and matching score works in selecting a proper tag from the tag library. We make an assumption that we have a populated tag library that contains input field labels for personal information for every form we test, and use common sense to decide whether we have a proper tag recommendations or not. We adopted two metrics, that captures the accuracy of tag recommendation at different aspects:

Success at rank k (S@k) represents the probability of finding a good descriptive tag among the top k recommended tags. For this evaluation, we used two values of k: S@1 and S@5.

Average Match Score(AMS) represents the match score of first relevant tag returned by the system, averaged over all input fields that found a relevant tag. An AMS value is calculated per organisation.

Table 6. Evaluation results for our tag recommendation strategy for personal details input fields

Organisation	Forms	Input fields	S@1	S@5	AMS
UNSW FIPRAS	6	31	90%	97%	0.81
NSW Licence	3	30	47%	73%	0.41
OSU HR	5	33	61%	84%	0.65
Ontario	3	42	64%	86%	0.66
BNZ	7	49	88%	98%	0.80
ARC	7	37	100%	100%	0.99
Average			76%	90%	0.68

Based on the evaluation result shown in Table 6, we observed that 76% of the time, the proper tag for each personal details input field received the highest score amongst a list of recommended tags, and 90% of the time, the proper tag was found amongst top 5 tag recommendations. The Average Match Score for input fields in 30 forms from 6 different organisations were 0.68. In observing input fields that did not retrieve a proper tag, we have found three main reasons why *EzyForms* were not able to generate a proper tag recommendation. First of all, it was due to *irregular* position of the text label. When it's difficult to determine which is the *regular* position for a certain organisation (i.e., positions of the text labels are not consistent), it is difficult for *EzyForms* to decide which tag is more relevant than the other when there are raw texts extracted from more than one position. Second, some input fields have rather descriptive text labels (e.g., Address (*licence address must be within NSW*)), and since input field tag recommendation strategy relies on the *Ngram* algorithm to compare the strings, more descriptive text label tends to hamper finding a proper tag recommendation. The third reason is somewhat relevant to the previous point. It is the size of the rectangle we use to extract raw texts. Since the length of the text labels is all different, the size of the rectangle could be too small that it did not extract enough information to determine a proper tag, or it could be too big that the size of the raw texts extracted is too big for accurate recommendation.

6   Related Work

Annotation of resources and its tooling support has been an active field of research. A recent study by Ames et al. [1] on the motivation for annotation in Flickr and the role of tag suggestion in the system has provided number of implications for the design of annotation system in general such as making the annotation pervasive and multi-functional, and not forcing the users to annotate. Their work also reveals that easy annotation features and relevant tag suggestions encouraged the tagging and gave users direction as to the sort of tags they should use. Marlow et al. [10] provides some insights for design decisions in architecting new tagging systems. They point out the importance of studying the incentives for driving participation, and the level of system support to embrace or limit these motivations.

[13] seeks to extract domain ontologies for Web Services from textual content attached to the services. Another work [14] proposes a large-scale annotation framework that does not rely on existing ontologies, but builds its vocabulary dynamically during the process. These are complementary to our work in that our initial annotation knowledge base construction follows similar approaches and these work can help during the initial bootstrapping phase of the *EzyForms* lifecycle.

We see the commercial work in this area as an important and relevant work we can learn from and contribute to. Microsoft SharePoint [15] suite provides a rapid development environment for, among other things, custom design forms and workflows. Its tool for managing forms allows the users to create forms and apply conditional formatting, actions and validations. Its workflow design tool is able to support creation and execution of sophisticated approval workflows. Liferay suite (e.g., Social Office and Portal [16]) focuses on the synchronisation of the work and communication between the team members. However, it has limited support for process automation. For example, its workflow engine Kaleo is used to control creation, approval and rejection by a person of one "asset" (e.g., a document), whereas a form-based process may involves multiple of such "asset". There are many other tools (e.g., Adobe LiveCycle ES2 [17], SAP Gravity [18]) in this category we can discuss, but the strong difference is that our focus is in supporting form-based processes in its "AS-IS" state (i.e., low cost, time and effort), which means we do not require the forms or user interfaces to be designed and created from scratch. We use the forms that are in situ and the users will always see them as the main interacting components for the processes.

7 Conclusions

EzyForms aims to improve the usability of form-based processes for both form owners and form users without heavy investment in IT infrastructure or human resource. In order to remove the complexity involved in the modelling of business processes, we proposed a new approach using form annotation. However, manual annotation is time-consuming and tedious. Therefore, in this paper, we first presented the results of the real-world form analysis which forms the foundation for the tag recommendation strategy presented in the second part of the paper. We then presented a prototype implementation and evaluation, which showed that the heuristics and algorithms incorporated in *EzyForms* works well in generating a tag recommendation for input fields, and lead us to believe that our approach can significantly contribute to the automation of form-based processes. We also presented the execution environment of *EzyForms* using an example of CSE travel request process which reuses the input data across multiple forms in a same process, and provides "AS-IS" look-and-feel environment to end-users by using a form as an main interaction object in the user-interface.

Our main idea for *EzyForms* is that, it will become an environment where the knowledge contributions from the users are reflected in the continual improvement of the system to support the long-tail processes. Hence, as for future work, we first plan to expand current KB by supporting additional form directives such as conditions or rules that may apply to the usage of the form, and email templates to support simplified email approval request. Secondly, we are investigating different types of matching relationship between the tags to improve data matching between the forms. And we are also looking for ways to utilise the annotations to semi-automate the form discovery/selection processes, and to find possible data matchings between the forms.

References

1. Ames, M., Naaman, M.: Why We Tag: Motivations for Annotation in Mobile and Online Media. In: Proceedings of the SIGCHI Conference on Human Factors in Computing Systems, CHI 2007, San Jose, California, USA, p. 971 (2007)
2. Cattuto, C., Loreto, V., Pietronero, L.: Semiotic dynamics and collaborative tagging. Proceedings of the National Academy of Sciences 104, 1461–1464 (2007)
3. Golder, S., Huberman, B.A.: The structure of collaborative tagging systems. Technical report, HP Labs (2005)

4. Nov, O., Ye, C.: Why do people tag? motivations for photo tagging. Communications of the ACM 53, 128 (2010)
5. Sen, S., Lam, S.K., Rashid, A.M., Cosley, D., Frankowski, D., Osterhouse, J., Harper, F.M., Riedl, J.: Tagging, communities, vocabulary, evolution. In: Proceedings of the 2006 20th Anniversary Conference on Computer Supported Cooperative Work, CSCW 2006, Banff, Alberta, Canada, p. 181 (2006)
6. Sigurbjörnsson, B., van Zwol, R.: Flickr tag recommendation based on collective knowledge. In: Proceeding of the 17th International Conference on World Wide Web, WWW 2008, Beijing, China, p. 327 (2008)
7. Girgensohn, A., Adcock, J., Cooper, M., Foote, J., Wilcox, L.: Simplifying the management of large photo collections. In: Proceedings of INTERACT 2003, pp. 196–203. IOS Press (2003)
8. Kuchinsky, A., Pering, C., Creech, M.L., Freeze, D., Serra, B., Gwizdka, J.: FotoFile: A Consumer Multimedia Organization and Retrieval System. In: Proceedings of the SIGCHI Conference on Human Factors in Computing Systems, CHI 1999, Pittsburgh, Pennsylvania, United States, pp. 496–503 (1999)
9. Shneiderman, B., Kang, H.: Direct annotation: a drag-and-drop strategy for labeling photos. In: IEEE International Conference on Information Visualization, London, UK, pp. 88–95 (2000)
10. Marlow, C., Naaman, M., Boyd, D., Davis, M.: HT06, tagging paper, taxonomy, flickr, academic article, to read. In: Proceedings of the Seventeenth Conference on Hypertext and Hypermedia, HYPERTEXT 2006, Odense, Denmark, p. 31 (2006)
11. Weber, I.M., Paik, H.Y., Benatallah, B., Gong, Z., Zheng, L., Vorwerk, C.: FormSys: Form-processing Web Services. In: Proceedings of the 19th International Conference on World Wide Web, WWW 2010, Raleigh, North Carolina, USA, p. 1313 (2010)
12. Larson, P.: BPM Suites and the Long Tail of Process Automation (2005), www.philiplarson.com/docs/bpm-longtail.pdf
13. Sabou, M., Wroe, C., Goble, C., Mishne, G.: Learning domain ontologies for web service descriptions. In: Proceedings of the 14th International Conference on World Wide Web, WWW 2005, Chiba, Japan, p. 190 (2005)
14. Kungas, P., Dumas, M.: Cost-Effective semantic annotation of XML schemas and web service interfaces. In: 2009 IEEE International Conference on Services Computing, Bangalore, India, pp. 372–379 (2009)
15. Microsoft SharePoint, http://sharepoint.microsoft.com
16. Liferay Social Office and Portal, http://www.liferay.com
17. Adobe LiveCycle EC2, http://www.adobe.com/products/livecycle
18. SAP Gravity, http://tiny.cc/gllur

A Web Services Variability Description Language (WSVL) for Business Users Oriented Service Customization

Tuan Nguyen, Alan Colman, and Jun Han

Faculty of Information and Communication Technology,
Swinburne University of Technology, Melbourne, Australia
{tmnguyen,acolman,jhan}@swin.edu.au

Abstract. To better facilitate business users in customizing Web services, customization options need to be described at a high level of abstraction. In contrast to related efforts that describe customization options at the technical level of service description, we propose a Web Services Variability description Language (WSVL) that facilitates the representation of such options at business level. The language has several advantages. Firstly, it does not require people, who perform customization, to have knowledge of Web service technologies. Thus, the language enables business users-friendly service customization. Secondly, the language captures not only what can be customized, but also how and where customization operations should happen in a service-oriented way. This self-described property removes the need for a separate procedure for governing service customization. Consequently, this property eases the adoption of the language. We elaborate the design of the language using a case study and describe its usages from both consumers and providers' viewpoints.

Keywords: Service variability, Service customization, Service description language, Feature model, Software Product Line (SPL).

1   Introduction

Services in Service Oriented Architecture (SOA) often contain many variants due to the variability in consumer requirements. In order to address such variability, service customization has proved to be an efficient approach [1-3]. Service customization refers to a *consumer-driven process* of deriving a *service variant*, which contains adequate service capability for a particular consumer, from a *super-service*, which contains a superset of all service capability required for all service variants.

In order to support service customization, customizable services need to be described. However, related work in the literature proposes approaches which are oriented toward IT professionals [2, 3]. In particular, all these approaches are concerned with expressing *variant service capabilities* in the service interface description (i.e. messages, operations, and data types) and exposing those variant service capabilities to service consumers for selection. Customizing services in these

A. Haller et al. (Eds.): WISE 2011 and 2012 Combined Workshops, LNCS 7652, pp. 321–334, 2013.

ways presumes that people who perform service customization have IT background and are very familiar with the technical description of services. And even for IT professionals, these approaches are still challenging because of the large number of variant service capabilities and dependencies among them [1].

In order to facilitate business professionals in customizing services, customization options need to be described at a high-level of abstraction. From business professionals' viewpoint, it is more important to know about what it is that a service variant will achieve, rather than how to technically invoke its capability. In other words, it is more beneficial for business professionals to be able to customize services at the problem space (i.e. business level service variability), rather than the solution space (i.e. technical realization of such variability). To this end, we are developing a feature-based service customization framework that captures and represents service variability at the business level so that it is much easier to customize services [1].

In this paper, we propose a language, namely Web Services Variability description Language (WSVL), for describing customizable services. WSVL adds variability description capability into WSDL, a de facto standard for describing services. Further, a WSVL document is *self-described* and captures not only the information on what customization options are, but also the information on how and where to perform such customization (i.e. the exchange of customization requests and responses) in a service-oriented way. The self-described property removes the need of defining a separate, informal service customization procedure which is likely to vary from one provider to another. Thus, it eases the adoption of the language. In addition to enabling service consumers to customize and consume one service variant, the WSVL language also allows service consumers to manage *variability inter-dependencies* between their applications and customizable partner services in case there are more than two service variants from the same customizable partner service involves.

The rest of the paper is structured as follows. Section 2 presents a case study that will be used throughout the paper to demonstrate the language. We elaborate motivations and designs of different aspects of WSVL in section 3. Section 4 presents the application of WSVL by demonstrating how consumers utilize WSVL documents in customizing services and how providers develop customizable services based on WSVL documents. Section 5 discusses related work and points to future work. We conclude the paper in section 6.

2   Case Study

Swinsure Insurance is an insurance company providing various types of building insurances to various consumers (e.g. insurance brokers, business users or personal users). In exposing its capability as a Web service (namely Swinsure WS), Swinsure Insurance has identified the following variations in its consumer requirements:

- Some consumers only need to get a quote and others will go on to purchase policies.
- Some consumers need to be able to view and update purchased policies.

- There are two policy types, namely residential insurance and business insurance, depending on the purpose of using a building.
- Consumers are able to include extra cover (i.e. accidental damage, fusion cover and extended third party liability) in their policies.
- When purchasing a policy, some consumers choose to use online credit card payment while others prefer to be issued a cover note and pay on invoice.

Note that these variations cannot be arbitrarily combined. Instead, there are dependencies among them. These dependencies come from both consumers' need and Swinsure Insurance's business policies. In particular, Swinsure Insurance has identified the following constraints:

- Those consumers who need to update policies also need to view policies.
- The extended third party liability extra cover is only available to business insurance policies.

In order to efficiently provide this Web service to consumers, Swinsure Insurance decides to develop a customizable service so that consumers can customize the service on their own while satisfying constraints imposed by the provider. In addition, Swinsure Insurance needs to describe its customizable service in a comprehensive and convenient way so that consumers can easily perform the customization. In the following section, we describe how to use WSVL for these purposes.

3   Web Service Variability Description Language (WSVL)

3.1   Overview

A WSVL document contains 4 different sections:

- *Feature Description* section describes the variability of the service by means of features.
- *Customization Description* section describes information related to customization operations.
- *Capability Description* section captures full service capability (i.e. superset of capability of all service variants).
- *Mapping Description* denotes the mapping between variant features and variant service capabilities.

Elements of the *Feature Description* section and the *Mapping Description* section are defined by our WSVL schema. Elements of the *Customization Description* extend the existing elements in WSDL due to semantic similarities. And elements of the *Capability Description* are existing WSDL elements. In the following subsections, we will describe each of these descriptions in detail.

3.2   Feature Description

One essential part of a service variability description is to express what can vary. This involves two types of information [5]. Firstly, what are variant service capabilities?

Secondly, what are dependencies among those variant service capabilities? Dependencies describe mutual inclusion and exclusion relationships among variant service capabilities. From consumers' viewpoint, it is also important that such variability is described at an appropriate level of abstraction for better comprehension. Therefore, in contrast to all related work focusing on capturing variability at the technical level, we support the description of variability at the business level.

To this end, the concepts of features and feature models from the Software Product Line (SPL) research domain well-suit our purpose [6]. SPL is a software engineering paradigm that aims to develop a family of software products from reusable core assets. In SPL, a feature model is used to capture the commonalities and differences among a family of software products [7]. Features are visible characteristics used to differentiate one family member from others. A feature model is a hierarchy of features with *composed-by* relationship between a parent feature and its child features. In addition, there are cross-tree *constraints* that typically describe inclusion or mutual exclusion relationships. A feature model is an efficient abstraction of variability and provides an effective means for communicating variability between different stakeholders. Therefore, the use of feature models in capturing business level service variability will provide an appropriate level of abstraction for service customization.

While there are many manifestations of feature modeling techniques, e.g. [8-11], in our work we exploit the cardinality-based feature modeling technique [12, 13]. The main reason for this choice is that the concepts of *feature cardinality* and *group cardinality* well suit the needs of service customization. A *feature cardinality*, associated with a feature, determines the lower bound and the upper bound of the number of the feature that can be part of a product. A *group cardinality*, associated with a parent feature of a group of features, limits the number of child features that can be part of a product when the parent feature is selected.

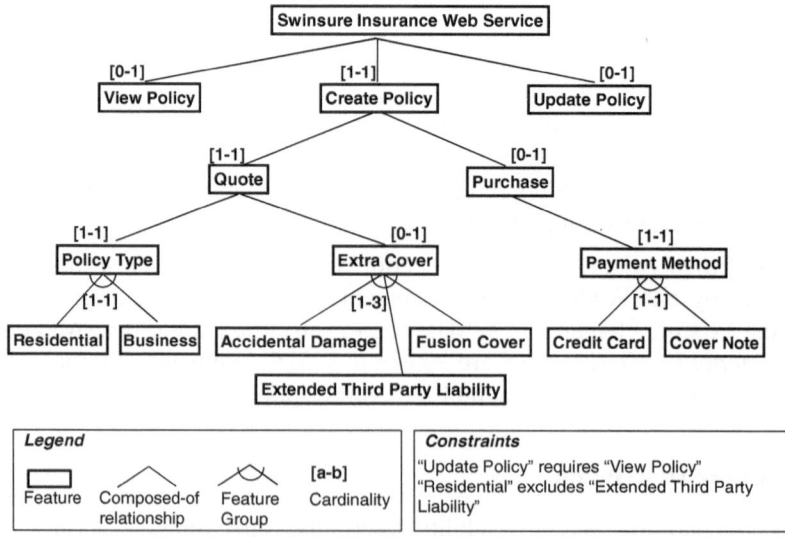

Fig. 1. Feature-based variability representation for the case study

Figure 1 demonstrates a feature model capturing variability of the case study. In the figure, cardinality represented above a feature is feature cardinality, while the one represented below a feature is group cardinality. In this feature model, feature *"Create Policy"* is mandatory (i.e. this feature is required by all consumers) while feature *"View Policy"* and *"Update Policy"* are optional (i.e. the necessity of these features are decided by consumers). Feature *"Create Policy"* is composed by a mandatory feature *"Quote"* and an optional feature *"Purchase"*. *"Policy Type"* is a mandatory feature and is also a feature group with two grouped features, namely *"Residential"* and *"Business"*. The group cardinality of this feature group is [1-1] which implies that consumers have to select one of these two alternative features for their service variants. All other features are described in a similar fashion.

Figure 1 also defines two constraints. These constraints represent dependencies between feature *"Update Policy"* and feature *"View Policy"*, as well as between feature *"Residential"* and feature *"Extended Third Party Liability"*. These constraints and the feature hierarchy precisely capture the variability of Swinsure WS.

The *Feature Description* section in a WSVL document represents the feature model. In particular, it contains description of feature hierarchy and description of feature constraints. Based on the feature description, consumers can specify a *feature configuration* and request a particular service variant. A feature configuration is a specialized form of a feature model in which all variability is resolved, i.e. all variant features are selected or removed.

Figure 2 presents an extracted XML from Swinsure WSVL for a partial feature hierarchy description, the feature constraint description, and a complete feature configuration. The feature configuration describes one consumer with minimum requirements. In this configuration, all optional features are not selected by the consumer when customizing the service.

```
<feature name="CreatePolicy" minCardinality="1"
         maxCardinality="1">
  <feature name="Quote" minCardinality="1"
           maxCardinality="1">
    <feature name="PolicyType" minCardinality="1"
             maxCardinality="1">
      <featureGroup minCardinality="1" maxCardinality="1">
        <feature name="Residential"/>
        <feature name="Business"/>
      </featureGroup>
    </feature>
  </feature>
</feature>        Partial feature hierarchy description
```

```
<featureDescription>
  <featureHierarchy>
    <feature name="SwinsureInsuranceWebService">
      <feature name="CreatePolicy">
        <feature name="Quote">
          <feature name="PolicyType">
            <feature name="Residential"/>
          </feature>
        </feature>
      </feature>
    </feature>
  </featureHierarchy>        A feature configuration
</featureDescription>
```

```
<featureConstraint>
  <constraint>
    <constraintDesc>if (//UpdatePolicy) then (//ViewPolicy) else true();</constraintDesc>
  </constraint>
  <constraint>
    <constraintDesc>if (//Residential) then not (//ExtendedThirdPartyLiability) else true();</constraintDesc>
  </constraint>
</featureConstraint>        Feature constraint description
```

Fig. 2. Extracted XML for feature model and feature configuration

3.3   Customization Description

In addition to the information about what can vary, a WSVL document needs to define how and where such variation can be requested. This is the information about how consumers should construct customization requests, where they should send those requests, and what should be expected as responses from service providers. The *Customization Description* section in a WSVL document is used for this purpose.

The *Customization Description* defines a set of customization operations that consumers may use to customize the service. One typical operation is the one that accepts a feature configuration as the request and returns a reference to WSDL description of a service variant as the response. However, there are many other possible customization operations. For instance, a customization operation might accept an incomplete feature configuration and return a revised WSVL document. This usage enables multi-stage service customization. Or service providers can group resolutions of several variant features into predefined packages. A service consumer can send one package as the request and the provider responses with an updated WSVL. Because of this variety, we have made the *Customization Description* generic, rather than confining it to describing only typical operations.

Customization operations can be described using conventional elements in WSDL due to the semantic similarity between the *Customization Description* and conventional service description. However, in order to clearly separate service operations for service customization and service operations for service consumption, we extend existing elements in WSDL to describe customization operations. Figure 3 presents an extract of Swinsure WSVL for the customization description. The customization operation *"swinsureCustomizationOperation"* is specified using element *<<wsvl:operation>>* which inherit the element *<<wsdl:operation>>*. Similarly, *<<wsvl:portType>>* is used to enclose all customization operations. Further, *<<wsvl:binding>>* specifies the binding of customization operations to transport and messaging protocols which are HTTP and SOAP respectively in the example. Lastly, *<<wsvl:port>>* specifies the endpoint to which a consumer can exchange customization requests and responses. As can be seen from the example, the *Customization Description* section provides sufficient information for governing service customization. Consequently, it makes WSVL documents self-described.

```
<wsvl:portType name="swinsureCustomizationPortType">
  <wsvl:operation name="swinsureCustomizationOperation"/>
</wsvl:portType>
<wsvl:binding name="swinsureBinding" type="swinsureCustomizationPortType">
  <soap:binding style="document" transport="http://schemas.xmlsoap.org/soap/http" />
  <wsvl:operation name="swinsureCustomizationOperation"/>
</wsvl:binding>
<wsdl:service name="swinsureCustomizationFrontend">
  <wsvl:port binding="swinsureBinding" name="swinsurePort">
    <soap:address location="http://localhost:9000/swinsureCustomizationFrontend" />
  </wsvl:port>
</wsdl:service>
```

Fig. 3. Extracted XML for customization description

3.4   Capability Description

One typical usage of customizable service descriptions is allowing service consumers to customize services. In such situation, the combination of feature description and customization description is sufficient. However, we also consider an advanced use of a customizable service in which the service is used in another customizable service and there are dependencies among their variants. In particular, the derivation of one variant of a customizable composite service requires a particular variant of a customizable partner services. We call such dependency as *variability inter-dependencies* [15]. To support variability inter-dependencies, WSVL needs to incorporate additional information explained in this section and the next one.

Let us consider an example in Figure 4. Swinbroker is building a customizable insurance quoting business process that reuses Swinsure WS. Depending on the type of policy its users request (i.e. residential or business), the Swinbroker will derive different business process variants that interact with different variants of Swinsure WS. In particular, for users requesting residential quoting, the Swinbroker will customize Swinsure WS with feature *"Residential"* enabled and feature *"ExtraCover"* disabled. The derived process variant will invoke the operation *getQuoting4Residential()* of the corresponding service variant. And for users requesting business quoting, the Swinbroker will customize Swinsure WS with feature *"Business"* enabled and feature *"ExtraCover"* disabled to be able to invoke the operation *getQuoting4Business()* of another service variant.

In such situation, it is necessary that Swinbroker knows not only the feature model of Swinsure WS, but also the consequence of customizing the service based on the feature model. That is, how they should interact with service variants which are the result of service customization. And such information should be made available before the service is actually customized (e.g. while Swinbroker models its business process). To this end, the WSVL document needs to describe full capability of a customizable service [16]. This full capability is the superset of capability of all service variants. We reuse WSDL elements for this purpose.

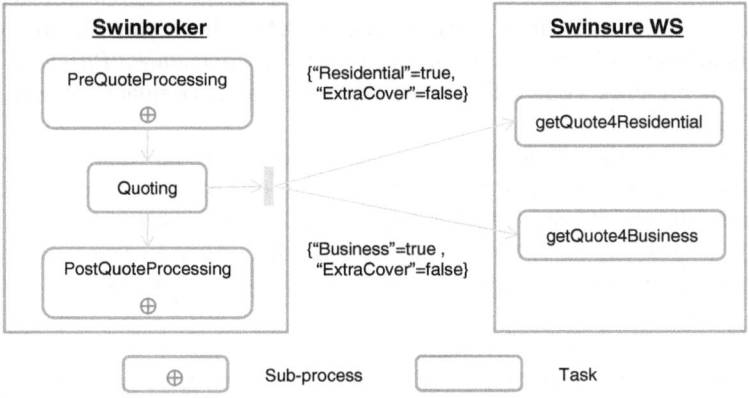

Fig. 4. Variability inter-dependencies between Swinbroker process and Swinsure WS

```
<wsdl:portType name="quotingPortType">
  <wsdl:operation name="getQuote4Residential"/>
  <wsdl:operation name="getQuote4Business"/>
</wsdl:portType>
<wsdl:portType name="purchasingPortType">
  <wsdl:operation name="purchasePolicyByCreditCard"/>
  <wsdl:operation name=" purchasePolicyByCoverNote"/>
</wsdl:portType>
```

Fig. 5. Extracted XML for service capability description

Figure 5 presents an excerpt of the capability description for Swinsure WS. There are two <<wsdl:portType>> of which the *"quotingPortType"* is available to all consumers while the *"purchasingPortType"* is only available if feature *"Purchase"* is selected. Among two operations of the *"quotingPortType"*, the operation *"getQuote4Residential"* (or *"getQuote4Business"*) is only available if the corresponding feature *"Residential"* (or *"Business"*) is selected and feature *"ExtraCover"* is disabled. Similarly, for two operations of the *"purchasingPortType"*, the operation *"purchasePolicyByCreditCard"* (or *"purchasePolicyByCoverNote"*) is only available if the corresponding feature *"CreditCard"* (or *"CoverNote"*) is selected.

3.5   Mapping Description

The full service capability description is the superset of the capability of all service variants. However, a WSVL document also needs to explicitly describe what capabilities are available if a feature is selected or disabled. We call a condition over a set of features from which we decide the existence of a variant service capability as *presence condition*. For example in Figure 4, the operation *"getQuote4Residential"* operation only exists in service variants if feature *"Residential"* is true (consequently feature *"Business"* is false because these two features are alternative) and feature *"ExtraCover"* is false. Or for the extracted XML in Figure 5, the portType *"purchasingPortType"* only exists in service variants if feature *"Purchase"* is true. Therefore, a presence condition can be expressed as a conjunction over a set of features involved.

A presence condition needs to be associated with relevant variant service capabilities. To this end, we introduce the concept of links as a mapping between a feature and relevant elements [17]. Each link has an additional attribute specifying whether feature should be selected or disabled for the presence of the elements. Since a presence condition is a conjunction over a set of features, the association of a presence condition and elements can be expressed as a set of links mapping relevant features and the elements. This approach does not require the use of a certain condition expression language for representing the presence condition in a WSVL document. Thus, we can avoid imposing more constraints on consumers in order to interpret the WSVL document.

```
<mappingInfo>
  <link name="LResidential">
    <featureRef ref="fd:Residential" presence="true"/>
    <serviceElementRef ref="tns:getQuote4Residential" target="operation" />
  </link>
  <link name="LExtraCover">
    <featureRef ref="fd:ExtraCover" presence="false"/>
    <serviceElementRef ref="tns:getQuote4Residential" target="operation" />
  </link>
  <link name="LPurchase">
    <featureRef ref="fd:Purchase" presence="true"/>
    <serviceElementRef ref="tns:purchasingPortType" target="portType" />
  </link>
</mappingInfo>
```

Fig. 6. Extracted XML for mapping description

Figure 6 presents an extracted XML from the mapping description section of Swinsure WSVL. The first two links associates feature *"Residential"* and feature *"ExtraCover"* with the operation *"getQuote4Residential"* operation with the *"presence"* attribute of *<<featureRef>>* elements are *"true"* and *"false"* respectively. These two links together specifies above mentioned presence condition that the operation *"getQuote4Residential"* is available only if *"Residential"* is true and *"ExtraCover"* is false. Similarly, the third link specifies that the portType *"purchasingPortType"* only exists if feature *"Purchase"* is true.

With regard to the types of element that will be objects of presence conditions, we consider only port types or operations. While there might be variants in message formats and data types that require presence conditions, we argue that those variabilities should be escalated to variabilities in operations (i.e. a separate operation for each variant message format/data types) and port types. This escalation enables simpler implementation of WSVL-based customizable services because in general each operation is transformed to a method in the service implementation [18]. As shown in Figure 6, types of elements are described using the attribute *"target"* of the corresponding *<<serviceElementRef>>*.

4   Application of WSVL

In order to illustrate the feasibility and applicability of WSVL we describe how service consumers and service providers utilize a WSVL document for service customization. While the mechanisms described in following subsections can be generalized, we just use them as one solution for validating the usefulness and the feasibility of the WSVL language. The explanation is based on Swinsure WSVL.

4.1   WSVL for Service Consumers

While WSVL uses XML which is a machine-readable format, many business professionals are not familiar with writing XML documents manually given an XML schema. Therefore, customizing services described in WSVL will be easier for

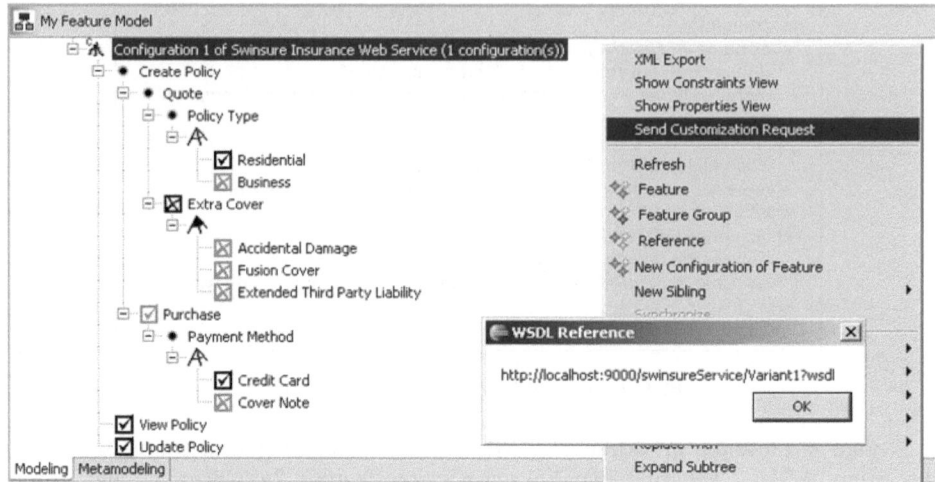

Fig. 7. Customizing services by business professionals

business professionals with the use of appropriate tools. We describe one such tool in this subsection. The tool is compatible with the WSVL schema to demonstrate the language's feasibility.

Figure 7 presents a screenshot of how business professionals can customize Swinsure WS in an intuitive and simple way. We extend the open source feature modeling tool to implement this plugin [19]. Information from the variability description section in the WSVL document is used to reproduce feature model on the consumer side. Feature model is then rendered to the business professionals through an interactive interface so that they can select features they need and remove features they do not need. As there exist feature constraints in the feature model, such constraints are used to validate business professionals' selection, as well as automatically propagate the selection through the feature configuration [14]. This helps to reduce the time and overhead during the customization process, as well as avoiding mistakes. For example, when a business professional selects feature "Residential", feature "Extended Third Party Liability" is automatically disabled due to the constraint between the two features (cf. Figure 1). Once the business professional finishes the customization, he can send a request for a service variant based on the generated feature configuration.

The feature configuration will be embedded in a SOAP message and be sent to the customization endpoint as specified in the customization description section of the WSVL document. As the result, the consumer will be issued an URL from which the consumer can retrieve the WSDL document of the generated service variant.

4.2   WSVL for Service Providers

We have implemented Swinsure WS as an atomic service. Figure 8 presents its software architecture. Swinsure WS has a customization Web service frontend that

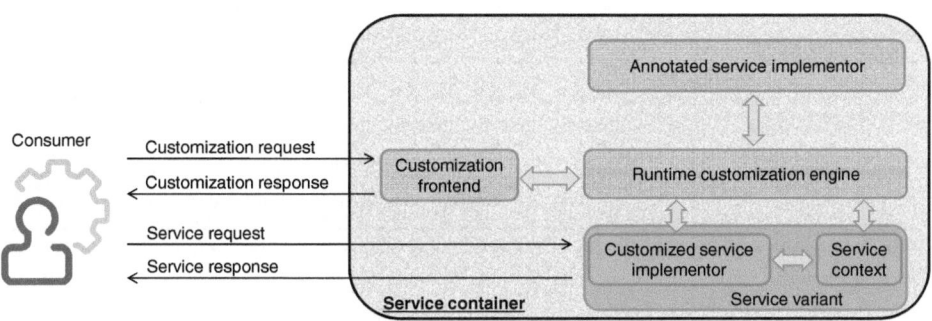

Fig. 8. Architecture for Swinsure WS

accepts a customization request (i.e. a feature configuration) from service consumers. The feature configuration is then passed to the *Runtime customization engine* which dynamically generates and deploys a service variant from an *Annotated service implementor* (described below). The reference to the WSDL of the generated service variant is returned to service consumers through the customization frontend. Consumers can then consume the service by invoking the service variant.

The *Customization frontend* is the Web service implementation of the customization description in the WSVL document. The *Annotated service implementor* is the implementation of service capability description in the WSVL document. The implementation contains additional annotations that specify presence conditions of variant portTypes and variant operations. An example of these annotations is shown with italic underline font in Figure 9. Note that these additional annotations are based on the information on the mapping description section of the WSVL document. The interface *PurchasingPortType* which implements the portType *"PurchasingPortType"* in the WSVL document is only available in a service variant if all features in the *enabledFeatureList* are selected and all features in the *disabledFeatureList* are removed. In this case, the interface is only available if feature *"Purchase"* is true. Similarly, the operation *"PurchasePolicyByCreditCard"* (or *"PurchasePolicyByCoverNote"*) is only available in a service variant if feature *"CreditCard"* (or *"CoverNote"*) is true. Based on these additional annotations, the runtime customization engine can derive and deploy a particular service variant that exposes appropriate service capability. Note that the *Service context* component of a

```
@WebService(name = "purchasingPortType")
@FeatureMapping(enabledFeatureList={"Purchase"}, disabledFeatureList={})
public interface PurchasingPortType {
    @WebMethod
    @FeatureMapping(enabledFeatureList={"CreditCard"}, disabledFeatureList={})
    public PurchasePolicyByCreditCardResponse purchasePolicyByCreditCard(PurchasePolicyByCreditCard request);

    @WebMethod
    @FeatureMapping(enabledFeatureList={"CoverNote"}, disabledFeatureList={})
    public purchasePolicyByCoverNoteResponse purchasePolicyByCoverNote(purchasePolicyByCoverNote request);
}
```

Fig. 9. Annotated service implementor

Fig. 10. Snapshots for published WSVL and WSDL documents

service variant stores information about what features are selected and what features are disabled for the service variant. The use of this component enables the correct business logic of the service variant during its execution in case dynamic switching between alternative behaviors for service variants is required.

Figure 10 presents screenshots of how a WSVL document for a customizable service (cf. left panel) and how the generated WSDL document for a particular service variant (cf. right panel) are published. The demonstrated service variant is the response for the feature configuration shown in Figure 7.

5   Discussion and Future Work

Describing customizability of a Web service has been a focus in a number of approaches [2, 3, 20]. Stollberg [2] makes explicit variants of service description elements (e.g. operations or data types) as well as constraints among them so that end users can select the appropriate variant. In contrast, Liang [3] defines customization policies that express which elements in the service description can be changed and what kind of changes is allowed. These customization policies are then used by consumers in customizing the service description. Tosic [20] lists all possible service variants from which consumers can select the most appropriate one. While these approaches are different in the way they represent customization options, they all focus on capturing variability at the technical level of service description. Consequently, the approaches cannot be used by business professionals who have a little knowledge of the underlying Web service technologies.

The work on WSVL is strongly influenced by the ideas of feature models from the SPL research domain [7, 12, 21]. And there are many variability description languages for representing feature models, e.g. [22-25]. Since our service customization framework utilizes the cardinality-based feature modeling technique [1, 12], we have develop the feature description part in the WSVL schema based on concepts of this technique. However, our contribution in developing WSVL is not about the representation of feature models in the WSVL Schema. Instead, we design the language so that it is self-described, comprehensive, and business users-friendly.

We have presented in the paper one way of developing a customizable service based on WSVL documents. We believe that such approach can be generalized to have a semi-automated process for developing customizable services using Model Driven Engineering (MDE) techniques. For example, additional annotations can be added as extension to JAX-WS [18] so that the skeleton of the service implementation (i.e. annotated service implementor) can be automatically generated from a WSVL document. Further, the software architecture for customizable services can be revised to have a middleware supporting customizable services.

6   Conclusion

In this paper, we define a Web Services Variability description Language (WSVL) that can be used to describe customizable services. The language facilitates business professionals to customize services on the consumer side by raising the level of abstraction at which customization options are described. To this end, we exploit the concept of feature models from SPL so that business professionals can reason about and customize services at the business level. The language is self-described because it describes not only what can be customized (i.e. feature description section), but also how and where such customization operations can be performed (i.e. customization description section). Furthermore, the language can be used to produce and consume one particular service variant, as well as support variability inter-dependencies between the service and other customizable service when more than one variant is involved (i.e. mapping description and capability description sections). The usage of the language is demonstrated thoroughly using a case study. We also describe how the language can be used by both consumers and providers with respect to service customization. As the future work, we plan to derive techniques for facilitating service providers in developing and deploying customizable services based on WSVL documents.

Acknowledgments. This research was carried out as part of the activities of, and funded by, the Smart Services Cooperative Research Centre (CRC) through the Australian Government's CRC Programme (Department of Innovation, Industry, Science and Research).

References

1. Nguyen, T., et al.: A Feature-Oriented Approach for Web Service Customization. In: IEEE International Conference on Web Services, pp. 393–400 (2010)
2. Stollberg, M., Muth, M.: Service Customization by Variability Modeling. In: Dan, A., Gittler, F., Toumani, F. (eds.) ICSOC/ServiceWave 2009. LNCS, vol. 6275, pp. 425–434. Springer, Heidelberg (2010)

3. Liang, H., et al.: A Policy Framework for Collaborative Web Service Customization. In: Proc. of the 2nd IEEE Int. Sym. on Service-Oriented System Engineering (2006)
4. Christensen, E., et al.: Web Services Description Language (WSDL) 1.1, March 15 (2001), http://www.w3.org/TR/wsdl
5. Schmid, K., et al.: A customizable approach to full lifecycle variability management. Science of Computer Programming 53(3), 259–284 (2004)
6. Pohl, K., et al.: Software Product Line Engineering: Foundations, Principles and Techniques. Springer-Verlag New York, Inc. (2005)
7. Kang, K.C., et al.: Feature-Oriented Domain Analysis (FODA) Feasibility Study, in Technical Report, Softw. Eng. Inst., CMU. p. 161 pages (November 1990)
8. Batory, D.: Feature Models, Grammars, and Propositional Formulas. In: Obbink, H., Pohl, K. (eds.) SPLC 2005. LNCS, vol. 3714, pp. 7–20. Springer, Heidelberg (2005)
9. Griss, M.L., et al.: Integrating Feature Modeling with the RSEB. In: Proceedings of the 5th International Conference on Software Reuse. IEEE Computer Society (1998)
10. Schobbens, P.-Y., et al.: Generic semantics of feature diagrams. Comput. Netw. 51(2), 456–479 (2007)
11. Kang, K.C., et al.: FORM: A feature-oriented reuse method with domain-specific reference architectures. Ann. Softw. Eng. 5, 143–168 (1998)
12. Czarnecki, K., et al.: Formalizing cardinality-based feature models and their specialization. Software Process: Improvement and Practice 10(1), 7–29 (2005)
13. Czarnecki, K., et al.: Cardinality-based feature modeling and constraints - a progress report. In: Proceedings of International Workshop on Software Factories, OOPSLA (2005)
14. Benavides, D., et al.: Automated analysis of feature models 20 years later: A literature review. Information Systems 35(6), 615–636 (2010)
15. Nguyen, T., et al.: Managing service variability: state of the art and open issues. In: Proc. of the 5th Int. Workshop on Variability Modeling of Software-Intensive Systems (2011)
16. Czarnecki, K., Antkiewicz, M.: Mapping Features to Models: A Template Approach Based on Superimposed Variants. In: Glück, R., Lowry, M. (eds.) GPCE 2005. LNCS, vol. 3676, pp. 422–437. Springer, Heidelberg (2005)
17. Didonet, M., et al.: Weaving Models with the Eclipse AMW plugin. In: Proceedings of Eclipse Modeling Symposium, Eclipse Summit Europe (2006)
18. Kotamraju, J.: JSR224 - The Java API for XML-Based Web Services (JAX-WS) 2.2, December 10 (2009), http://jcp.org/aboutJava/communityprocess/mrel/jsr224/index3.html
19. Czarnecki, K.: Feature Modeling Plug-in, http://gsd.uwaterloo.ca/projects/fmp-plugin/
20. Tosic, V., et al.: WSOL — Web Service Offerings Language. In: Web Services, E-Business, and the Semantic Web, pp. 57–67 (2002)
21. Chen, L., et al.: Variability management in software product lines: a systematic review. In: Proc. of the 13th Int. Software Product Line Conf. (2009)
22. Boucher, Q., et al.: Introducing TVL, a Text-based Feature Modelling Language. In: Proc. of the 4th Int. Workshop on Variability Modeling of Soft. Intensive Syst., pp. 159–162 (2010)
23. Mendonca, M., et al.: S.P.L.O.T.: software product lines online tools. In: Proceeding of the 24th ACM SIGPLAN Conference Companion on OOPSLA. ACM (2009)
24. Benavides, D., et al.: On the Modularization of Feature Models. In: First European Workshop on Model Transformation (2005)
25. Cechticky, V., et al.: XML-Based Feature Modelling, in Software Reuse: Methods, Techniques and Tools, pp. 101–114 (2004)

Towards a Service Framework for Remote Sales Support via Augmented Reality

Ross Brown and Alistair Barros

Faculty of Science and Technology
Queensland University of Technology
2 George St, Brisbane, Qld, 4000, Australia
{r.brown,alistair.barros}@qut.edu.au

Abstract. Real-time sales assistant service is a problematic component of remote delivery of sales support for customers. Solutions involving web pages, telephony and video support prove problematic when seeking to remotely guide customers in their sales processes, especially with transactions revolving around physically complex artefacts. This process involves a number of services that are often complex in nature, ranging from physical compatibility and configuration factors, to availability and credit services. We propose the application of a combination of virtual worlds and augmented reality to create synthetic environments suitable for remote sales of physical artefacts, right in the home of the purchaser. A high level description of the service structure involved is shown, along with a use case involving the sale of electronic goods and services within an example augmented reality application. We expect this work to have application in many sales domains involving physical objects needing to be sold over the Internet.

Keywords: Augmented Reality, Service User Interfaces, Customer Service Support.

1 Introduction

The explosion of online retail sales has brought with it much convenience for consumers of goods, in particular, digital goods have benefited from their ability to be easily sampled and then purchased, especially in the case of music and video [6]. However, some items remain problematic to sell remotely, due to their inherently spatial nature, requiring physical interaction in order to be sold effectively (for example, cars, musical instruments, clothing, amongst many others). Such physical qualities are difficult, if not impossible, to transmit with the level of immersion available on a flat screen home computer. In such cases, the consumer is forced to visit an actual store to test the artefact they wish to purchase.

Recently, the emergence of the computer game industry has driven the development of graphics and multi-core processing systems, to the extent that high end processing capabilities are available on mobile devices such as phones or tablets [16]. With such computing power comes the possibility to create more immersive environments using Augmented Reality (AR) in order to allow the customer to experience interactively,

A. Haller et al. (Eds.): WISE 2011 and 2012 Combined Workshops, LNCS 7652, pp. 335–347, 2013.

convincing representations of physical products in their own home, where the products will be juxtaposed with their home environment in order to assist goods selection. Such an AR approach can be positioned as a *Purchase Simulation Service,* providing the user with the ability to perform in-situ analysis of purchases in their home.

AR has been used to provide further synthetic information, overlaid on video feeds of real scenes in order to provide insight into the physical characteristics of information [12], and to assist in remote collaboration [18, 9, 2]. Of particular interest is its ability to provide remote representations that are registered with the physical world, and so facilitate analysis of synthetic representations of objects at the location of the customer. In addition, such environments can be augmented even further with representations of remotely connected sales assistants, who can assist in the explanation and process of explaining the physical device and services needed in the process of the sales transaction [8, 9].

Fig. 1a. shows a typical web-page that displays electronics goods available for sale on the Internet, via an online shop front. We note in this example, that the devices sold are large screen TVs with an online sales option.

While such web page approaches are able to present large physical items via images, a number of limitations become evident. Firstly, the customer is unable to

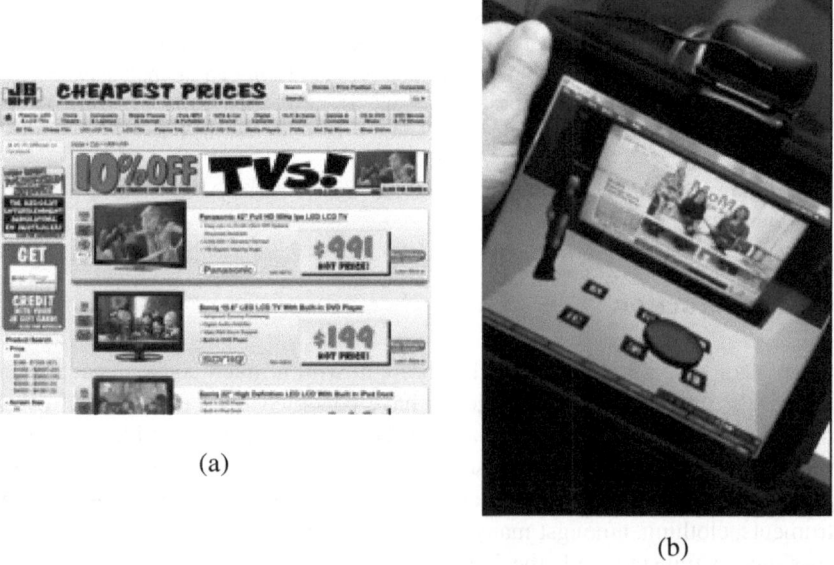

(a)

(b)

Fig. 1. Image (a) shows a typical vendor website selling home electronics and related services. Image (b) shows our addition of a spatial context to these representations in order to facilitate decision making by purchasers at home using mobile devices. The image shows a preliminary prototype running an AR application, with a remotely connected avatar demonstrating the features of the TV in the home of the customer.

examine the product from any angle other than presented in the web page and is unable to see if the product complements the interior décor of their home. In addition, they are unable to determine if it is compatible with other audio or computer devices by inspection; usually this process requires examination of manuals or the advice of a shop assistant in a store. In addition, other services, while easily presented on the web page, may be quite complex to consume (for example insurance) and thus need advice to be presented by a sales assistant. Typical web collaboration tools, such as audio and video environments [14], suffer from problems with a lack of insight into the actual context of actions by the remotely connected collaborator (in this case a sales assistant). The customer may struggle with understanding the actual artefact that is being discussed by the sales assistant, and a lack of relevant gestures and spatial organisation makes the process of service provision by sales staff problematic.

From this analysis we see two main problems with such service provision, the first is the lack of an inherently spatial representation of the product and services, the second is the lack of ability for remote sales staff to adequately interact with customers in order to provide assistance with the product and other aligned services.

In this paper, we present a potential solution to these problems as a novel augmented reality customer sales assistance service framework that provides additional information to a remotely connected customer to assist them with analysis and configuration of their purchase. The key innovation in this paper is the use of a remote representation in a *human-in-the-loop* manner to assist with service discovery and aggregation for a customer engaged in a sales transaction. We show examples from a preliminary AR system we have developed to support these services. We present the framework, the key services offered, interactions within a typical session and illustrate with a use case of a large screen television sale using such a framework. We then conclude with discussion of how this preliminary work may be further developed.

2   Virtual World and Augmented Reality for Sales Services

In order to provide the functionality required for a remote sales service framework, we have built a test system to provide illustrations of the interface components, and to show how a remotely connected avatar can assist with provision of sales services of synthetic representations of physical products. Two major components required are an AR component, to provide the product visualisations, and a Multi-User Virtual World (MUVE) to provide a remote avatar service to the consumer.

The MUVE used in this work was Open Simulator[1], which is an open-source server that allows users to create and deploy virtual worlds over a network. Users are able to connect to Open Simulator servers using a client called a viewer. We use the Second Life viewer[2], which is compatible with Open Simulator. Once the viewer has established a connection with an Open Simulator server, the user receives a figure called an avatar, which the user can personalise and control (see Fig. 1). The avatar serves as a proxy for each user inside the MUVE.

---

[1] Open Simulator - www.opensimulator.org
[2] Second Life – www.secondlife.com

Augmented Reality is an interface mode in which the user is immersed in a world that is real but contains computer generated augmentations. As such it falls in between reality, in which the surrounding environment is completely real, and virtual reality, in which the surrounding environment is completely computer generated [15].

AR systems aim to combine real and virtual world seamlessly in three dimensional space and allow for real-time interactions [1]. They provide information that is not readily available to human senses to facilitate tasks [1]. The video-see-through approach to AR, which we use in our prototype, works by capturing an image from real space then adding the virtual objects and displaying it on some kind of screen, often on a mobile device (refer to Fig. 1). For this method a camera that is positioned and oriented approximately similar to the display in space captures an image of the real space as seen from the perspective of the user. Virtual objects are then drawn into the image with 3D graphics rendering approaches.

To enable the illusion of virtual objects existing in the real space, a consistent registration between virtual and real space is required [1]. Tracking is used to register the positions and orientations of the user and relevant real-world objects in the mixed reality space [24]. We use an icon based tracking approach, as shown in Fig. 5, to perform this registration.

It has been found that AR can help to overcome many of the shortcomings of current user interfaces and allows for more natural interactions and communication between users. This happens because they merge the task, communication and work spaces into a whole communication channel. The position of participants relative to each other is the same in the real and the virtual world. Due to these spatial relationships, related communication behaviours, such as pointing and gazes, are supported [3]. Since AR includes images of the real world, users can see each other, which supports body-language communication better [11, 19]. Kiyokawa [11] demonstrated the importance of users seeing each other for collaboration and the position of task space in relation to communication space for task performance. Support of these communication cues decreases interruptions and communication overlap.

Since each user has his own view (7, 19) AR interfaces can also use space more efficiently to place service information around the user and allow private display of information to relevant people. This also allows the use of peripheral senses to perceive information that may contribute to a task, without interrupting the user's focus through data overload, as described by Ishii and Ulmer [10]. In the MagicBook project Billinghurst, Kato and Poupyrev [4] demonstrated collaborative exploration of spaces in a story across multiple reality modes with similar results. All these features have shown benefits to computer supported collaboration in co-located work settings.

In addition, such capabilities are beneficial to remote collaboration as well. Regenbrecht et al. [19] and Uva, Cristiano, Fiorentino and Monno [23] have applied AR to augment co-located as well as remote product configuration and development. Lee, Rhee and Park [12] have applied it to prototyping and evaluation of product design. Billinghurst, Kato and Poupyrev [4] and Regenbrecht, Haller, Hauber and Billinghurst [20] demonstrated the merging of virtual and augmented reality to facilitate both co-located and remote collaboration.

Similar work has been carried out in the space of AR for use in related areas to sales, including sales recommendation systems [8], IT technical help desk systems [9] and ubiquitous house service simulations [13]. In each case, some services are proffered within the application, but what is lacking is the ability of a remotely connected collaborator to discover, orchestrate and aggregate the digital services in real time. This has not been done before to our knowledge, and we believe the use of AR Avatars will facilitate the ability of remote service aggregation by humans in a much more intuitive fashion than standard 2D interfaces in such a remote collaborative setting.

Media Rich Theory promotes concepts of immersion and engagement [5] brought about by the closeness of the representations to objects required for the task at hand. We can then postulate that customer purchase intentions may increase in strength by use of immersive technology, due to the closeness of representation providing the ability to support remote decision making in purchases [22]. In addition, due to the digital nature of the presentation, other digital services can be presented in-situ, thus potentially motivating the customer to consume them, as the sense of immersion, on presentation of the service, is not broken.

Therefore, there is a need to provide the ability to present synthetic versions of physical objects. Digital sales of products such as music and magazines, are easily assimilated into online delivery channels (for example, iTunes). Other physical object, such as musical instruments, cars, houses, amongst others, do require, in part, a physical intimacy with the object in order to ascertain its suitability to the purchaser.

However, it can be argued that some home products, such as TVs, sound systems, whitegoods, kitchen devices and furniture, fall into a category that may be readily supported by AR sales services. Each of these can be evaluated more for their appearance and functionality in the home of the purchaser, and are less reliant on requiring a physical interaction to be sold. Key to this presentation approach is a number of factors, first the presentation of the product in the home of the user, the explication of the product features and then aligned services to be presented to the consumer. All of these have been utilised in person by other home sales companies (for example, Amway, Avon, Tupperware) but such methods can be expensive and require costly travel and the inconvenience of sales site setups.

The product features can be presented via animations and videos, but the major contribution here is the use of a remotely connected sales assistant to answer questions and engage with the consumer to establish rapport and support the consumer in their purchasing decisions. Our readings indicate that by applying an AR interface to remote services in a MUVE, it should be possible to create a tool that is intuitive to use and supports a high quality, immersive form of remote collaboration to provide services at the point of sale for physical artefacts. We now show a preliminary service framework and AR implementation to support this approach.

3   Prototype AR Service Interaction and Aggregation Implementation

The AR sales service framework consists of the following key components of services, context management and major interactions involved in product configurations. We focus on the product configuration in this paper, as it is the most novel component of our work, and we consider transactional components can be implemented using service structures that have already been developed [17]

3.1   Service Manifest

We have implemented a prototype using a MUVE with added Augmented Reality features in order to facilitate the in-situ discovery and aggregation of services in a 3D graphical manner for point of sale services.

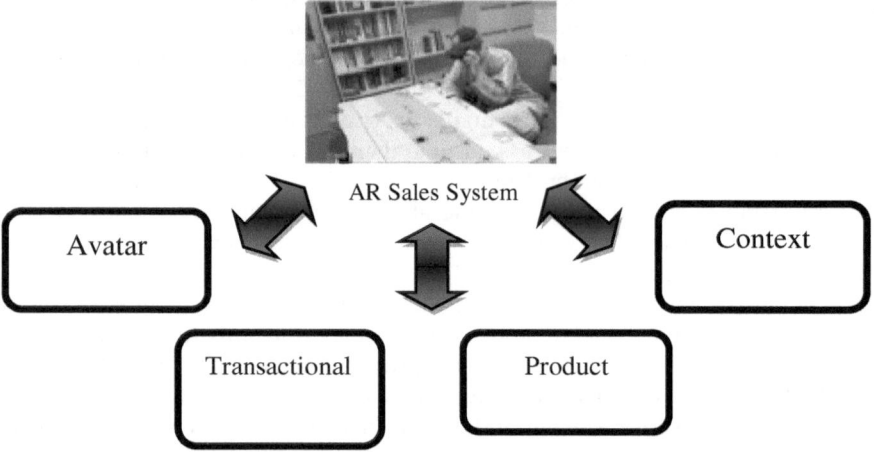

Fig. 2. Illustration of the major components of our AR Service Framework, all of these services are accessed from within the AR Sales System that is used by a customer

The AR sales service framework contains four major high-level services: Avatar, Product, Context and Transactional, as illustrated in Fig. 2.

- Avatar Service – chat bot or human controlled [21], providing a graphical representation of a sales assistant providing information about product capabilities, service interactions and compatibilities;
- Product – product information, in particular, 3D representations to be sent to the AR application and compatibility information with other available products;
- Context – management of customer account information, including physical location information, relative contexts at the customer location and stored product configurations in previous transactions.
- Transactional – other allied services to the main sales transaction, such as, best price services, credit services, insurance, amongst others;

These services can be offered as a human mediator via remote connected sales personnel, or an assisting chat bot. Each of the services integrates to provide a simulation of purchase, which we believe may increase intent to purchase [22], and thus could be a potent remote sales delivery concept.

3.2   Context Management

Absolute Context. The absolute context for data in this framework is derived from the customer location. This is the main spatial point from where the transactions occur within the wider geospatial context (see Fig. 3). In the context of our use case shown later, such an absolute context can be considered to be the home address of the person making the purchases, so we name it as a *Customer Context*. This absolute context contains information regarding customer physical location and the set of customer transaction histories and relative contexts within the house.

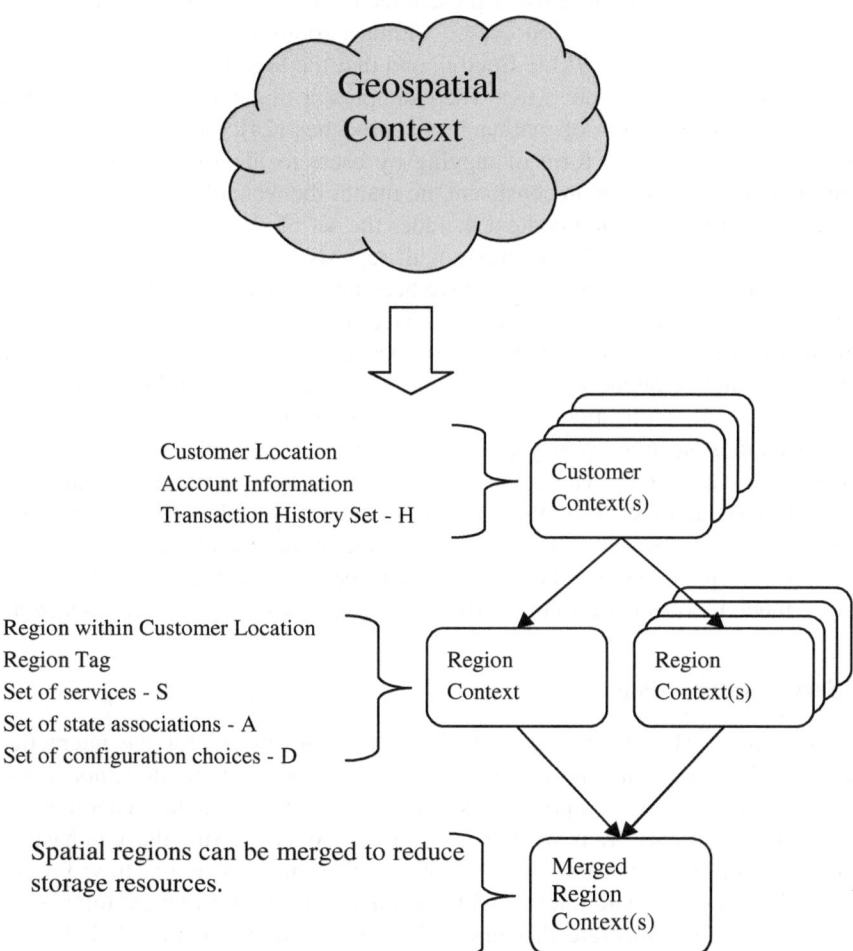

Fig. 3. Illustration of the main components used in context management, viz. an Absolute Context (Customer) and Relative Context (Region). Multiple customers can be allocated to physical locations, with multiple regions used within a house. We control the generation of contexts, by providing a merging capability, to integrate similar relative contexts.

Relative Context. Our relative context approach is an adaptive spatial subdivision of the house, known as a *Region Context*. This structuring is derived from the observation that the internals of the home of a customer are divided up into rooms and form a natural context for purchases within a house using an AR system. Instantiation of this context occurs when the AR marker pattern is laid on the ground, wall or roof of the room and is annotated with a text tag, for example, "Lounge Room East Wall." Any finer subdivision of the house would be problematic, due to unnecessary redundancy, and follows the structure of other research work in this regards [12].

In addition, there is no real way to identify the exact GPS location down to fine resolutions inside a typical house using present technology. We show, however, that users may tag multiple registration events within a room via a text term, such as, "Lounge-TV-Wall," "Lounge-Table-Centre" and that the high level model allows for merged regional contexts within a tag. The examples in this paper use marker-based AR. Even with the adoption of marker less approaches [24], such a service framework will still require some form of tagging by users to identify a session, so the information requirements remain consistent, no matter the technology used.

The data stored in the region context includes the set of services for the configuration, associations to other region contexts with regards to merging operations and a set of product configuration choices that have been previously created by customers.

Such contexts are made available via a service to allow for a number of possible configuration modalities. For example, a customer may have already enquired about a product using a mobile phone in a shop, but wishes to check the configuration in their home using the AR application. The contexts are therefore available from other software modalities, such as web-systems and normal mobile applications that are not augmented reality in nature. This allows for the user to load up previous enquiries in the AR application and then view them in-situ, to determine if they should be actually purchased. Our use case will show an AR configuration example only, but it is in principle possible for other methods of enquiry to be integrated into this framework, providing a much more general solution to consumer services using this AR application.

3.3   Service Interactions

In previous work, [21] a five stage model has been developed for the interactions between virtual sales agents and customers. However, their model does not include product configuration as a component, as only single items are sold. Therefore, the interaction model in this framework has been developed to suit the emphasis on product configuration that we wish to describe. The intention is to allow for both software and human agents to interact with customers in the environment to provide a general solution. We illustrate this interaction model below, in Fig. 4. It has two major activities, the creation of the appropriate contexts, related to the marker pattern laid down by the customer in their home, and the process of configuring the product required by the customer, which may be a combination of a number of products in a spatial organization.

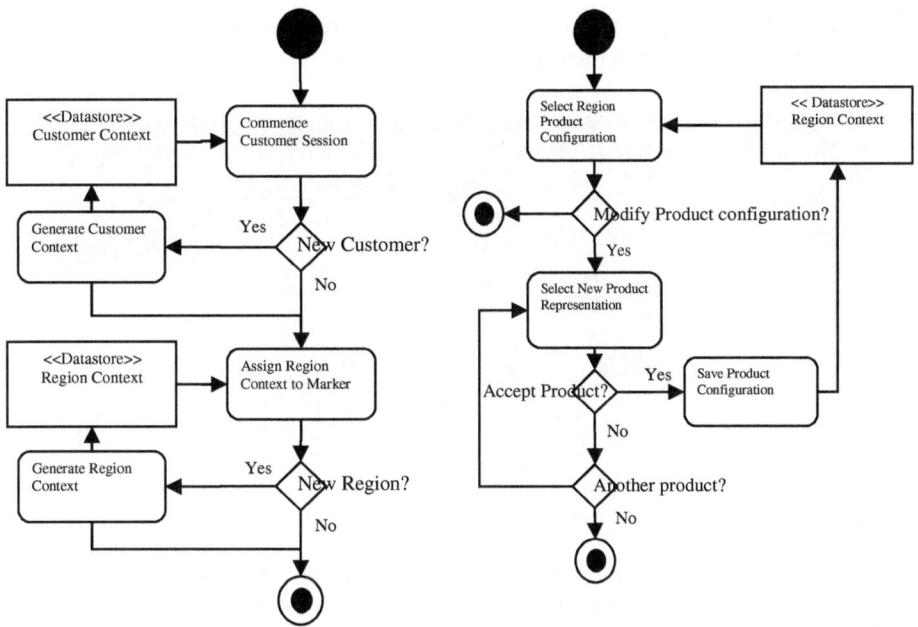

Fig. 4. Service activity diagrams for the use case illustrated, the diagrams illustrate the context creation step (left) and the configuration phase for our particular sales domain (right)

In the above interactions, it should be noted that the avatar is able to support the key decision points in the interactions, either by being a live human directing the customer as a remotely logged in avatar, or by a chat bot service that provides feedback information from a product knowledge base, typically implemented as an expert system [21].

4   Use Case

In this section we use the sale of audio goods to provide examples of the interactions and service aggregations used in guiding the user through a sales transaction, based upon the previously illustrated modelling.

4.1   Scenario

Our example customer is intending to purchase a large screen TV, potentially with related audio equipment and furniture. We present the customer with a front-end selling environment for electrical goods as an application on their mobile device. The customer is at home and places an augmented reality registration pattern onto the ground (see Fig. 5), where the actual television is to be installed. A dialog appears requesting customer information (creating a new *Customer Context* if needed for the customer address) and then asks for a text tag to be applied to the marker location (creating a new *Region Context* if needed). The user tags the marker with "Lounge

Floor East." A sales representative is alerted to the creation of a session, and logs in from the other end to assist with any questions.

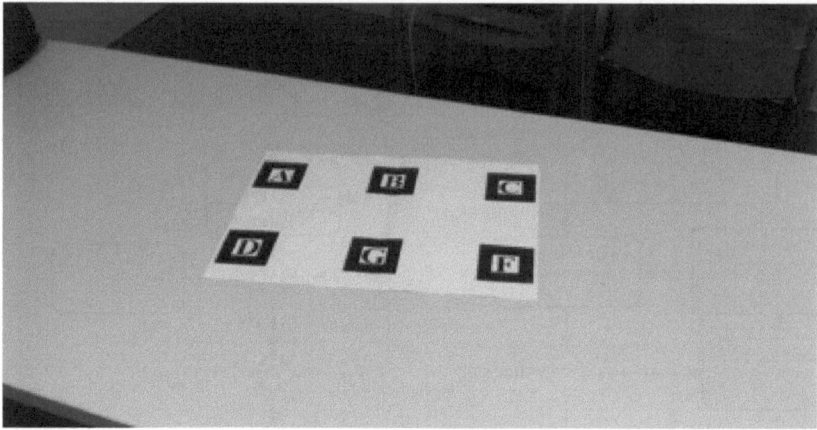

Fig. 5. Illustration of the coordinate system registration marker, which anchors the position of the objects to be sent to the viewer. This marker is placed at a location in the home close to the location of the desired objects and is tagged by the user to identify a Region Context.

The user requests to view large screen televisions from a specified brand available from this seller, along with appropriate entertainment furniture. A large screen television of a brand of interest is brought into the display, which the user can see in their room through the AR application. The remotely connected avatar then demonstrates the features of the new television, see Fig. 6.

Fig. 6. Illustration of an avatar showing the key features of a product; in this case showing the remote control features on a television set

The sales assistant asks if the customer wishes to see some related furniture and audio equipment, to which the customer gives assent. A selection is provided, that are compatible with the large screen television, and in this case the furniture is compatible to the weight of the television. The audio equipment is inspected for compatibility (such as fiber optic audio) and is also checked via web services shown with the AR view (see Fig. 7).

Fig. 7. In this image the avatar is accessing information services for the customer regarding furniture for the TV using an invoked web service within the application

Fig. 8. Illustration of an avatar displaying the Best Price service for large screen televisions

The user then asks to see where the best prices are for this device, which is shown on the side of the view as a web page from a best price service (see Fig. 8). In addition, credit services can be displayed displayed along with insurance tie-ins for this company. The purchase is made via linked application for credit, and the session is the ended, with the information saved to a sales history for the *Customer Context*.

5   Conclusion

In this paper we have outlined a high-level framework for services in customer transactions, mediated by AR technology. A number of novel elements have been introduced and developed. Firstly, the use of AR technology has been explored, with regards to assisting users to simulate their purchases in their own home. Secondly, we have introduced the concept of using an avatar to guide the user through the process of purchase remotely, assisting with interactions with other services that are required in the process. Finally, we have outlined the data requirements to support such an approach, providing high-level definitions for the services, absolute and relative context information and relevant product configuration service interactions. A preliminary implementation of the interface concept was shown to illustrate the service framework via a use case regarding the purchase of electronic entertainment goods.

All the examples in this paper have been shown on a desktop system, linked to a tablet via VNC. Future work will include the implementation of this framework within an appropriate mobile tablet device or smartphone and the full integration of the sales support services. Furthermore, we intend that this framework will be augmented with a purpose built scripting language, which will, in a similar vein to level editing systems in virtual worlds and games, enable the easy creation of sales service scenarios that can be embedded within AR applications.

References

1. Azuma, R.: A survey of augmented reality. Presence-Teleoperators and Virtual Environments 6, 355–385 (1997)
2. Azuma, R., Baillot, Y., Behringer, R., Feiner, S., Julier, S., MacIntyre, B.: Recent advances in augmented reality. IEEE Computer Graphics and Applications 21, 34–47 (2001)
3. Billinghurst, M., Kato, H.: Collaborative augmented reality. Communications of the ACM 45, 64–70 (2002)
4. Billinghurst, M., Kato, H., Poupyrev, I.: The MagicBook: a transitional AR interface. Computers & Graphics 25, 745–753 (2001)
5. Daft, R.L., Lengel, R.H.: Organizational information requirements, media richness and structural design. Management Science 32(5), 554–571 (1986)
6. Elberse, A.: Should You Invest in the Long Tail? Harvard Business Review 86, 88–97 (2008)
7. Fuhrmann, A., Löffelmann, H., Schmalstieg, D., Gervautz, M.: Collaborative visualization in augmented reality. IEEE Computer Graphics and Applications 18, 54–59 (1998)

8. Güven, S., Oda, O., Podlaseck, M., Stavropoulos, H., Kolluri, S., Pingali, G.: In: IEEE International Conference on Pervasive Computing and Communications, PerCom 2009, pp. 1–3 (2009)
9. Güven, S., Podlaseck, M., Pingali, G.: Exploring Co-Presence for Next Generation Technical Support. In: Proceedings of IEEE VR 2009 (Virtual Reality), pp. 103–106 (2009)
10. Ishii, H., Ullmer, B.: Tangible bits: towards seamless interfaces between people, bits and atoms. In: Conference on Human Factors in Computing Systems, Atlanta, GA, pp. 234–241. ACM Press, New York (1997)
11. Kiyokawa, K., Billinghurst, M., Hayes, S., Gupta, A., Sannohe, Y., Kato, H.: Communication behaviors of co-located users in collaborative AR interfaces. In: International Symposium on Mixed and Augmented Reality, Darmstadt, Germany, pp. 135–144. IEEE Press, Washington (2002)
12. Lee, J., Rhee, G., Park, H.: AR/RP-based tangible interactions for collaborative design evaluation of digital products. The International Journal of Advanced Manufacturing Technology 45, 649–665 (2009)
13. Lee, J.Y., Rhee, G.W., Seo, D.W., Kim, N.K.: Ubiquitous home simulation using augmented reality. In: Proceedings of the 2007 Annual Conference on International Conference on Computer Engineering and Applications, pp. 112–116. World Scientific and Engineering Academy and Society (WSEAS) (2007)
14. Lanubile, F., Ebert, C., Prikladnicki, R.: Collaboration tools for global software engineering. IEEE Software 27(2), 52–55 (2010)
15. Milgram, P., Takemura, H., Utsumi, A., Kishino, F.: Augmented reality: A class of displays on the reality-virtuality continuum. In: Proceedings of Telemanipulator and Telepresence Technologies, SPIE 2351-34, pp. 282–292 (1994)
16. Morrison, A., Oulasvirta, A., Peltonen, P., Lemmela, S., Jacucci, G., Reitmayr, G., Naanen, J., Juustila, A.: Like bees around the hive: a comparative study of a mobile augmented reality map. In: Proceedings of the 27th International Conference on Human Factors in Computing Systems, CHI 2009, pp. 1889–1898 (2009)
17. Papazoglou, M., van den Heuvel, W.-J.: Service oriented architectures: approaches, technologies and research issues. The VLDB Journal 16, 389–415 (2007)
18. Poppe, E., Brown, R.A., Recker, J.C., Johnson, D.M.: A prototype augmented reality collaborative process modelling tool. In: Proceedings of the Demo Track of the Ninth Conference on Business Process Management 2011. CEUR Workshop Proceedings, vol. 820 (2011)
19. Regenbrecht, H., Wagner, M., Baratoff, G.: Magicmeeting: A collaborative tangible augmented reality system. Virtual Reality 6, 151–166 (2002)
20. Regenbrecht, H., Haller, M., Hauber, J., Billinghurst, M.: Carpeno: Interfacing remote collaborative virtual environments with table-top interaction. Virtual Reality 10, 95–107 (2006)
21. Mumme, C., Pinkwart, N., Loll, F.: Design and Implementation of a Virtual Salesclerk. In: Ruttkay, Z., Kipp, M., Nijholt, A., Vilhjálmsson, H.H. (eds.) IVA 2009. LNCS, vol. 5773, pp. 379–385. Springer, Heidelberg (2009)
22. Underhill, P.: Why we buy: The Science of Shopping. Simon & Schuster, New York (1999)
23. Uva, A., Cristiano, S., Fiorentino, M., Monno, G.: Distributed design review using tangible augmented technical drawings. Computer-Aided Design 42, 364–372 (2008)
24. Zhou, F., Duh, H.B.-L., Billinghurst, M.: Trends in augmented reality tracking, interaction and display: A review of ten years of ISMAR. In: 7th IEEE and ACM International Symposium on Mixed and Augmented Reality, Cambridge, England, pp. 193–202. IEEE Press, Washington (2008)

Author Index